G-W PUBLISHER

Print Reading for HVACR

Write-In Text with Large Prints

Eugene Silberstein

Introduction to reading and interpreting professional HVACR prints through the study of key HVACR and print reading topics.

G-W PUBLISHER EduHub®

Be Digital Ready on Day One with EduHub

EduHub provides a solid base of knowledge and instruction for digital and blended classrooms. This easy-to-use learning hub delivers the foundation and tools that improve student retention and facilitate instructor efficiency. For the student, EduHub offers an online collection of eBook content, interactive practice, and test preparation. Additionally, students have the ability to view and submit assessments, track personal performance, and view feedback via the Student Report option. For instructors, EduHub provides a turnkey, fully integrated solution with course management tools to deliver content, assessments, and feedback to students quickly and efficiently. The integrated approach results in improved student outcomes and instructor flexibility.

Michael Jung/Shutterstock.com

eBook

The EduHub eBook engages students by providing the ability to take notes and highlight key concepts to remember.

Objectives

Course objectives at the beginning of each eBook chapter help students stay focused and provide benchmarks for instructors to evaluate student progress.

eAssign

eAssign makes it easy for instructors to assign, deliver, and assess student engagement. Coursework can be administered to individual students or the entire class.

Monkey Business Images/Shutterstock.com

Assessment

Self-assessment opportunities enable students to gauge their understanding as they progress through the course. In addition, formative assessment tools for instructor use provide efficient evaluation of student mastery of content.

Reports

Reports, for both students and instructors, provide performance results in an instant. Analytics reveal individual student and class achievements for easy monitoring of success.

Print Export

Score	Items	
100%	●	●
80%	●	●
100%	●	●
80%	●	●
100%	●	●
100%	●	●

Instructor Resources

Instructors will find all the support they need to make preparation and classroom instruction more efficient and easier than ever. Lesson plans, answer keys, and PowerPoint® presentations provide an organized, proven approach to classroom management.

Guided Tour

The instructional design includes student-focused learning tools to help students succeed. This visual guide highlights the features designed for the textbook.

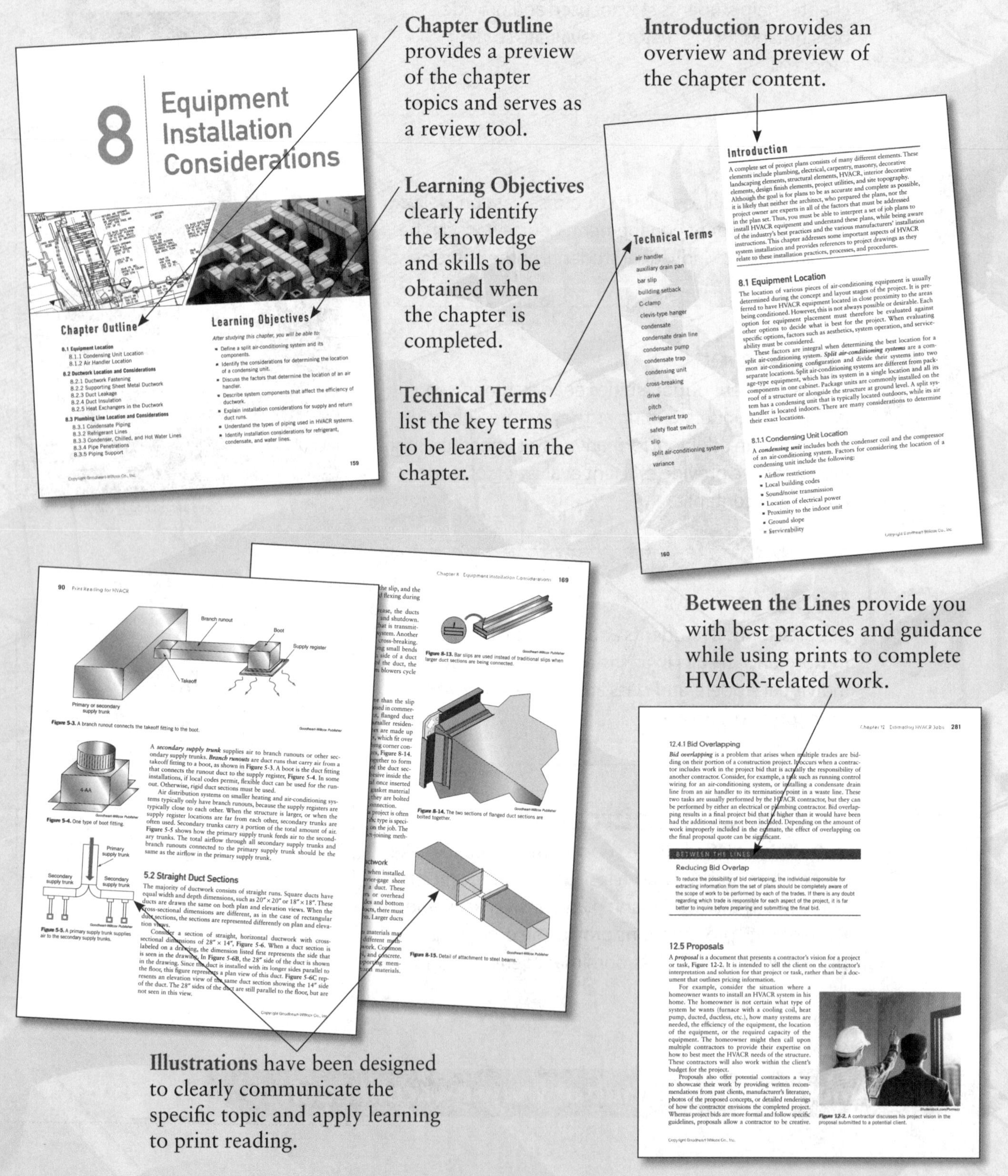

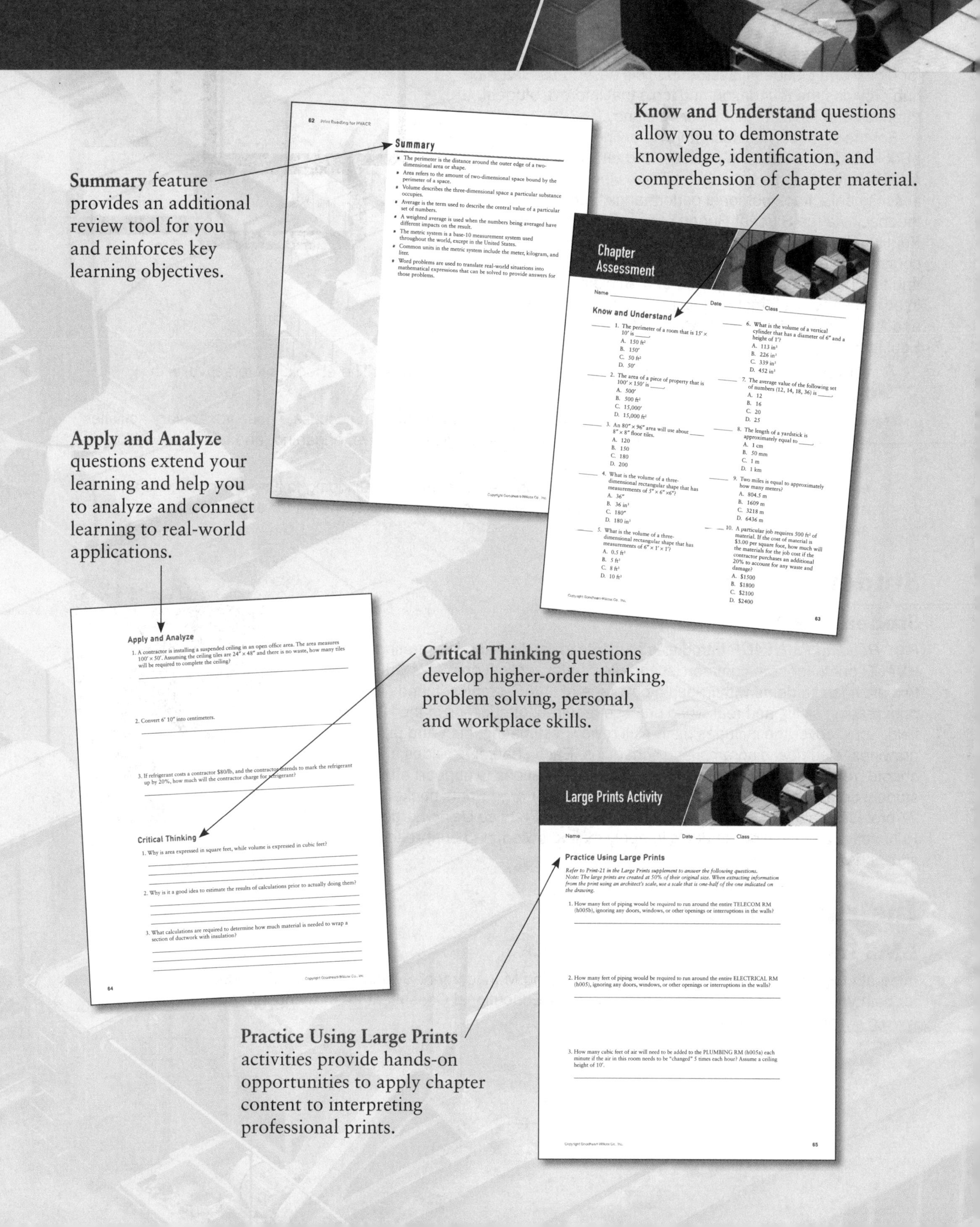
Summary feature provides an additional review tool for you and reinforces key learning objectives.
Know and Understand questions allow you to demonstrate knowledge, identification, and comprehension of chapter material.
Apply and Analyze questions extend your learning and help you to analyze and connect learning to real-world applications.
Critical Thinking questions develop higher-order thinking, problem solving, personal, and workplace skills.
Practice Using Large Prints activities provide hands-on opportunities to apply chapter content to interpreting professional prints.
62 Print Reading for HVACR
Summary
Chapter Assessment
Know and Understand
Apply and Analyze
Critical Thinking
Large Prints Activity
Practice Using Large Prints

TOOLS FOR STUDENT AND INSTRUCTOR SUCCESS

EduHub

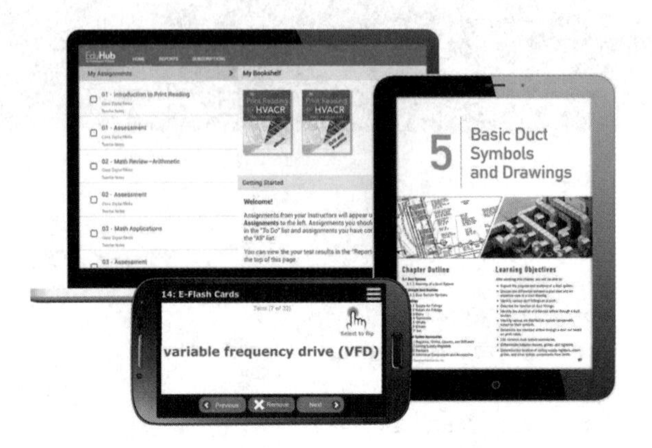

EduHub provides a solid base of knowledge and instruction for digital and blended classrooms. This easy-to-use learning hub provides the foundation and tools that improve student retention and facilitate instructor efficiency.

For the student, EduHub offers an online collection of eBook content, interactive practice, and test preparation. Additionally, students have the ability to view and submit assessments, track personal performance, and view feedback via the Student Report option. For the instructor, EduHub provides a turnkey, fully integrated solution with course management tools to deliver content, assessments, and feedback to students quickly and efficiently. The integrated approach results in improved student outcomes and instructor flexibility. Be digital ready on day one with EduHub!

- **eBook content.** EduHub includes the textbook in an online format. The eBook is interactive, with highlighting, magnification, and note-taking features.
- **Vocabulary activities.** Learning new vocabulary is critical to student success. These vocabulary activities, which are provided for all key terms in each chapter, provide an active, engaging, and effective way for students to learn the required terminology.
- **eAssign.** In EduHub, students can complete online assignments—including text review questions and assessments—as specified by their instructor. Many activities are autograded for easy class assessment and management.

Student Tools

Student Text

Print Reading for HVACR introduces students to reading and interpreting HVACR prints for both commercial and residential applications. This write-in text provides students with in-depth coverage of HVACR topics, foundational print reading skills, and real-world practice required for success in the HVACR field. Topic coverage includes math-skill review, estimating, duct and piping drawings, electrical diagrams, and schedules. Each chapter provides a chapter review and three levels of assessment questions to support effective student comprehension. This text is bundled with a Large Prints packet of 27 professional HVAC prints. Select chapters offer Practice Using Large Print activities that allow students to apply their knowledge to print-reading situations.

Instructor Tools

LMS Integration

Integrate Goodheart-Willcox content in your Learning Management System for a seamless user experience for both you and your students. Contact your G-W Educational Consultant for ordering information or visit www.g-w.com/lms-integration.

Instructor Resources

Instructor Resources provide all the support needed to make preparation and classroom instruction easier than ever. Available in one accessible location, you will find Instructor Resources, Instructor's Presentations for PowerPoint®, and Assessment Software with Question Banks. These resources are available as a subscription and can be accessed at school, at home, or on the go.

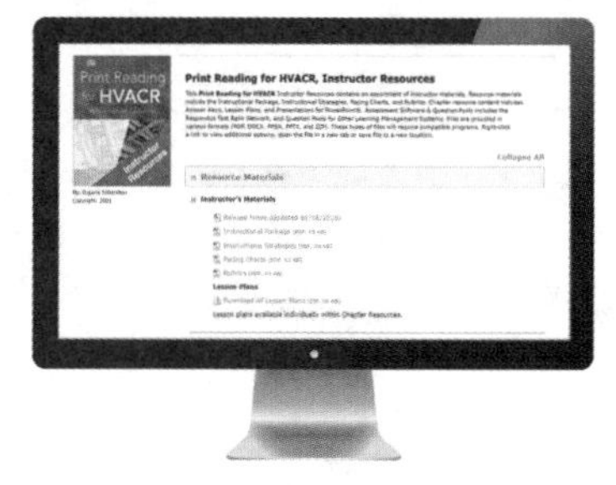

Instructor Resources One resource provides instructors with time-saving preparation tools such as answer keys, chapter outlines, editable lesson plans, and other teaching aids.

Instructor's Presentations for PowerPoint® These fully customizable, richly illustrated slides help you teach and visually reinforce the key concepts from each chapter.

Assessment Software with Question Banks Administer and manage assessments to meet your classroom needs. The following options are available through the Respondus Test Bank Network:

- A Respondus 4.0 license can be purchased directly from Respondus, which enables you to easily create tests that can be printed on paper or published directly to a variety of Learning Management Systems. Once the question files are published to an LMS, exams can be distributed to students with results reported directly to the LMS gradebook.
- Respondus LE is a limited version of Respondus 4.0 and is free with purchase of the Instructor Resources. It allows you to download question banks and create assessments that can be printed or saved as a paper test.

G-W Integrated Learning Solution

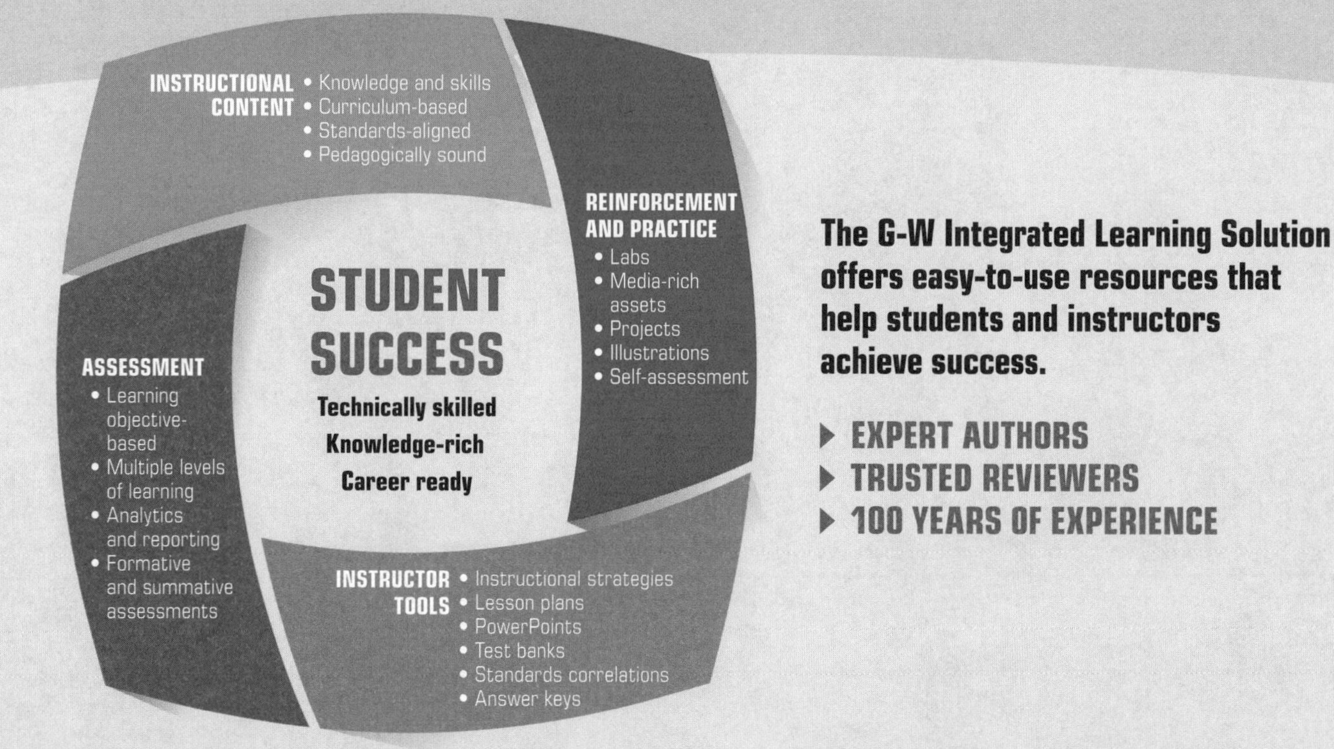

The G-W Integrated Learning Solution offers easy-to-use resources that help students and instructors achieve success.

- ▶ **EXPERT AUTHORS**
- ▶ **TRUSTED REVIEWERS**
- ▶ **100 YEARS OF EXPERIENCE**

EMPLOYABILITY SKILLS · TECHNICAL SKILLS · ACADEMIC KNOWLEDGE · INDUSTRY RECOGNIZED STANDARDS

Print Reading for HVACR

Write-In Text with Large Prints

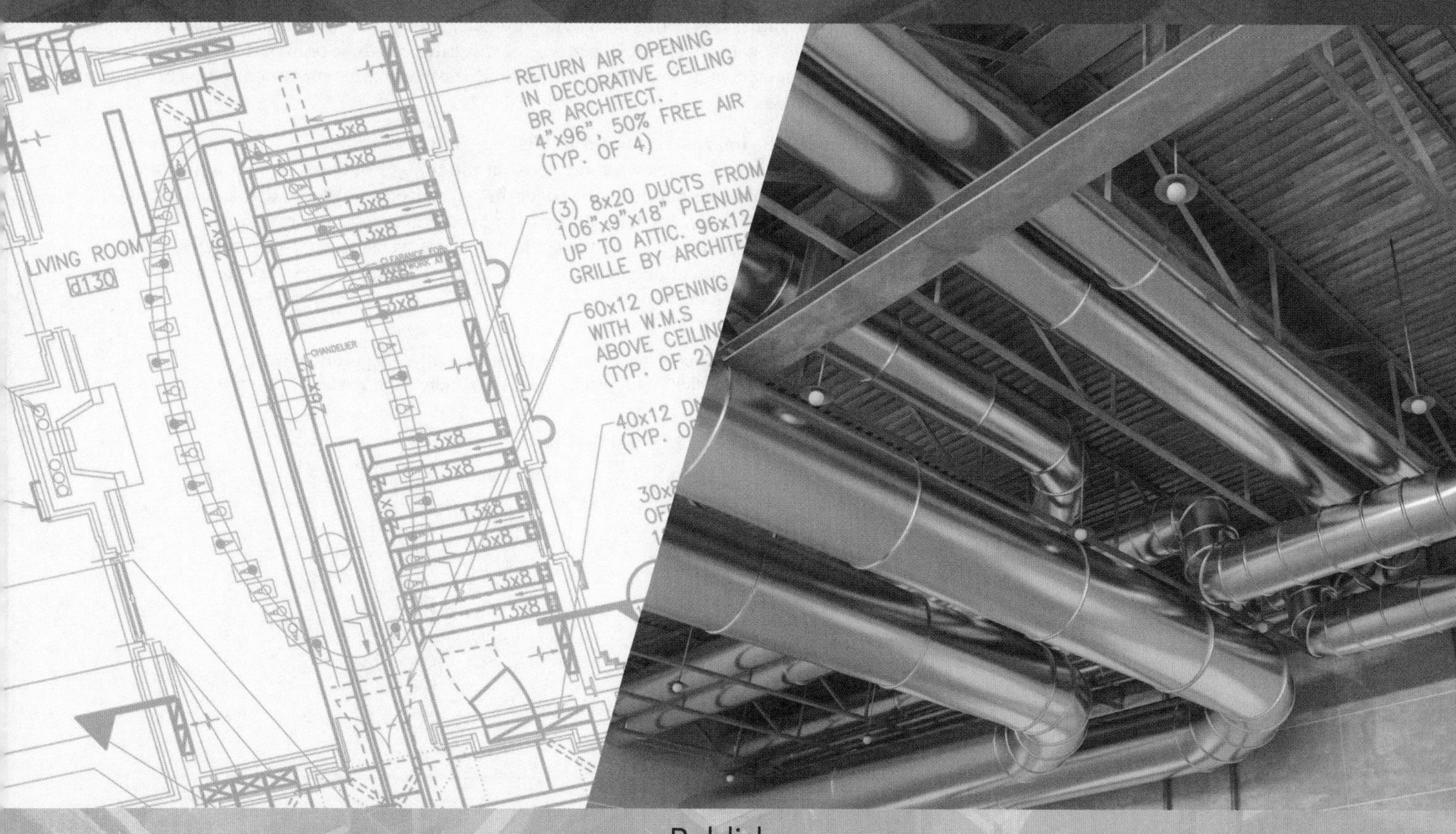

Publisher
The Goodheart-Willcox Company, Inc.
Tinley Park, IL
www.g-w.com

Manufactured in the United States of America.

Library of Congress Catalog Card Number 2019010801

ISBN 978-1-63563-882-0

1 2 3 4 5 6 7 8 9 – 21 – 24 23 22 21 20 19

Image Credits. Front cover: (background image) Tang Yan Song/Shutterstock.com, (left image) Lizardos Engineering Assoc., P.C., (right image) vchal/Shutterstock.com. Chapter opener images: (left) Lizardos Engineering Assoc., P.C., (right) Tang Yan Song/Shutterstock.com

Library of Congress Cataloging-in-Publication Data

Names: Silberstein, Eugene, author.
Title: Print reading for HVACR / Eugene Silberstein.
Description: Tinley Park, IL : The Goodheart-Willcox Company, Inc., [2021] |
Includes index.
Identifiers: LCCN 2019010801 | ISBN 9781635638820
Subjects: LCSH: Heating--Equipment and supplies--Drawings. |
Ventilation--Equipment and supplies--Drawings. | Air
conditioning--Equipment and supplies--Drawings. | Refrigeration and
refrigerating machinery--Equipment and supplies--Drawings. | Blueprints.
Classification: LCC TH7331 .S55 2021 | DDC 697.0028/4--dc23 LC record available at https://lccn.loc.gov/2019010801

Preface

The key to success for any construction or renovation project is the ability of tradespeople to effectively and efficiently read, interpret, and understand written instructions and guidelines. ***Print Reading for HVACR*** provides students with the foundation for learning these key skills to later excel in the HVACR field. To aid in the learning process, a set of professional Large Prints accompany the text that aligns with the chapter material and exercises to give the reader applicable, on-the-job HVAC experiences.

As an HVACR professional with over 40 years in the HVACR industry, I wrote this text as a comprehensive guide that provides the student with the tools required to read and understand project plans, prints, tables, and charts, while gaining insight into the methods used to create such drawings. ***Print Reading for HVACR*** is a valuable resource to those just starting to learn about print reading, as well as those with previous knowledge in the subject area. The information in this book has been organized in a logical, progressive format, allowing the reader to work through the content at an appropriate, individualized pace. Throughout the book, the reader will encounter a feature titled "Between the Lines," which offers best practices and applicable insight on the subject matter.

Each chapter starts with a list of objectives that outlines what the student can expect to learn in the pages that follow and concludes with a detailed summary of the material covered. End-of-chapter review questions are also presented in three different formats pertaining to a specific level of Bloom's taxonomy. "Know and Understand" questions evaluate the level of the reader's retained knowledge from the chapter. "Analyze and Apply" questions provide the reader with the opportunity to demonstrate a deeper level of understanding of the concepts covered in the chapter. When appropriate, these questions require the reader to perform calculations and reach conclusions about the chapter content. "Critical Thinking" questions provide the reader with challenging, open-ended questions that encourage subjective responses to print-related topics and elements.

Select chapters also offer "Practice Using Large Prints" exercises, which provide opportunities for the reader to manipulate and draw information from the Large Prints provided with the text. These exercises strive to help students complete real-world exercises that are performed on a regular basis when working with job prints.

I truly hope that you enjoy reading and learning from this book as much as I have enjoyed writing it.

Eugene Silberstein

About the Author

Eugene Silberstein, M.S., CMHE, BEAP, entered the HVACR industry in 1980. He is currently the Director of Technical Education and Standards at the ESCO Institute since retiring from his tenured professorship position at Suffolk County Community College in 2015. Eugene earned his Bachelor's degree from The City College of New York and his Masters of Science in Energy and Environmental Systems from Stony Brook University.

Eugene brings 25 years of experience teaching air-conditioning and refrigeration at several secondary, postsecondary, and vocational institutions. While teaching, he was selected as one of the top three HVACR instructors in the country for the 2005, 2006, and 2007 academic school years by the Air-Conditioning, Heating and Refrigeration Institute (AHRI) and the Air Conditioning, Heating and Refrigeration (ACHR) News. Eugene earned his Certified Master HVACR Educator (CMHE) credential from HVAC Excellence and the ESCO Group, as well as ASHRAE's BEAP credential, which classifies him as a Building Energy Assessment Professional. Eugene continues to serve as a subject matter expert as an author and coauthor of numerous HVACR textbooks and workbooks.

Reviewers

The author and publisher wish to thank the following industry and teaching professionals for their valuable input into the development of *Print Reading for HVACR*.

Senobio Aguilera
Riverside City College
Riverside, CA

Ed Burns
Harrisburg Area Community College
Harrisburg, PA

Steven Gutsch
Chippewa Valley Technical College
Eau Claire, WI

John Holley
Calhoun Community College
Decatur, AL

Gary Larson
Ozarks Technical College
Springfield, MO

Dr. Christopher Molnar
Porter and Chester Institute
Rocky Hill, CT

Jeffrey M. Muschlitz
Middle Bucks Institute of Technology
Jamison, PA

Richard L. Paznik
Lincoln Technical Institute
Union, NJ

Acknowledgments

The author and publisher would like to thank Lizardos Engineering Assoc., P.C. for their contribution of professional HVAC large prints in the development of *Print Reading for HVACR*.

Brief Contents

Contents

Feature Contents

BETWEEN THE LINES

1 Introduction to Print Reading

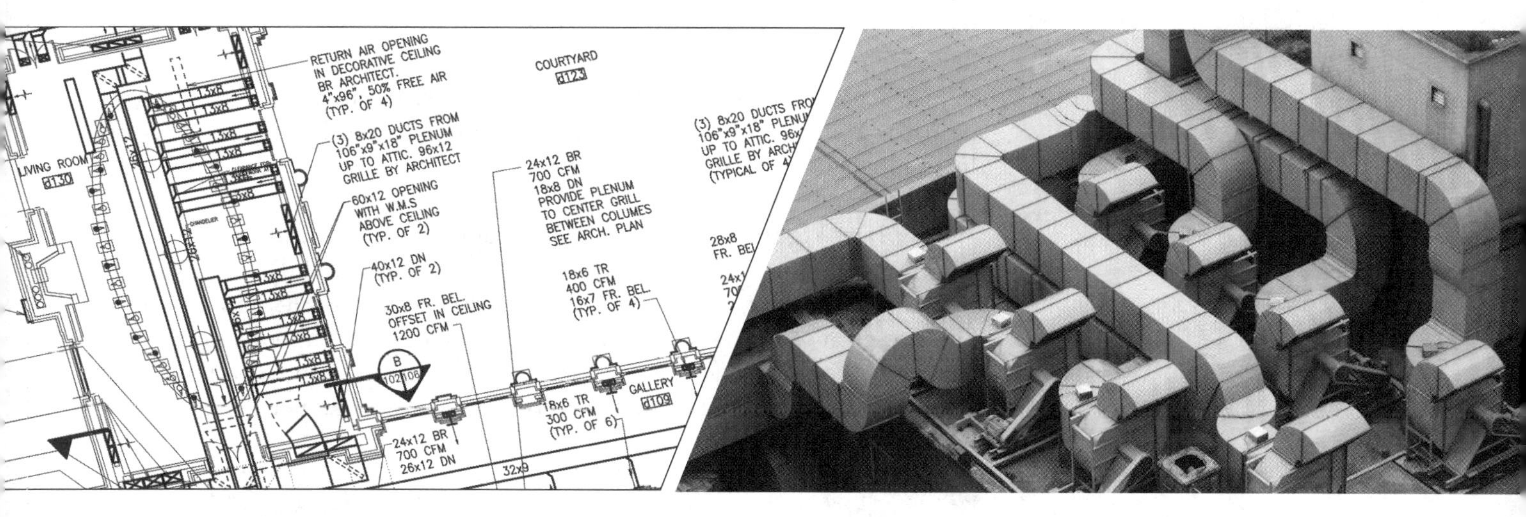

Chapter Outline

1.1 What Is a Print?

1.2 How Are Prints Created and Used?

1.2.1 Drawings and Prints

1.3 Why Are Prints Needed?

1.3.1 Visual Representation
1.3.2 Uniform Depiction
1.3.3 Integration of Individual Trades
1.3.4 Calculating Job Costs
1.3.5 Printed Record of Work

1.4 Construction Drawings

1.4.1 Floor Plans
1.4.2 Site Plans
1.4.3 Foundation Plans
1.4.4 Mechanical Plans
1.4.5 Framing Plans
1.4.6 Structural Plans
1.4.7 Sections
1.4.8 Details

1.5 Schedules

Learning Objectives

After studying this chapter, you will be able to:

- Discuss prints and the history of prints.
- Distinguish the components of a print.
- Explain the purpose of prints and why they are needed in the building trades.
- Explain how prints are created.
- Understand the purpose of floor plans and site plans.
- Differentiate between a plan view and an elevation view.
- Discuss the types of prints used in the HVACR field and other trades.
- Define a schedule and its purpose.

Introduction

A well-executed construction project relies on an accurately produced set of prints or plans to guide the project. It also requires well-trained individuals to read them. This chapter introduces prints as they apply to specific trades projects and the types of prints used on the job. By applying this learning to later chapters, you will better understand the importance of prints and their role in the HVACR field.

Technical Terms

details
electrical plan
elevation view
equipment specifications
floor plan
foundation plan
framing plan
legend
material list
mechanical plan
plan
plan view
plumbing plan
print
schedule
sections
site plan
structural plan
vicinity map

1.1 What Is a Print?

Imagine trying to construct a house, install an HVACR system, or renovate an office building with only a set of written instructions. These instructions might consist of countless pages of text and descriptions of the various tasks to perform. This causes a number of problems. First, individual details and instructions can easily be overlooked when reading the material. Second, the interpretation of the material varies greatly depending on the individual reading it. Third, a large amount of time is required to create these written instructions. Finally, there is often no uniformity between two sets of instructions written by two individuals. In an effort to resolve these issues, we turn to the old adage, "A picture is worth a thousand words."

A ***print***, **Figure 1-1**, is a graphical representation of an architect's or engineer's design. It is used to convey the ideas to the skilled worker performing the construction work. Prints use a standardized set of symbols and notations to convey this information in a concise and accurate manner. This allows the workers to precisely duplicate in the physical world what originally existed in the mind of the project designer.

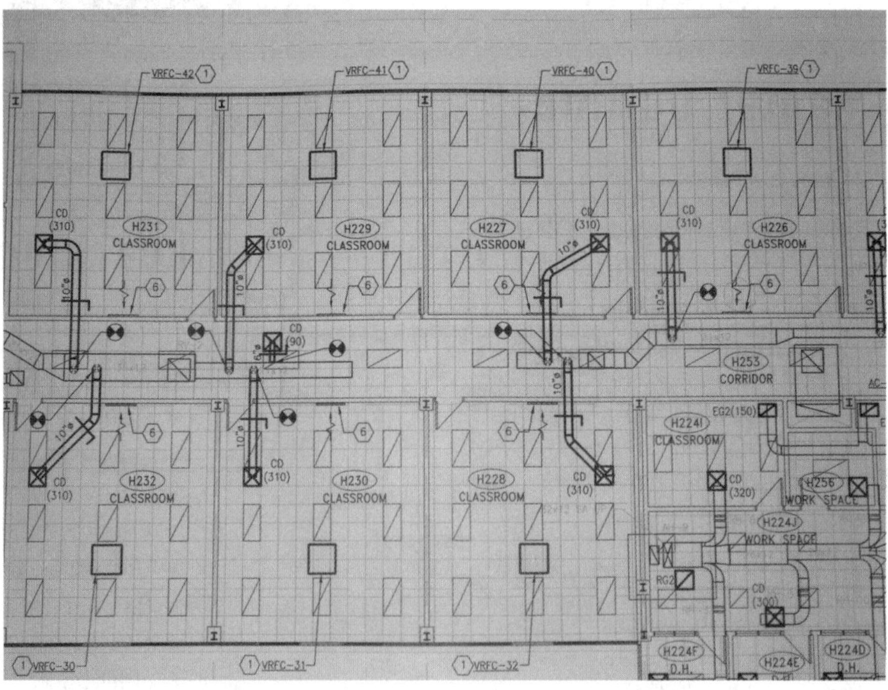

Goodheart-Willcox Publisher

Figure 1-1. A portion of an air-conditioning print.

It would be unpractical and confusing to put all the information from a complete set of prints into a single print. There are individual drawings for each of the building trades, including electricians, plumbers, carpenters, and HVACR installers, where each contractor can focus on the jobs for which they are responsible. As part of the construction process, all of the contractors communicate with each other on a continuous basis to ensure all the construction pieces fit together. Failure to communicate effectively can result in having to redo portions of the work, causing project delays that can be costly.

1.2 How Are Prints Created and Used?

Blueprints are the result of a long design process. In the construction industry, the process may begin with a company executive deciding that the company's office space needs a new look or with a newly married couple thinking over plans to build their dream house. Although these individuals have general ideas regarding the appearance or design of the new space, an architect must become involved to put these ideas down on paper. Architects are experts in construction techniques and creating spaces that are aesthetically pleasing while focusing on the needs of its occupants. The architect meets with the owners, or their representatives, and generates rough sketches representing possible designs for the finished project.

Once a final design is established and approved, the architect prepares original construction drawings that become the guidelines to be followed during all phases of the project. Copies of these prints or plans are submitted to a general contractor, commonly known as the GC, who will oversee and supervise all of the individual trades working on the job.

Sets of plans are distributed to each of the trades as well. Each trade receives a set of prints related directly to the specific tasks that the trade is to complete. It is from these prints that the individual contractors can price the job and submit bids. For example, a plumbing contractor receives a set of plumbing prints, while the HVACR contractor receives the air-conditioning and ventilation prints. A complete set of drawings should be on the jobsite at all times for reference and is usually located on the desk of the GC.

1.2.1 Drawings and Prints

In most cases, construction drawings are created using CAD (computer-aided design) software. Drawings exist as electronic files. A print is created when an electronic drawing file is printed.

In the past, drawings were hand-drawn. They would then be copied using various methods, and the copies were distributed to the construction workers. Prints back then consisted of white lines on blue paper and were called "blueprints." More recently, drawings were copied as blue lines on white paper and were called "blueline drawings" or "blueline prints." See **Figure 1-2.** Although these methods are no longer common, "prints" may still be referred to as "blueprints" or "blueline prints."

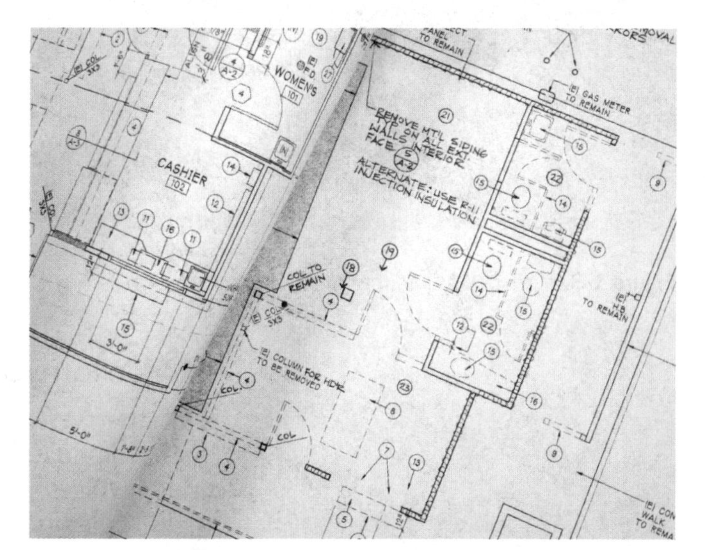

Chad McDermott/Shutterstock.com

Figure 1-2. Sample of a traditional blueprint, or blueline drawing.

Today, prints are created by printing the electronic drawing file from a computer. Advances in technology have also helped to reduce the need for paper prints altogether. Many contractors and workers rely on their tablet computers and other electronic devices, which store the drawings needed for the particular project. Because blueprints are a thing of the past, the terms "drawings," "plans," and "prints" will be used interchangeably throughout this book.

1.3 Why Are Prints Needed?

There is a great deal of information contained in a print, which are rather small. Prints are used for a number of purposes:

- Visual representation of the proposed desired result, thereby reducing any ambiguities that may arise from written descriptions and/or instructions.
- Uniform depiction that can be easily understood by those trained to read prints.
- Means by which the work of all related trades can be integrated.
- Scaled drawing from which individual contractors can accurately calculate job costs, including materials, labor, profit, and time requirements.
- Printed record of what must be accomplished, which aids in the resolution or clarification of any discrepancies that may arise during the construction process.

1.3.1 Visual Representation

It is much easier to determine what must be accomplished if a worker can see what needs to be done. A print provides this visual image in a uniform manner, reducing the chances of incorrect assumptions by field workers. Consider a suspended, dropped ceiling in an office area, **Figure 1-3**. Ceiling real estate may include supply registers, air-conditioning return grilles, lighting fixtures, surveillance cameras, fire sprinkler heads, and audio-visual equipment. The print provides detailed information regarding what goes where when the suspended ceiling grid components are located and mounted. There is then no confusion as to where a particular supply register or light fixture needs to go.

Goodheart-Willcox Publisher

Figure 1-3. A suspended ceiling in an office space.

1.3.2 Uniform Depiction

A print serves as a road map that, when properly followed, leads to the desired destination. Thus, prints must be presented in a manner understood by all involved parties to properly guide the work to the correct end result.

All construction drawings use a similar set of elements and devices to represent the structure being built. Symbols, abbreviations, and line types are relatively consistent in all drawings. See **Figure 1-4**. This consistency

Line Types Commonly Used on Construction Drawings	
Property line	Identifies the boundaries of a building lot.
Border line	Thick, continuous line used around the border of the drawing.
Object line	Thick, continuous line identifies objects such as walls, windows, doors, foundations, and other physical components of the building.
Hidden line	Thin, dashed line identifies the edge of an object located behind another object in the view shown in the drawing.
Center line	Thin line of alternating long and short dashes used to mark the locations of centers of circular features, column lines, and lines of symmetry.
Dimension line **Extension line** Extension line 12'-4" Dimension line	Thin lines used with dimension values to show the "real life" distance between two points.
Section cutting line B A4	Identifies the place in the drawing from which a section drawing is viewed, and identifies the location of the section view in the set of plans.
Break lines	Thick or thin lines identifying a break in an object.

Goodheart-Willcox Publisher

Figure 1-4. Construction drawings use a standard "alphabet of lines" to effectively describe a structure. Line types differ in thickness and in the use of various patterns and lengths of dashes.

allows the construction drawings for any project to be read easily by an experienced builder.

It is sometimes difficult to identify items represented or determine what one needs to do. However, just as a road map has a legend that includes symbols denoting points of interest, a legend may be included to describe the symbols and conventions in a particular drawing. A ***legend***, **Figure 1-5**, is the portion of the print that tells the reader what each symbol in the drawing represents, without having to label or describe them multiple times. It allows the print reader to identify components and easily pull information from a print, **Figure 1-6**.

1.3.3 Integration of Individual Trades

There are often multiple contractors and subcontractors on a construction or renovation project. These can include designers, carpenters, electricians, plumbers, masons, painters, and HVACR installation personnel. The needs of the other trades are not always known or considered, so project plans and prints keep the project moving forward with limited obstacles.

KEY TO SYMBOLS

DUCTWORK
EXISTING DUCTWORK TO REMAIN
EXISTING DUCTWORK TO BE REMOVED
NEW DUCTWORK
DUCT UNDER POSITIVE PRESSURE (SUPPLY AIR OR FAN DISCHARGE)
DUCT UNDER NEGATIVE PRESSURE (RETURN, EXHAUST OR OUTSDIE AIR)
(400) SUPPLY DIFFUSER, (AIR QUANTITY IN CUBIC FEET PER MINUTE)
(400) RETURN REGISTER, (AIR QUANTITY IN CUBIC FEET PER MINUTE)
RECTANGULAR DIFFUSER WITH BLANKING PLATE
S.R. S.G. SUPPLY REGISTER SUPPLY GRILLE
R.R. R.G. RETURN REGISTER RETURN GRILLE
LOUVER & SCREEN
LINEAR DIFFUSER
10x6 TR (150) 10" BY 6" TOP REGISTER 150 CFM SUPPLY AIR
10x6 TR(TG) (150) 10" BY 6" TOP REGISTER (TOP GRILLE) 150 CFM RETURN AIR
10x6 BR(BG) (150) 10" BY 6" BOTTOM REGISTER (BOTTOM GRILLE) 150 CFM RETURN AIR

PIPING
EXISTING PIPING TO REMAIN
EXISTING PIPING TO BE REMOVED
NEW PIPING
C CONDENSATE PIPING
HWS HOT WATER SUPPLY
HWR HOT WATER RETURN
CHWS CHILLED WATER SUPPLY
CHWR CHILLED WATER RETURN
RFGT REFRIGERANT PIPING
GAS GAS PIPING
DIRECTION OF FLOW
C.O. CLEANOUT
F&T TRAP
LOW PRESSURE DRIP
UNION
STRAINER
HB HOSE BIBB
FLOOR DRAIN, F.D.
FLEXIBLE CONNECTION
AUTOMATIC AIR VENT

VALVES
TRIPLE DUTY VALVE

INSTRUMENTATION
T THERMOSTAT
H HUMIDISTAT
F FIRESTAT
G GAS DETECTOR
S REMOTE TEMPERATURE SENSOR
DD DUCT DETECTOR
RM REFRIGERANT LEAK DETECTOR
TH TEMPERATURE AND HUMIDITY RECORDER
LD LIQUID DETECTOR
PRESSURE GUAGE
F FLOW SWITCH
THERMOMETER
EP ELECTRIC PNEUMATIC SWITCH
PE PNEUMATIC ELECTRIC SWITCH
ESAP ENVIRONMENTAL SENSOR ALARM PANEL
P MANUAL POTENTIOMETER

EQUIPMENT
BASEBOARD RADIATION
CABINET HEATER
DUCT MOUNTED UNIT HEATER
HORIZONTAL UNIT HEATER
PUMP
COILS
CENTRIFUGAL FAN OR PUMP
ELECTRIC HEATER
CEILING EXHAUST FAN
CENTRIFUGAL ROOF FAN
VARIABLE FREQUENCY DRIVE

ANNOTATION
CONNECT TO EXISTING
LIMIT OF REMOVAL

A

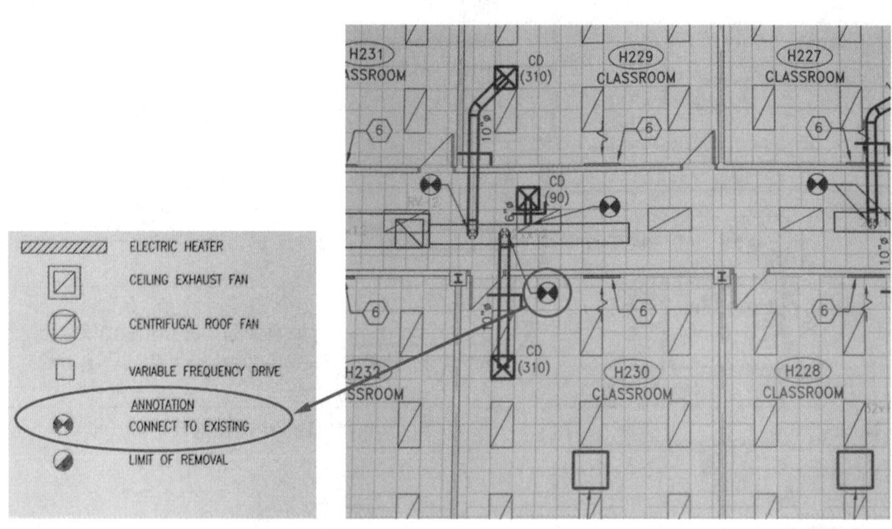

B

Goodheart-Willcox Publisher

Figure 1-5. A—Portion of a legend from an HVACR plan. B—The symbol in the legend that indicates the new duct section must be connected to the existing duct.

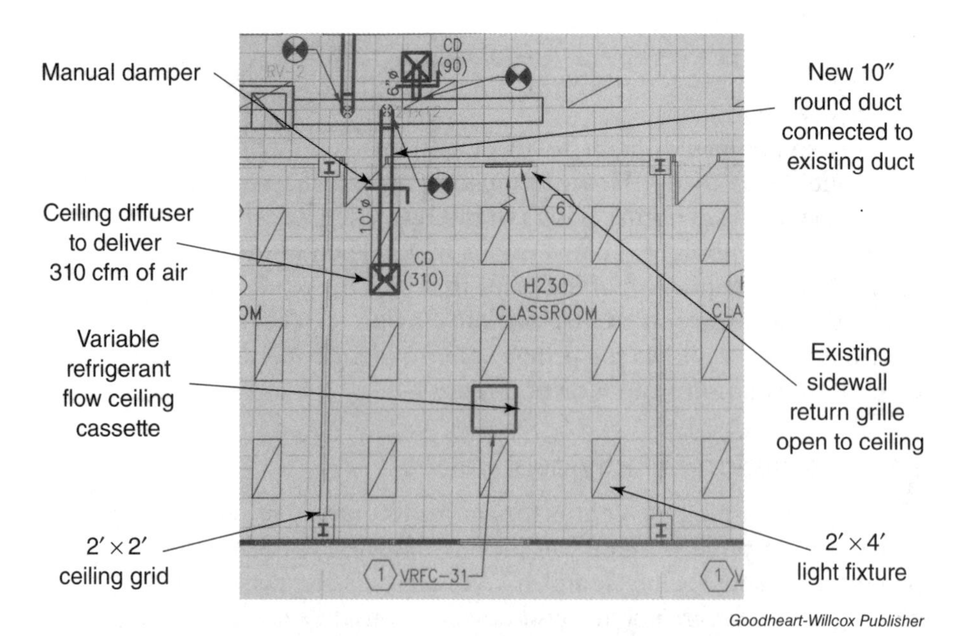

Goodheart-Willcox Publisher

Figure 1-6. Various pieces of information contained in the plan.

A print is a useful tool to help foresee and avoid potential problems that may arise during the construction process. For example, consider a 12″-high duct section from an air-conditioning system passing through a ceiling with a 14″ clearance. It is unlikely the electrical contractor can mount a recessed light fixture in a space to be occupied by the air duct. Either the duct or the light fixture must be relocated. The general contractor, who is the authority on the jobsite, must ultimately decide which of the two is moved.

1.3.4 Calculating Job Costs

Contractors rely on prints to estimate the total job cost, which is the total sum of all project-related expenses that must be paid by the contractor to complete the project as described. This is different from the quoted price that is presented to the customer.

The print provides information regarding the complexity of the job and the work conditions. For example, the process of running a branch duct or installing a ceiling diffuser can be easy or involved. It depends on how the work needs to be performed, the building materials used in the structure, and where the duct or diffuser must be located. Is the ceiling diffuser installed in a suspended 8′ ceiling, or located in an angled, sheetrock ceiling 18′ in the air? The print provides this information.

Another important element of estimating costs is determining the amount of material and equipment needed to successfully complete the task. Consider the example of registers, which are decorative grilles that serve as supply air outlets from an air-conditioning system. With a print, the HVACR job estimator can simply count the number of each supply register type needed to determine the costs involved.

Some prints include material lists and schedules. ***Material lists*** provide information about materials or items required to complete each project area. Each material type or item is represented by an identifying mark or label. See **Figure 1-7.** By providing these abbreviations, the material list eliminates the need for repetitive labeling on the print, which would result in clutter. Schedules are similar to material lists, but schedules are used to provide information, such as model numbers, capacities, flow rates, and electrical criteria, about the specific pieces of equipment used on the job.

1 24 GAUGE GALVANIZED SHEET METAL

2 10" ROUND ATD COLLAR

3 8" ROUND ATD COLLAR

Goodheart-Willcox Publisher

Figure 1-7. A portion of a material list on an HVACR plan. The number in the diamond is the identifying symbol or label to represent each material

BETWEEN THE LINES

Project Items

Using a combination of the plan, material list, and schedules, the estimator, or the individual responsible for pricing, can determine what is needed to execute the project.

1.3.5 Printed Record of Work

A working print resolves any discrepancies that may arise during construction. Unless any approved and documented changes have been made to the original set of prints, individual contractors are responsible for performing the work as outlined by the print without deviation.

Some changes made to the original plan or scope of the project may require additional labor or materials to complete. These revisions must be explained in detail on official documents referred to as *change orders*. Change orders include all of the pertinent information regarding the modifications, as well as information pertaining to any additional payments made to the contractor related to the change.

1.4 Construction Drawings

Buildings are constructed using a set of construction drawings, which completely describe the finished structure. An architect or a builder prepares the drawings. Often, the construction drawings must be submitted as part of a building permit application and must be approved before construction can begin.

A set of construction drawings for a small residence may consist of only a few prints, while a set for a large commercial building may include hundreds of prints. The number of prints included in the set is determined by the complexity of the project. **Figure 1-8** lists some common prints that may be included in a set of drawings. Commercial projects may include multiple prints in each category. For example, the set of drawings for a multistory office building will likely include a floor plan, electrical plan, mechanical plan, and structural plan for each building floor.

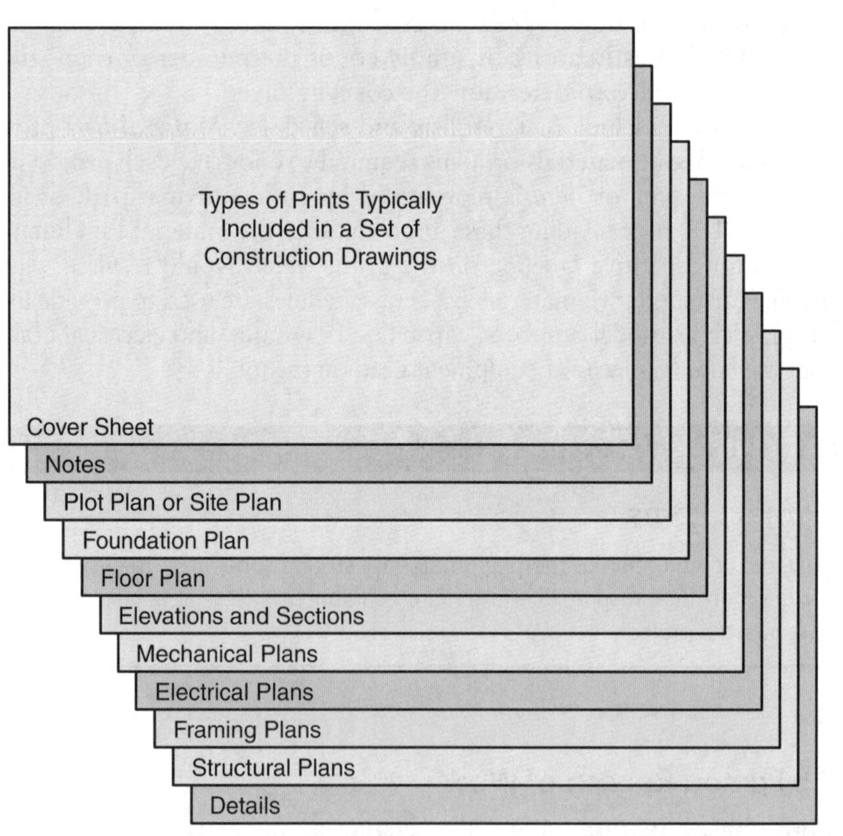

Figure 1-8. A set of construction drawings may consist of only a handful of prints or hundreds of prints. The complexity of the project determines the number of prints required to fully describe the structure.

A type of drawing called a ***plan*** shows features of a structure from above. Plans are common on most prints. There are many types:

- Floor plans
- Site plans
- Foundation plans
- Mechanical plans
- Framing plans
- Structural plans
- Sections
- Details

1.4.1 Floor Plans

A ***floor plan***, **Figure 1-9**, is a scaled diagram showing all of the parts of the finished project. Floor plans show the location of the walls, plumbing fixtures, electrical fixtures, windows, and doors of a project on a print. For larger prints, floor plans are drawn for each of the building trades. In this case, an HVACR contractor can use a set of prints that includes only the air-conditioning and heating portion of the job, while the plumbing contractor uses a set that includes only the plumbing portion. The general contractor possesses a set of prints on the jobsite that incorporates the tasks of all building trades.

Floor plans are created using a ***plan view***, which is a view from the top of the project. In this type of view, the overall footprint of the structure is shown, while only the tops of the structure's walls are drawn. An ***elevation view***, **Figure 1-10**, is created to see the structure from the side.

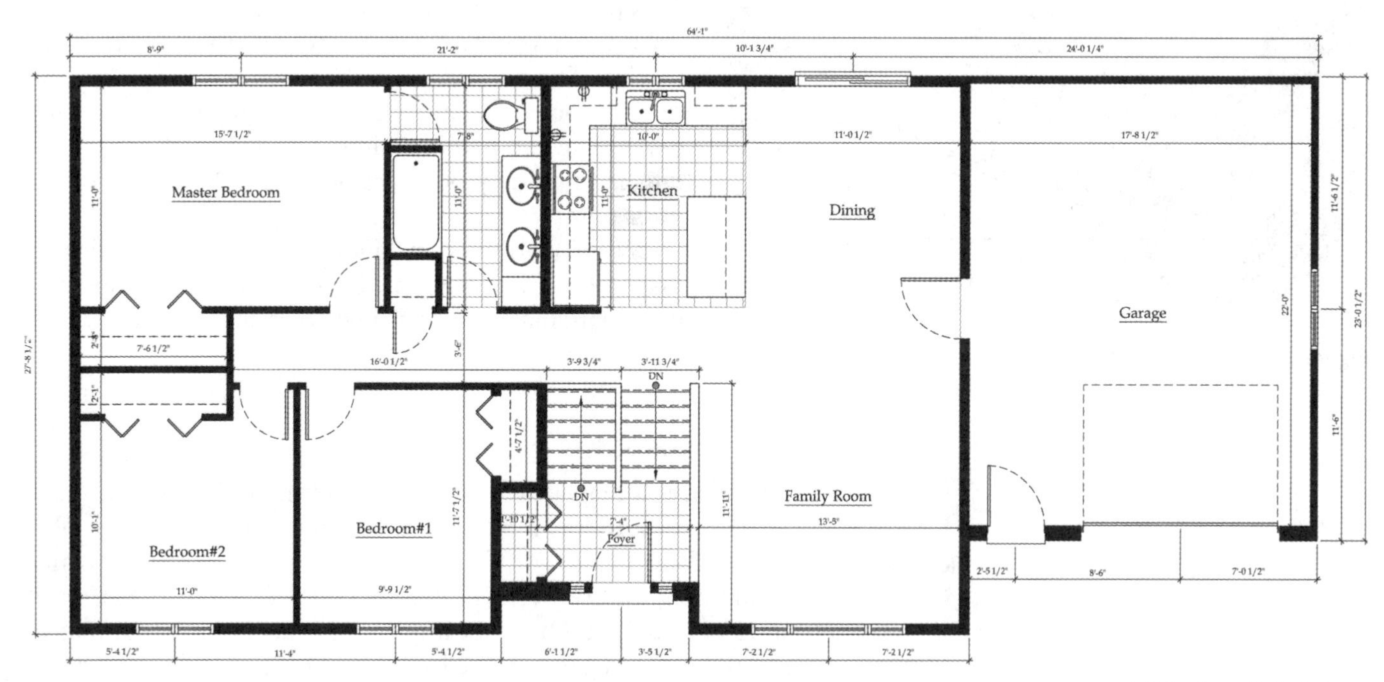

Evita Van Zoeren/Shutterstock.com

Figure 1-9. Sample floor plan.

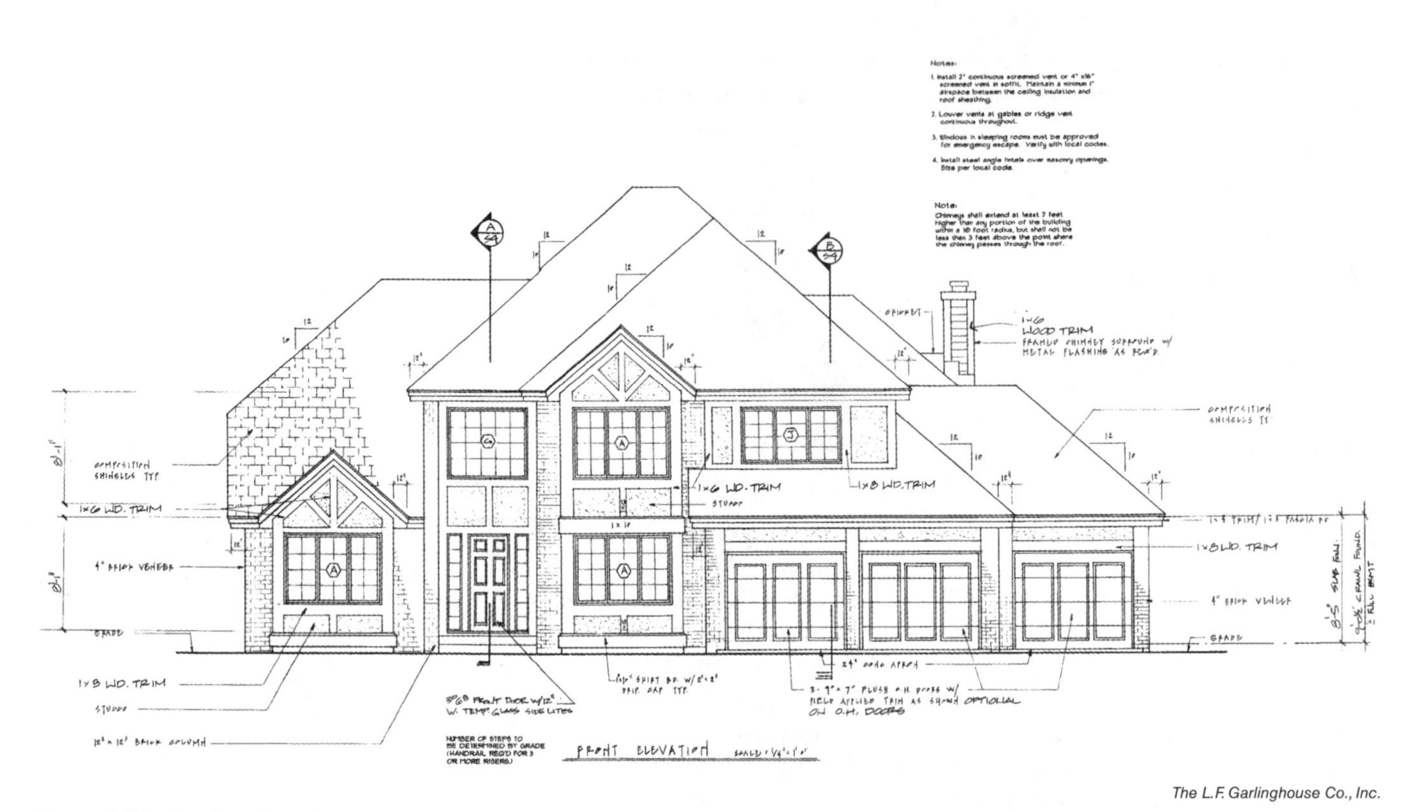

The L.F. Garlinghouse Co., Inc.

Figure 1-10. An elevation view.

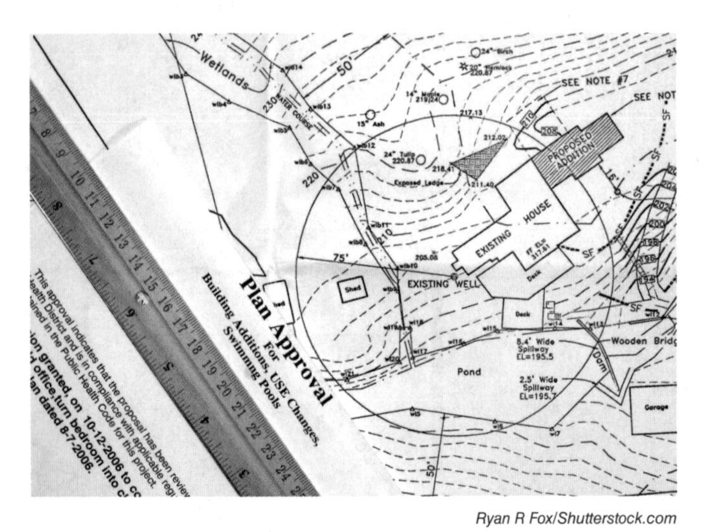

Ryan R Fox/Shutterstock.com

Figure 1-11. A site or plot plan.

1.4.2 Site Plans

When a construction project involves the structure itself and the surrounding grounds and property, a ***site plan***, or ***plot plan***, is prepared. See **Figure 1-11**. This plan includes additional details, such as the trees, driveways, sewage lines, electrical service, and water service. The topography, shape of the property, and orientation of the structures, are also part of most site plans. The site plan provides a scaled rendition of the finished property but does not show the details of the floor plans.

A site plan often includes a ***vicinity map***. A vicinity map shows the location of the project with respect to major landmarks, such as highways, in the area in which it is located.

1.4.3 Foundation Plans

A ***foundation plan*** is a plan-view drawing of a structure's foundation. In cases where the structure has a basement, the foundation plan will also include a detailed floor plan of the basement with all partition walls, doors, and plumbing fixtures. These plans usually include the location of any mechanical equipment. One very important aspect of the foundation plan is the information regarding the locations of structural support elements, such as piers, columns, and the perimeter walls of the foundation.

Foundation plans include any penetrations and openings needed in the foundation walls to facilitate water or electrical power into the structure, or to install windows, vents, or doors.

1.4.4 Mechanical Plans

Mechanical plans are project drawings that provide information about the mechanical equipment and systems. Mechanical plans typically include the heating, ventilation, air-conditioning, and refrigeration (HVACR) systems. On larger projects, these plans can include other large elements, such as elevators and escalators.

These plans can provide a set of drawings that detail electrical or plumbing work that falls under the mechanical umbrella, such as gas, water, and electricity. ***Electrical plans*** show electrical wiring, fixtures, lighting plans, and ceiling plans. ***Plumbing plans*** illustrate the hot- and cold-water system, sewage disposal, and fixture locations. In other cases, individual sets of plans are created for each of these areas.

1.4.5 Framing Plans

Framing plans are used by carpenters and show the details of structural and support elements. The plan also includes information regarding the support beams, joists, roofing materials, trusses, and rafters. Framing plans are also useful in determining which walls in a structure are classified as *load-bearing*—those that support the structure.

1.4.6 Structural Plans

A ***structural plan*** is a drawing, or set of drawings, that provides valuable information about how a building is constructed. Information in a structural plan often addresses structural calculations, structural weight loads, foundation information, structural material limitations, safety factor calculations, and structural footing data. In many instances, the HVACR contractor installs equipment in an already-built structure, but the information helps the contractor plan out and price a project.

1.4.7 Sections

Sections or cross-section drawings provide a view of components not seen from the outside. They show a structure as if it were cut through to see what occurs on the inside. Details, such as how construction elements arc layered or assembled, are included. This type of cross-sectional view provides a better understanding of the assembly.

1.4.8 Details

Details provide an exploded view of particular building components that require more uncommon assembly. These components may include arches, cornices, or retaining walls. Such drawings are particularly useful to show how project elements are to be assembled, connected, positioned, or otherwise constructed. The drawings are often large and not drawn to scale because they are intended to show how a particular completed connection or assembly looks.

1.5 Schedules

On a typical print, information regarding the equipment and materials to be used is included on the drawings themselves. This information is referred to as the equipment specifications. ***Equipment specifications*** often include information such as the product description, manufacturer, and model.

Schedules list the required components for a project and provide information on each component, such as the item, size, and number required. This gives print readers a comprehensive understanding of the project. The different types of schedules can include door schedules, lighting fixture schedules, and duct fitting schedules. Schedules will be discussed in more detail in Chapter 11, *HVACR Schedules*.

A typical duct-fitting schedule, **Figure 1-12** is a table that identifies the different types of takeoff fittings used on a particular job, as well as important information regarding them. For example, duct takeoffs identified as #1 on the plan is of the 6″ airtight with damper (ATD) variety and made from galvanized sheet metal. There will be 15 of these fittings required for the job. A number of different schedules are used on the job, and they help individuals tasked with gathering and compiling data from the prints. Schedules commonly found on an HVACR print include those for air handlers, condensing units, packaged units, supply air registers, return air grilles, thermostats, and sensors.

Duct Takeoff Fitting Schedule				
Plan Number	**Material**	**Size**	**Quantity**	**Note**
1	Galvanized	6″ ATD	15	Air Tight w/Damper
2	Galvanized	8″ ATD	9	Air Tight w/Damper
3	Galvanized	10″ ATD	7	Air Tight w/Damper
4	Galvanized	6″ CLINCH	4	Clinch w/Damper
5	Galvanized	8″ CLINCH	4	Clinch w/Damper

Goodheart-Willcox Publisher

Figure 1-12. A duct take-off fitting schedule.

Summary

- Prints are graphical representations of an architect's or engineer's design that conveys information on the outlined work. Prints use standardized symbols and notations to convey information.
- In most cases, construction drawings or prints are now created using CAD (computer-aided design).
- Prints provide a visual representation and uniform depiction of a project, thus reducing the possibility of errors. A legend is also included in prints to give further instruction.
- A print can integrate individual trades and help determine potential problems during the construction process.
- Contractors rely on prints to properly estimate job costs and determine the amount of material and equipment needed to successfully complete a project.
- A material's list provides information about which materials are needed to complete each project area.
- A working print resolves any discrepancies that arise during the project. Any revisions to be made must be explained in detail.
- Floor plans show the location of walls, windows, doors, and other features of the building. Elevation views show a side view of a building's exterior.
- Site or plot plans show the property on which the structure will be situated. These plans often include a vicinity map, which shows the property's location with respect to cross roads, highways, and other local landmarks.
- Foundation plans are shown as a plan view and are used to determine a structure's foundation.
- Mechanical plans are projects that provide information about mechanical equipment and systems, especially HVACR systems. Electrical plan and plumbing plan components may be included in these based on the size of the project.
- Framing plans are used by carpenters to show the details of structural and support elements.
- Structural plans are drawings that provide valuable information on how a building is constructed.
- Sections are cross-sectional and show components not seen from the outside, while details provide an exploded view of uncommon building components.
- Schedules are used to summarize equipment specifications.

Chapter Assessment

Name ________________________ Date ____________ Class ________________

Know and Understand

_______ 1. The print is a graphical representation of the _____.

A. work a tradesperson wishes to perform

B. designs an engineer wishes to convey to the tradesmen

C. work the engineer will personally perform

D. All of the above are correct.

_______ 2. The legend is the portion of the print that tells _____.

A. what each symbol on the print represents

B. the tradesmen how much material the job will use

C. Both A and B are correct.

D. Neither A nor B is correct.

_______ 3. Prints are needed for all of the following reasons except providing a _____.

A. visual representation of the desired completed project

B. period within which the work must be completed

C. means to coordinate all of the individual trades

D. printed record of what the job must accomplish

_______ 4. The person who oversees the individual trades on the job is the _____.

A. architect

B. design engineer

C. general contractor

D. None of the above.

_______ 5. All of the following will likely be found on a vicinity map except _____.

A. nearby highways

B. a compass

C. the topography of the property

D. cross streets bordering the property

_______ 6. Which of the following will most likely show what a structure will look like from the side?

A. Site map

B. Plan view of the structure

C. Elevation view of the structure

D. Floor plan

_______ 7. A schedule will most likely *not* include which of the following pieces of information?

A. The total cost of the equipment

B. The manufacturer of the equipment

C. The total number of pieces of equipment required

D. The model numbers of the equipment

Apply and Analyze

1. Why are detail drawings provided as part of a complete set of project drawings?

2. Why are project drawings provided as opposed to written descriptions of the work to be performed?

3. How does the legend benefit the print reader?

4. Describe three types of plan-view drawings used in the trades.

Critical Thinking

1. How can prints be used to avoid potential jobsite problems that might arise between different trades involved?

2. When would a site plan prove to be more useful than a floor plan?

Large Prints Activity

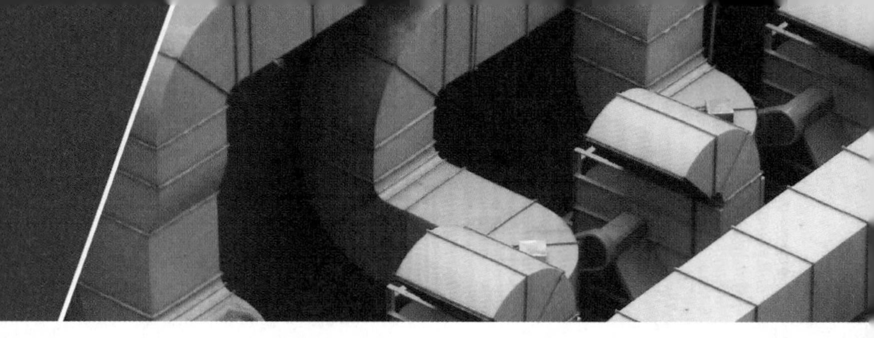

Name ______________________ Date ____________ Class ______________

Practice Using Large Prints

Use the prints in the Large Prints supplement to answer the following questions. Note that the sheet number of a print is found in the lower-right portion of the title block along the right edge of the print.

1. Find an example of a schedule on one of the prints. Write the title of the schedule and the sheet number of the print.

   ____

   ____

2. Find an example of a legend on one of the prints. Write the title of the legend and the sheet number of the print.

   ____

   ____

3. Find an example of a detail on one of the prints. Write the title of the detail and the sheet number of the print.

   ____

   ____

Math Review—Arithmetic

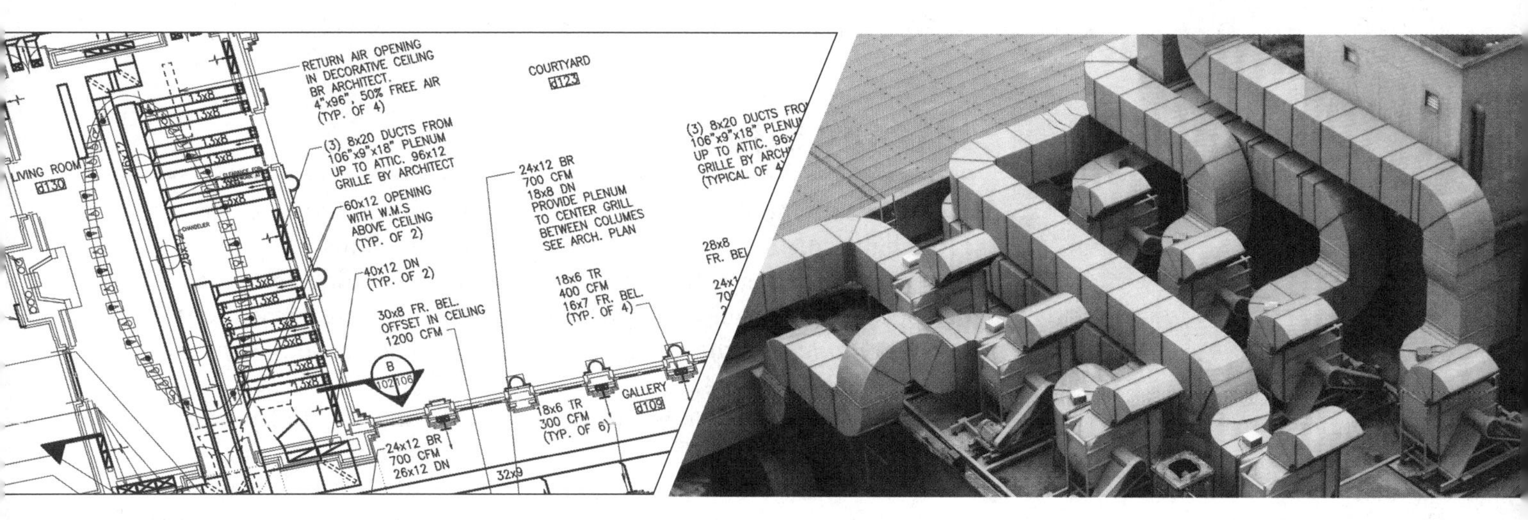

Chapter Outline

Technical Terms

addend
airflow
borrowing
cube
denominator
difference
dividend
divisor
exponent
factor
grouping
improper fraction
inverse
lowest common denominator
mixed number
numeracy
numerator
percentage
product
quotient
ratio
reducing
remainder
root
rounding
square
sum

Learning Objectives

After studying this chapter, you will be able to:

- Explain the importance of having a working knowledge of basic mathematical operations.
- Perform basic mathematical operations of addition, subtraction, multiplication, and division with whole numbers.
- Perform mathematical calculations using radicals and exponents.
- Perform basic mathematical operations of addition, subtraction, multiplication, and division with fractions.
- Find the lowest common denominator between fractions.
- Reduce fractions to their lowest terms.
- Identify improper fractions and mixed numbers and convert between the two.
- Convert fractions to their decimal equivalents.
- Convert decimals to their fraction equivalents.
- Explain mathematical relationships in terms of ratios and percentages.

Introduction

Working with numbers is a valuable skill for an individual to possess. From balancing a checking account to building a house, a solid understanding of numbers and their relationships will help to make complicated problems easier to solve. There are strong correlations between ***numeracy***, the ability to work effectively with numbers, and employability.

The ability to work with mathematical concepts is also needed in the building trades. Tradespeople must often extract or interpret information from a print using math. This chapter provides an overview of mathematical operations, along with some of their industry-related applications. This will allow you practice working effectively with numbers and prints.

2.1 The Importance of Math

Individuals in the building trades must be able to read prints and extract information. This often requires a basic mathematical knowledge. A few examples where math is used to interpret or read prints include:

- A contractor installing wood flooring in a given space must be able to determine how much material is required to cover the floor, as well as how many feet of molding are needed along the edge of the room.
- An electrician running an electrical conduit around a shop area for lighting and receptacle circuits must know how much wire and conduit is required.
- A painter must know how much paint and primer will be needed to cover the walls and ceilings of a given space, as rendered in the project plan.

- An air-conditioning system installer must know how far apart the indoor and outdoor units of a split air-conditioning system should be to determine how long the refrigerant line set will be, and whether this length falls within the manufacturer's guidelines.

These four examples provide a few scenarios that occur on a typical construction site. A tradesperson must be able to work with simple mathematical formulas to determine various parameters that may present themselves both on the jobsite and off.

Consider the real-life situation where a replacement refrigerator for a home needs to be purchased. A number of factors might need to be considered:

- Will the new refrigerator fit in the space occupied by the old appliance?
- Will the new refrigerator fit through the existing doors and passageways?
- Are the internal volumes of the refrigerator/freezer compartments comparable to those of the older appliance?
- How much energy will be saved by replacing the appliance?
- How much will the new appliance cost, taking into account any discounts or other incentives the store might be offering?

In the HVACR field, mathematical calculations must be made on a daily basis. It can range from determining how much sheet metal is required to fabricate a run of ductwork to calculating the final cost for an air-conditioning system installation after all discounts and rebates have been accounted for.

2.2 Common Myths about Math

Many people are uncomfortable with math, or even have "math anxiety." This is prevalent especially when using math in our daily lives or on the job. The way to improve your confidence in math is by learning and practicing the basic concepts until you become proficient. You can then apply your learning to the next concept.

Before you begin to practice math, first remove any misconceptions you may have about the topic. The following section serves to disprove some common myths about math.

2.2.1 Myth #1: Every Problem Has Only One Solution

There is often only one correct answer, but the path to that solution may be different for each individual. We all approach problems differently but often arrive at the same answer. For example, when you and your friend drive from your school to the store, you might take different routes. You both arrive at the same destination, but the path to the result was different. Who is right? Who is wrong? When it comes to math, there is no right method or wrong method as long as the method is logical and leads to the correct answer.

2.2.2 Myth #2: You Have to Be a Genius to Know Math

Many people say they are not math people when given a problem that requires using math. They may also believe they must be a genius to work with numbers in a coherent manner, or that being intelligent requires excelling in math. Although Charles Darwin, Michael Faraday, Alexander Graham Bell, and Thomas Edison are considered geniuses of their era, each struggled with math.

These individuals, as well as many others, succeeded by:

- Appreciating the power of mathematics
- Knowing what math can accomplish
- Practicing mathematical basics to become knowledgeable
- Asking many questions
- Looking at the big picture—what goal are you trying to reach?

2.2.3 Myth #3: Math Is All about Memorization

Because math has one correct answer, we are trained to believe math must be memorized. Nothing could be further from the truth. Because every problem or situation is unique, memorization is not the most efficient route. It is very important to learn from previous challenges. So, when faced with a similar situation, the road to the solution will likely be far less bumpy. By looking at the challenge ahead and thinking about the best way to solve it, the solution will be far easier to find.

2.2.4 Myth #4: You Must Follow the Rules as Presented in Books

Many say, "Rules were made to be broken." Quite often, a solution method presented in a book, including this one, is one of many possible approaches that can be taken. One way to improve in math is to read problems and try to come up with alternate ways of solving the same problem. Challenge yourself by asking, "How would I have approached this problem if it were presented to me?" By coming up with your own way of tackling problems, you take ownership of them and are more likely to master the methods used.

2.3 Whole Number Operations

Becoming proficient in the use and manipulation of more involved mathematical expressions starts with becoming proficient in the four basic mathematical operations: addition, subtraction, multiplication, and division.

2.4 Addition

Addition is helpful when determining the cumulative quantity of a substance, material, product, or service that will be required to perform a task. The formal term used to describe the numbers being added is ***addends***. The term used to describe the answer obtained from the addition process is ***sum*** or *total*.

Many applications require this mathematical operation, some of which are used in our everyday lives. Some common uses for addition include:

- Determining the total amount of money needed to pay our monthly bills.
- Determining the total cost of materials needed to complete a project.
- Adding up the loose change in your pocket.
- Adding up a company's outstanding invoices to determine how much revenue the company can expect to receive.
- Adding multiple linear measurements together to determine the amount of piping or similar material needed for an HVACR installation.

2.4.1 Rounding

Rounding, or *rounding off*, is the process of altering a number to a higher or lower value for more convenient, but less exact calculations. For instance, if the total cost of eight pieces of equipment needed to complete a job needs to be determined, it would not be a difficult task to simply add the numbers together in your head or use the calculator on your phone. However, it may be helpful for the contractor to initially round the numbers to get an estimate of the equipment cost.

Consider a contractor who needs to purchase eight pieces of equipment. In this scenario, rounding can be used to help get a rough estimate of the equipment cost. The individual cost for each piece of equipment is as follows:

System 1: $600
System 2: $500
System 3: $800
System 4: $200
System 5: $700
System 6: $400
System 7: $300
System 8: $400

The first step in estimating the total is to round, either up or down, the numbers being added together. Because the addends range from $200 to $800, values of $0, $500, and $1000 are best to work with:

- Values that are equal or less than $200 → Rounded to $0
- Values from $300 to $700 → Rounded to $500
- Values that are equal to or greater than $800 → Rounded to $1000

Next, round the equipment prices according to the preceding guidelines. The rounded cost for the equipment is provided here:

System 1: $500
System 2: $500
System 3: $1000
System 4: $0
System 5: $500
System 6: $500
System 7: $500
System 8: $500

Once you have rounded each value, add them together to determine the estimated sum. Add the numbers together.

$500 + $500 + $1000 + $0 + $500 + $500 + $500 + $500 = $4000

From this calculation, we find that the cost of the equipment required to complete this job is $4000. To compare this total to the precise total, add together the actual equipment costs.

$600 + $500 + $800 + $200 + $700 + $400 + $300 + $400 = $3900

The total cost for the equipment required to complete this job is $3900. By estimating the total first, and noting that they are relatively close to each other, we can be confident that the result of our actual calculation is correct.

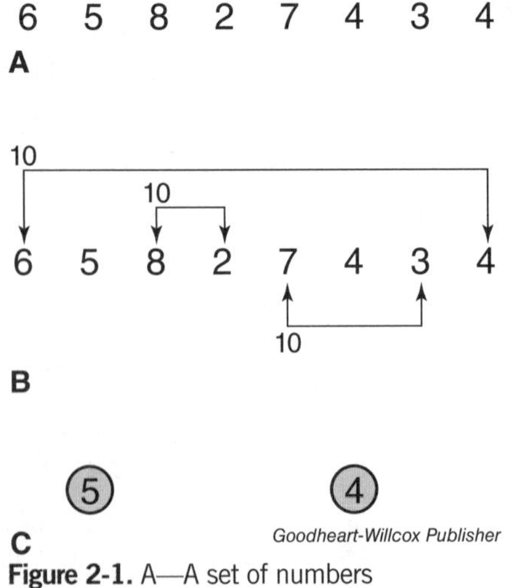

Goodheart-Willcox Publisher

Figure 2-1. A—A set of numbers representing the number of gallons of paint required for each area. B—The numbers have been grouped to form three bundles of 10. C—The ungrouped numbers.

2.4.2 Grouping

In addition to rounding off, ***grouping*** is a similar technique that can be used to aid in the addition process. The process of grouping involves organizing the numbers in smaller groups that each add up to a number easier to work with. Although common group values are 10, 100, or 1000, any group value can be used. The number of groups is then multiplied by the group value to arrive at the result. Consider the original numbers that represent the number of gallons of paint required for each of the areas in the office building, **Figure 2-1**.

These numbers can be grouped together in bundles of 10 to make the addition process easier. Three groups of 10 have been formed. The ungrouped numbers, namely 5 and 4, are added separately. By grouping and forming these bundles, we can readily determine the amount of paint needed:

10 + 10 + 10 + 9 = 39 gallons of paint

2.4.3 Determining Total Airflow

An air-conditioning technician needs to determine the total amount of airflow introduced into the offices on one side of a building. ***Airflow*** is the amount of air per unit of time that flows through a component and is measured in cubic feet per minute (cfm). The eight areas being served have the following individual airflow rates:

Area 1: 410 cfm
Area 2: 380 cfm
Area 3: 275 cfm
Area 4: 525 cfm
Area 5: 605 cfm
Area 6: 490 cfm
Area 7: 525 cfm
Area 8: 490 cfm

We can also round off our numbers to get an estimate on the total airflow. In this case, rounding off to the nearest 50 cfm, or even 100 cfm, would be much quicker and would still provide useful information.

In this example, the numbers are rounded to the nearest 100 cfm value. Thus the rounded-off airflow values will be expressed as:

Area 1: 400 cfm
Area 2: 400 cfm
Area 3: 300 cfm
Area 4: 500 cfm
Area 5: 600 cfm
Area 6: 500 cfm
Area 7: 500 cfm
Area 8: 500 cfm

By grouping the values into bundles of 1000 cfm, we can further simplify the addition process, **Figure 2-2**. There are three groups of 1000 cfm, leaving 700 cfm ungrouped. This gives a total estimate:

1000 cfm + 1000 cfm + 1000 cfm + 700 cfm = 3700 cfm

This estimate is very close to the actual airflow rate of 3800 cfm.

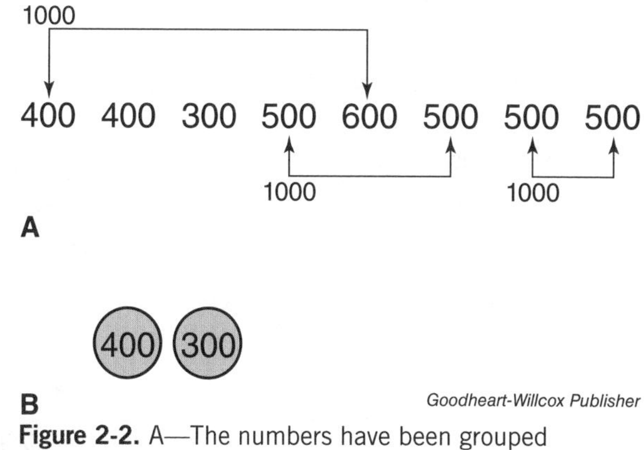

Figure 2-2. A—The numbers have been grouped in bundles, each representing 1000 cfm. B—The ungrouped numbers.

2.5 Subtraction

Subtraction involves removing number values from another to find the ***difference***. When subtracting one number from another, we are actually determining the difference between two or more numbers. We can thus express a mathematical expression that involves subtraction as, "How far is X from Y?"

Rounding for a simple estimation can be used with subtraction. For example, if you were asked to subtract 197 from 498, first subtract 200 from 500. This math helps you learn that the actual result is close to 300.

The first method is to break down the problem into multiple pieces. For example, assume that the difference between 152 and 37 is needed. Mathematically, this would be expressed as 152 – 37 = X. However, this problem can be broken down into two parts:

- Part 1: How far is 37 from 100?
- Part 2: How far is 100 from 152?

Each of these two parts is easier to solve than the original problem.

$$100 - 37 = 63$$

$$152 - 100 = 52$$

For the result:

$$63 + 52 = 115$$

2.5.1 Borrowing

Borrowing is another strategy that is used when subtracting. ***Borrowing*** is the process of subtracting 1 from a place value, and then adding 10 to the next lower place value.

Consider the following subtraction problem:

$$\begin{array}{r} 9418 \\ \underline{-6749} \end{array}$$

In the ones column, the 9 is greater than 8. Because 9 cannot be subtracted from 8 without leaving a negative number, the 8 is increased by 10 (from 8 to 18), and the tens column (the second column from the right) is reduced from 1 to 0.

$$\begin{array}{r} {\scriptstyle 0\;18} \\ 9418 \\ \underline{-6749} \end{array}$$

Next, in the tens column, the 4 is greater than the 0. Since 4 cannot be subtracted from 0 without leaving a negative number, the 0 is increased by 10 (from 0 to 10), and the hundreds column (the third column from the right) is reduced from 4 to 3.

$$\begin{array}{r} {\scriptstyle 3\;10\,18} \\ 9418 \\ \underline{-6749} \end{array}$$

We will examine the hundreds column next. In this column, the third column from the right, the 7 is greater than the 3. Since 7 cannot be subtracted from 3 without leaving a negative number, the 3 is increased by 10 (from 3 to 13), and the thousands column (the leftmost column) is reduced from 9 to 8.

$$\begin{array}{r} {\scriptstyle 8\;13\,10\,18} \\ 9418 \\ \underline{-6749} \\ 2669 \end{array}$$

At this point, the top number has an 8 in the thousands column, a 13 in the hundreds column, a 10 in the tens column, and an 18 in the ones column. These numbers are all greater than the corresponding values in the bottom number. Now, to finish the problem, the bottom values in each of the columns must be subtracted from the corresponding values in the top number.

2.6 Multiplication

Multiplication is the process of adding like numbers numerous times. Consider a painter who needs to paint six rooms that each require three gallons of paint. To calculate the total number of gallons of paint required to complete this job, one could simply add three six times, or 3 + 3 + 3 + 3 + 3 + 3 to get a total of 18 gallons. In this case, we can simply multiply 6 by 3 to obtain 18 as the ***product***. The product is the result of numbers, called ***factors***, which are multiplied together. The expression can be stated as six rooms at three gallons each (6 × 3), or three gallons of paint for each of six rooms (3 × 6).

Multiplication is commutative, meaning the order in which numbers are multiplied together does not matter. In general, multiplication problems are often graphically represented in rectangular form, **Figure 2-3**.

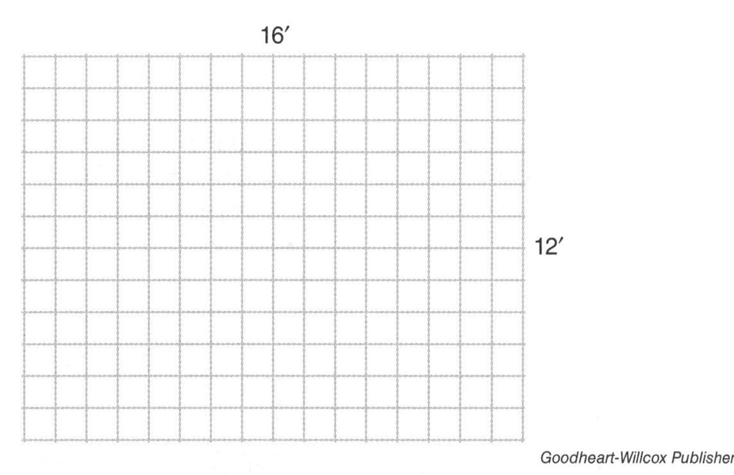

Goodheart-Willcox Publisher

Figure 2-3. This drawing represents a room that is 12′ wide and 16′ long. The grid can be depicted as either 12 rows of 16 ft², or 16 rows of 12 ft².

All multiplication problems are broken down into a series of single-digit multiplication calculations. A basic multiplication chart is provided in **Figure 2-4**.

This chart provides the products for all combinations of single-digit factors. To use the chart, locate one of the factors across the top row of the chart and the other factor in the left-hand column of the chart. The selected row and column will intersect at the product of the two factors. Because the multiplication process is commutative, the order of the numbers is unimportant. Multiplying 8 × 4 will yield the same result as multiplying 4 × 8. See **Figure 2-5**.

Memorizing the multiplication chart can help with multiplying single-digit numbers, as well as performing multiplication with multiple-digit numbers.

$$\begin{array}{r} 16 \\ \underline{\times 12} \end{array}$$

To multiply 12 and 16 together, we start out by multiplying the rightmost number in the bottom factor (in this case, 2) by each of the

X	1	2	3	4	5	6	7	8	9
1	1	2	3	4	5	6	7	8	9
2	2	4	6	8	10	12	14	16	18
3	3	6	9	12	15	18	21	24	27
4	4	8	12	16	20	24	28	32	36
5	5	10	15	20	25	30	35	40	45
6	6	12	18	24	30	36	42	48	54
7	7	14	21	28	35	42	49	56	63
8	8	16	24	32	40	48	56	64	72
9	9	18	27	36	45	54	63	72	81

Goodheart-Willcox Publisher

Figure 2-4. A multiplication table used to complete single-digit multiplication problems.

X	1	2	3	4	5	6	7	8	9
1	1	2	3		5	6	7		9
2	2	4	6		10	12	14		18
3	3	6	9		15	18	21		27
4								32	36
5	5	10	15		25	30	35	40	45
6	6	12	18		30	36	42	48	54
7	7	14	21		35	42	49	56	63
8				32	40	48	56	64	72
9	9	18	27	36	45	54	63	72	81

Goodheart-Willcox Publisher

Figure 2-5. The multiplication process is commutative.

numbers in the top factor. Multiplying 2 × 6 gives us a product of 12, so the number 2 goes underneath the line and the 1 is written over the 1 in 16.

$$\begin{array}{r} {\scriptstyle 1} \\ 16 \\ \times 12 \\ \hline 2 \end{array}$$

This indicates that 1 needs to be added to the tens column. Then, the 2 is multiplied by the 1 in the top factor, giving us a result of 2. This result should then be entered in the second column under the solution line. Mentally, we add the remainder of 1 to 2, to get 3.

$$\begin{array}{r} 16 \\ \times 12 \\ \hline 32 \end{array}$$

The next step is to multiply the number in the tens place in the lower factor by each of the numbers in the top factor. Because the 1 in the bottom factor is in the second column from the right, it is important that the entries in the solution area start in the same column. For this reason, a zero, which is acting as a placeholder, is added in the first column.

$$\begin{array}{r} 16 \\ \times 12 \\ \hline 32 \\ 0 \end{array}$$

Multiplying 1 and 6 together yields a product of 6, which is written to the left of the 0. In a similar fashion, the product of 1 multiplied by 1 is 1. So a 1 is written to the left of the 6:

$$\begin{array}{r} {\scriptstyle 1} \\ 16 \\ \times 12 \\ \hline 32 \\ 160 \end{array}$$

Finally, the columns in the solution area are added together to provide the result of the multiplication operation, which is 192.

$$\begin{array}{r} 16 \\ \times 12 \\ \hline 32 \\ +160 \\ \hline 192 \end{array}$$

2.6.1 Exponents

The ***square*** of a whole number is the number multiplied by itself. This can be expressed as Y × Y, or Y^2, where the 2 is shown as a superscript. A superscript indicates the number of times that Y is expressed as a factor.

In general, the superscripted number, that indicates the power to which the number is raised, is referred to as an ***exponent***.

Some examples of exponents are shown in **Figure 2-6**.

A number squared is often expressed as a square that has two adjacent sides of the same length. Multiplying these lengths together yields square units (such as square inches or square feet). If a number is used as a factor three times, such as $Y \times Y \times Y$, the mathematical expression Y^3 is used. This is the ***cube*** of a whole number, and can be called Y to the third power, or simply "Y-cubed."

Exponents of Five			
Exponent	Power	Equation	Example
5^1	First power	5	$2 \times 5^1 = 10$
5^2	Second power	5×5	$2 \times 5^2 = 50$
5^3	Third power	$5 \times 5 \times 5$	$2 \times 5^3 = 250$
5^4	Fourth power	$5 \times 5 \times 5 \times 5$	$2 \times 5^4 = 1250$
5^5	Fifth power	$5 \times 5 \times 5 \times 5 \times 5$	$2 \times 5^5 = 6250$

Goodheart-Willcox Publisher

Figure 2-6. This table illustrates the powers of 5.

2.6.2 Radicals

Radicals, or roots, are the opposite of exponents. The mathematical symbol for radical or root is $\sqrt{}$, which closely resembles a checkmark. The ***root*** of a number is a number that, when multiplied by itself a certain number of times, yields a product equal to the original number. The square root of a number is a number that, when used as a factor twice, will yield the original number. For example, the square root of 25 is 5, since $5 \times 5 = 25$.

$$\sqrt{25} = 5$$

Since roots and exponents are mathematical opposites of each other, they cancel each other out. For example, if you take the square root of a squared number, the result is the original number.

$$\sqrt{5^2} = 5$$

$$(\sqrt{25})^2 = 25$$

2.7 Division

Division is the process of separating numbers into smaller parts or groups. Multiplication is often referred to as repeated addition, whereas division is often described as repeated subtraction.

To practice basic division, assume that an air-conditioning installer has an 8′ length of copper pipe that needs to be cut into four equal pieces without any waste. Expressing this problem in other terms, we could ask ourselves, how many times does 4 go into 8?

$$8 \div 4 = 2$$

Each of the four pieces of pipe must be 2′ in length.

The number being divided into, in this case 8, is referred to as the ***dividend***. The number that divides the dividend, in this case 4, is called the ***divisor***. The answer to the problem, or 2 in this example, is the ***quotient***.

Now consider that the installer took an 8′ piece of copper pipe and needed to cut 2′ sections from it. This can be expressed mathematically by placing the dividend below a calculation line with the divisor on the left side of the dividend. The quotient to be solved is then placed on the top of the calculation line.

In this case, the 8 is the dividend, while the 2 is the divisor.

$$2\ \overline{)\,8\quad}$$

Since it is known that 2×4 is 8, then $8 \div 2$ is 4. This means the installer can cut four 2′ sections from the material. For any division problem, the quotient can be checked by multiplying the quotient times the divisor. This answer should be the dividend.

However, consider the case where a system installer needs to cut 3′ sections of pipe from an 8′ length.

$$3\ \overline{)\,8\quad}$$

When a divisor does not equally go into the dividend, first determine how many times the divisor can go into the dividend without going over.

$$\begin{array}{rl} & \;\;2 \\ 3 & \overline{)\,8\quad} \end{array}$$

In this case, 3 can go into 8 twice. Thus there are 2′ of material left over after the 3′ sections are cut. The amount left over, in this case 2′, is referred to as the ***remainder***. This example could be rephrased mathematically as 8 divided by 3 equals 2, with a remainder of 2.

$$\begin{array}{rl} & \;\;2\quad R2 \\ 3 & \overline{)\,8\qquad\;} \\ & \underline{-6} \\ & \;\;2 \end{array}$$

Consider a multiple-digit division problem.

$$414 \div 23$$

The first step is to determine how many times 23 can go into 4, which is the first number in the dividend. Since 23 cannot go into 4, we then include the next digit in the dividend, which becomes 41. Determine how many times 23 goes into 41 without going over 41. Since $23 \times 2 = 46$, which is greater than 41, 23 will only go into 41 one time. Therefore, 1 is placed directly over the rightmost digit in 41.

$$\begin{array}{rl} & \quad\;\; 1 \\ 23 & \overline{)\,414} \\ & \underline{-23} \end{array}$$

The number 23 ($23 \div 1 = 23$) is then entered immediately below the 41 and is subtracted from 41, leaving 18.

$$\begin{array}{rl} & \quad\;\; 1 \\ 23 & \overline{)\,{}^{3}\not{4}\,1\,{}^{1}4} \\ & \underline{-23} \\ & \;\;18 \end{array}$$

The 4 from the leftmost column, because it was not used in this step, is brought down under the solution line. It must then be determined how many times 23 goes into 184 without going over 184.

```
       18
23 |34̸ 1'4
   -23↓
    184
```

It turns out that 23 goes into 184 exactly eight times, leaving a remainder of zero.

2.8 Fractions

The ability to work with fractions ensures the accuracy of the HVACR work performed. A fraction is a value expressed as a portion of a whole. Fractions consist of two numbers. The ***denominator*** is the bottom number in a fraction. The ***numerator*** is the top number in a fraction. The line separating the two numbers indicates division. For example, in the fraction 1/7, the denominator 7 divides the numerator 1. Fractions commonly used in the HVACR industry are halves, quarters, eighths, and sixteenths.

2.8.1 Adding and Subtracting Fractions

When adding or subtracting fractions, the denominator of the fractions being added or subtracted must all be the same. Once the denominators are the same, the numerators are added together. For example, if we need to add 1/8 and 3/8 together, the denominator will be 8 and the numerator will be 4, which is the sum of 1 and 3.

$$\frac{1}{8} + \frac{3}{8} = \frac{1+3}{8} = \frac{4}{8}$$

This result can then be reduced. ***Reducing*** fractions means lowering the numerator and denominator to the lowest term, while maintaining the same ratio between the fractions. This requires determining whether a numerator and denominator are both divisible, or able to be divided by the same number. Both the numerator and denominator are divisible by 4, so both can be divided to get our final answer, which is 1/2.

$$\frac{4 \div 4}{8 \div 4} = \frac{1}{2}$$

When adding fractions with different denominators, the fractions must be adjusted to be expressed with the same denominator. For example, if we need to add 1/8 and 1/4 together, they each have different denominators. Based on simple multiplication, we also know that 4 can be multiplied by 2 to equal 8. Thus, the numerator and denominator of 1/4 can be multiplied by 2 to have the same denominator. Once the denominators are the same, we can then simply add the numerators to obtain the sum:

$$\frac{1}{8}+\frac{1}{4}=\frac{1}{8}+\frac{2}{8}=\frac{1+2}{8}=\frac{3}{8}$$

2.8.2 Lowest Common Denominator

When adding or subtracting fractions, first identify the ***lowest common denominator***. This is the lowest number that one or both of the denominators can be changed to in order to make the denominators the same.

To determine the lowest common denominator, first list each denominator's multiples, or a number that can be divided by another number without a remainder. Then, identify the lowest multiple that both the denominators have in common. For example, if the fractions 2/9 and 1/6 were to be added together, the lowest common multiple of 6 and 9 would need to be identified.

Multiples of 6: 6 12 **18** 24 30
Multiples of 9: 9 **18** 27 36 45

The lowest common multiple of both 6 and 9 is 18. In this case, both the 2/9 and 1/6 will be converted to fractions, with 18 as their denominators. To change the 2/9 into a fraction that has 18 as the denominator, both the numerator and the denominator must be multiplied by 2 in order to maintain the correct ratio:

$$\frac{2}{9}=\frac{2\times 2}{9\times 2}=\frac{4}{18}$$

Multiplying and dividing a number by the same number does not change the value of the number. As a result, our original fraction of 2/9 can now be expressed as 4/18. In a similar manner, 1/6 can be expressed as 3/18.

$$\frac{1}{6}=\frac{1\times 3}{6\times 3}=\frac{3}{18}$$

Having obtained the lowest common denominator, the fractions can then be added to obtain the total of 7/18.

$$\frac{4}{18}+\frac{3}{18}=\frac{4+3}{18}=\frac{7}{18}$$

Like addition, when we subtract fractions, the denominator of each fraction must also be the same. Instead of adding the numerators together, one is subtracted from the other. For example, if we subtract 1/16 from 5/16, the result is 4/16. This result can be reduced to give us 1/4.

$$\frac{5}{16}-\frac{1}{16}=\frac{5-1}{16}=\frac{4}{16}$$

If we need to subtract 1/8 from 5/16, the lowest common denominator, as in the case of addition, must be found. Since 8 is a multiple of 16, the fraction 1/8 can be expressed as 2/16.

$$\frac{1 \times 2}{8 \times 2} = \frac{2}{16}$$

The fraction 2/16 can then be subtracted from 5/16.

$$\frac{5}{16} - \frac{2}{16} = \frac{5-2}{16} = \frac{3}{16}$$

2.8.3 Multiplying Fractions

It is also necessary to multiply fractions together on the job. For example, an HVACR duct fabricator may need to multiply fractions to determine the amount of duct-lining material necessary to complete a particular job. When multiplying fractions, the numerators and denominators are multiplied together.

If we wanted to multiply 1/2 and 1/4, we would multiply the numerators, giving us 1 times 1, which equals 1. The denominator of the answer would be 2 times 4, which equals 8. The answer to this problem is then 1/8.

$$\frac{1}{2} \times \frac{1}{4} = \frac{1 \times 1}{2 \times 4} = \frac{1}{8}$$

The final fraction should always be reduced to its simplest terms.

2.8.4 Dividing Fractions

When dividing fractions, the operation is changed to that of multiplication. We do this by first obtaining the ***inverse***, where the numerator and denominator of the fraction that is being divided by are interchanged. Then, the fractions are multiplied together.

For example, if we wish to divide 1/4 by 1/8, we invert the 1/8 to give us 8/1.

$$\frac{1}{4} \times \frac{1}{8} = \frac{1}{4} \times \frac{8}{1}$$

Then, we multiply.

$$\frac{1}{4} \times \frac{8}{1} = \frac{8}{4} = 2$$

This gives us a result of 2.

2.8.5 Converting Improper Fractions to Mixed Numbers

An ***improper fraction*** is a fraction that has a numerator larger than its denominator. This means the value of the fraction is larger than 1. Improper fractions are commonly converted to ***mixed numbers***. A mixed number is a value consisting of a whole number and a fraction.

Division is used to change an improper fraction into a mixed number. For example, 11/8 is an improper fraction because the numerator, 11, is larger than the denominator, 8. To convert 11/8 to a mixed number, we must first determine how many times the denominator can be divided into the numerator. We know 8 will go into 11 one time and there will be a remainder of 3. The number 8 cannot be divided into the remainder, so the remainder is written as the numerator over the original denominator, 8.

$$\frac{11}{8} = 1\frac{3}{8}$$

Adding and subtracting mixed numbers is often difficult. So, in the case where mixed numbers need to be manipulated, it is often desirable to convert the mixed numbers, or at least part of them, to improper fractions first.

2.8.6 Converting Mixed Numbers to Improper Fractions

While improper fractions must be divided to create a mixed number, mixed numbers are converted to improper fractions by multiplication. First, multiply the denominator of the fraction by the whole number. Then, add the numerator of the fraction to this result. This number becomes the new numerator, and the denominator is kept the same as in the original fraction.

Consider the mixed number 5 1/4. To convert this mixed number to an improper fraction, first multiply the whole number, 5, by the denominator, 4:

$$5 \times 4 = 20$$

Then, add the product to the numerator:

$$20 + 1 = 21$$

The number 21 will now be the numerator for the improper fraction while the denominator remains the same as the one in the mixed number.

$$\frac{21}{4}$$

So, 5 1/4 expressed as an improper fraction is 21/4.

2.9 Decimals

Decimals, like fractions, represent a portion of a whole. A decimal is expressed with place values to the right of a decimal point. Digits to the left of a decimal point indicate a whole number. Decimals are indicated by the numbers to the right of a decimal point. The fraction 1/2 and the decimal 0.50 both represent one-half of the whole.

The place value of each digit is determined by the location in the decimal number. See **Figure 2-7**.

Place Values								
Thousands	Hundreds	Tens	Ones		Tenths	Hundredths	Thousandths	Ten Thousandths
3	1	8	4	.	9	8	2	6

Goodheart-Willcox Publisher

Figure 2-7. This chart depicts the place values of a number with a decimal.

Using the decimal 3184.9826, the first number to the right of the decimal point, in this case the 9, represents the tenths column, and the numbers become smaller in value the farther they are from the decimal. The number directly to the left of the decimal, in this case the 4, is the smallest value of the numbers on that side, and the numbers grow in value the farther left of the decimal they are.

2.9.1 Converting Fractions to Decimal Equivalents

A fraction is expressed in decimal form by dividing the numerator by the denominator. For instance, the fraction 3/4 can be stated as being 3 divided by 4. Because 3/4 is less than 1, we know our decimal equivalent is less than 1.00. The result is 0.75. We can then see that 3/4 is equal to 0.75. Using a calculator yields the same result.

Many devices in the HVACR industry, such as electrical test instruments, utilize decimals to display data. Any fraction can be converted into decimal form. A decimal equivalent chart can be referred to for quick conversions. See **Figure 2-8.**

A decimal such as 0.125 is expressed as 125 one thousandths, or 125/1000, since there are three numbers to the right of the decimal point and the third column to the right of the decimal represents thousandths. The decimal 0.75 is expressed as 75/100, since there are two numbers to the right of the decimal point and the second column to the right of the decimal represents hundredths.

Fraction and Decimal Equivalents			
Fraction	Decimal	Fraction	Decimal
1/32	0.03125	1/2	0.5
1/16	0.0625	9/16	0.5625
3/32	0.09375	21/32	0.65625
1/8	0.125	11/16	0.6875
3/16	0.1875	3/4	0.75
7/32	0.21875	13/16	0.8125
1/4	0.25	27/32	0.84375
5/16	0.3125	7/8	0.875
3/8	0.375	15/16	0.9375
7/16	0.4375	31/32	0.96875
15/32	0.46875	1	1.0

Goodheart-Willcox Publisher

Figure 2-8. A fraction and decimal equivalents chart.

2.9.2 Converting Decimals to Their Fraction Equivalents

Converting a decimal to a fraction is relatively easy since a decimal is a fraction with a power of ten. For instance, the decimal 0.2 is two tenths. To convert this decimal to a fraction, simply use the digits to the right of the decimal as the numerator, in this case 2. The denominator is a 1 followed by the number of zeroes that is equivalent to the number of digits in the numerator, which is 1. Thus, the fraction equivalent of 0.2 is 2/10.

Other examples of decimal-to-fraction conversion are:
0.02 = 2/100
0.002 = 2/1000
0.0002 = 2/10,000

0.46 = 46/100
0.897 = 897/1000
0.0392 = 392/10,000

When converted from decimals, fractions are not always in their lowest terms. For example, 0.2 converted to 2/10 can be reduced to 1/5.

2.10 Ratios

Ratio refers to the proportional relationship between two quantities. Ratios are often used when comparing manufacturers' data on equipment performance. Consider a belt-driven blower assembly where the motor turns at a speed of 3500 rotations per minute, and the blower is turning at a speed of 700 rotations per minute. The ratio of the motor speed to the blower speed can be expressed as 3500:700, or 5:1. The 5:1 ratio indicates that, for every five rotations of the motor, the blower rotates one time. This is the same as the original 3500:700 ratio. When both of the numbers are divided by 700, the ratio 5:1 is obtained.

A common application of ratios is to describe the layout, or relative "squareness" of a room. The aspect ratio is used to compare the length to the width of a room. For example, if a room has an aspect ratio of 1:1, the room is a square. If the aspect ratio is 2:1, this means the room is twice as long as it is wide. Given a room that is 30 feet long and 20 feet wide, we can determine the aspect ratio of the room is 3:2 or 1.5:1.

2.11 Percentages

Percentages are a portion of a whole, or similar to a ratio that is compared to the base value of 100. The term 100 percent, denoted as 100%, is used to describe a situation when all of the conditions are met for a certain event or occurrence. For HVACR applications, percentages can be used to calculate the fuel efficiency for furnaces or boiler burners, the amount of outside air brought into a structure, or job pricing, such as profits or losses.

Fractions are often expressed as percentages. For example, the fraction 3/5, can be described as three parts out of five, or represented as 60%. So, if an HVACR crew completed the installation of three out of five air handlers on a particular day, it can be stated that 60% of the air handlers were installed. In order to convert a fraction to a percentage, the fraction must first be converted to a decimal, and then the decimal must be multiplied by 100. From the example just presented, the conversion from 3/5 to 60% is shown here:

$$(3 \div 5) \times 100 = 0.60 \times 100 = 60\%$$

In many industries, including HVACR, many suppliers will allow contractors to discount their purchases if they are paid for within a certain period of time. The method to calculate the discount and determine a product's cost after a discount has been applied is shown next:

$$\text{discount (in dollars)} = \text{percentage in decimal form} \times \text{original price of item}$$
$$30\% \times \$200 = 0.30 \times \$200$$
$$= \$60$$

This is the discount, not the final purchase price of the item. This discount must be subtracted from the original purchase price to determine the discounted final price of the item. In general terms, the final price of a discounted item can be calculated as follows:

$$\text{discounted price} = \text{original price} - (\text{percentage discount} \times \text{original price})$$

In this equation, the percentage discount is expressed as a decimal. So, the final price of our \$200 item, when discounted by 30%, can be calculated directly as:

$$\$200 - (0.30 \times \$200) = \$200 - \$60$$
$$= \$140$$

2.11.1 Interpreting Percentages

Percentages can also be used in conjunction with ratios. Consider the 30′ × 20′ room that was discussed earlier. It was determined that the aspect ratio of that room was 3:2, or 1.5:1, meaning that, for every foot of room width, there was 1.5′ of room length. It can be stated, in terms of percentage, that the length of the room is 150% that of the width.

Some confusion can arise when interpreting percentages due to a major difference between the following two statements:

1. The actual cost of the job is 200% of what it was when it was originally estimated.
2. The actual cost of the job increased by 200% from the time it was originally estimated.

The key phrases in the two statements are the following: "is 200%" in the first statement, and "increased by 200%" in the second. The word "is" is often used mathematically as the equivalent of an equal (=) sign, whereas the phrase "increased by" is used as the equivalent of the plus (+) or addition sign. Examining a numerical example will help clarify the difference between the two statements.

Assume that the original estimate for the job was $5000. Statement #1 can be rephrased as "the actual job cost is 200% of $5000." Since 200% is the same as twice, the statement can, once again, be reworded to state that "the actual job cost is twice $5000," or $10,000. Statement #2 can be rephrased as "there was a 200% increase in the cost of the job" or "the cost of the job increased by 200%." Since the cost of the job was originally $5000 and 200% of the job cost is $10,000, statement #2 can be calculated as $5000 + $10,000, or $15,000. A summary of the two statements is provided:

1. actual cost = 200% × original estimate = 2 × $5,000 = $10,000
2. actual cost = original estimate + (200% × original estimate) = $5,000 + $10,000 = $15,000

The interpretation of the statement can lead to different assumptions, so it is very important to clarify the statement's intentions before moving forward with a project.

Summary

- Math problems can often be solved differently, while still arriving at the same solution.
- Before performing calculations, an estimate of the solution should be made first.
- It is good practice to double check all calculations and results, in addition to comparing the results with any estimates or predictions.
- Addition is the process of combining numbers to obtain a sum, or total.
- Rounding off numbers can help provide a quick prediction of what the actual solution will be.
- Airflow is the amount of air per unit of time that flows through a component. It can be determined by addition.
- Subtraction involves removing number values from another number to determine the difference. Borrowing is a strategy used to evaluate a problem as a whole.
- Multiplication can be described as repeated addition, and is the process of adding like numbers numerous times.
- The product is the result of two numbers, called factors, which are multiplied together.
- A multiplication chart can be used to help multiply single-digit and multiple-digit numbers.
- An exponent refers to the number of times that a particular number is multiplied by itself.
- Radicals or roots refer to the number that, when multiplied by itself a certain number of times, yields the original number.
- Division can be described as repeated subtraction and is used to determine how many times one number can fit into another number.
- Portions of a whole number can be expressed as either fractions or decimals.
- The decimal equivalent of a fraction is obtained by dividing a fraction's numerator by its denominator.
- The fraction equivalent of a decimal is obtained by using the number to the right of the decimal point as the numerator and 1 followed by the number of zeroes equivalent to the number of digits in the numerator as the denominator. When adding or subtracting fractions, the lowest common denominator must be found.
- A ratio refers to a proportional relationship that exists between two quantities.
- Percentages are portions of a whole as they relate to a base value of 100.

Chapter Assessment

Name ______________________ Date ____________ Class ______________

Know and Understand

_______ 1. A close estimation of 11 + 31 + 39 + 22 is _____.

A. 70

B. 80

C. 100

D. 150

_______ 2. After paying out $15, $20, and $35, how much money remains if the initial amount was $95?

A. $20

B. $25

C. $30

D. $35

_______ 3. 492 – 176 = _____

A. 248

B. 298

C. 316

D. 324

_______ 4. The expression 12 + 12 + 12 + 25 + 25 +25 can be written as _____.

A. $12^3 + 25^3$

B. $(12 \times 3) + (25 \times 3)$

C. $(12 + 25)^3$

D. $(12 + 3) \times (25 + 3)$

_______ 5. The expression 5^3 can be written as _____.

A. 5 + 5 + 5

B. $3 \times 3 \times 3 \times 3 \times 3$

C. $5 \times 5 \times 5$

D. 3 + 3 + 3 + 3 + 3

_______ 6. The correct value of 5^3 is _____.

A. 15

B. 125

C. 243

D. 342

_______ 7. The square root of 36 is _____.

A. 6

B. 9

C. 18

D. 27

_______ 8. The fraction 4/16 can also be expressed as _____.

A. 1/4

B. 1/8

C. 3/8

D. 1/2

_______ 9. The decimal 0.625 can be expressed as _____.

A. 5/16

B. 3/8

C. 5/8

D. 3/4

_______ 10. 3/4 + 1/8 = _____

A. 1/4

B. 3/8

C. 5/8

D. 7/8

_______ 11. 3/4 – 5/8 = _____

A. 1/8

B. 1/4

C. 3/8

D. 3/4

Apply and Analyze

1. It takes Crew A three days to install a split air-conditioning system, and it takes Crew B five days to install the same system. Working together, how many days will it take to install the same type of air-conditioning system?

____

2. If a piece of equipment costs $1000, how much will a contractor need to bill the customer if 40% profit is needed on the equipment?

____

3. A contractor receives a 20% discount on an equipment purchase from an HVACR supplier. The contractor will also receive an additional discount of 10% if the invoice is paid within 14 days. What is the contractor's final cost if a $5000 purchase is paid for within the two-week period?

____

4. If an HVACR installation crew must install 10 complete systems in five days, what percentage of the units will be installed after two days, assuming the crew works at a consistent rate?

____

Name ______________________________ Date ____________ Class ____________________

5. The installation of a particular model of air-conditioning system requires 45′ of 1/2″ copper tubing, 45′ of 1/4″ copper tubing and 20′ of 7/8″ rigid copper pipe. If five of these systems must be installed, how many feet of piping and tubing material, in total, will be required?

____

6. The installation of a particular model of air-conditioning system requires 15′ of 1/2″ copper tubing. If the copper tubing is supplied in 50′ rolls, how many systems can be installed using just this one roll of tubing? Assuming no waste, how many feet of tubing will be left over once these systems have been installed?

____

7. Each linear foot of a particular duct run weighs 8.5 lb. If the complete duct run is 50′ long, how many duct support straps will be needed, at a bare minimum, if each strap can support 25 lb?

____

Critical Thinking

1. Why is multiplication often equated to adding like numbers multiple times? Create a drawing that supports your answer.

____

____

____

2. Why is a discount of 25%, with an additional discount of 10%, not equal to a total discount of 35%?

____

____

____

Name ______________________________ Date ____________ Class ____________________

3. When adding multiple numbers together, why does rounding off result in an answer that is quite often very close to the actual sum? Create a simple example of where rounding off the addends would result in an unacceptably inaccurate estimate.

____

____

____

4. If efficiency is represented as 100 × (output/input), how would you describe the efficiency of a process where the numerator (output) is greater than the denominator (input)?

____

____

____

3 Math Applications

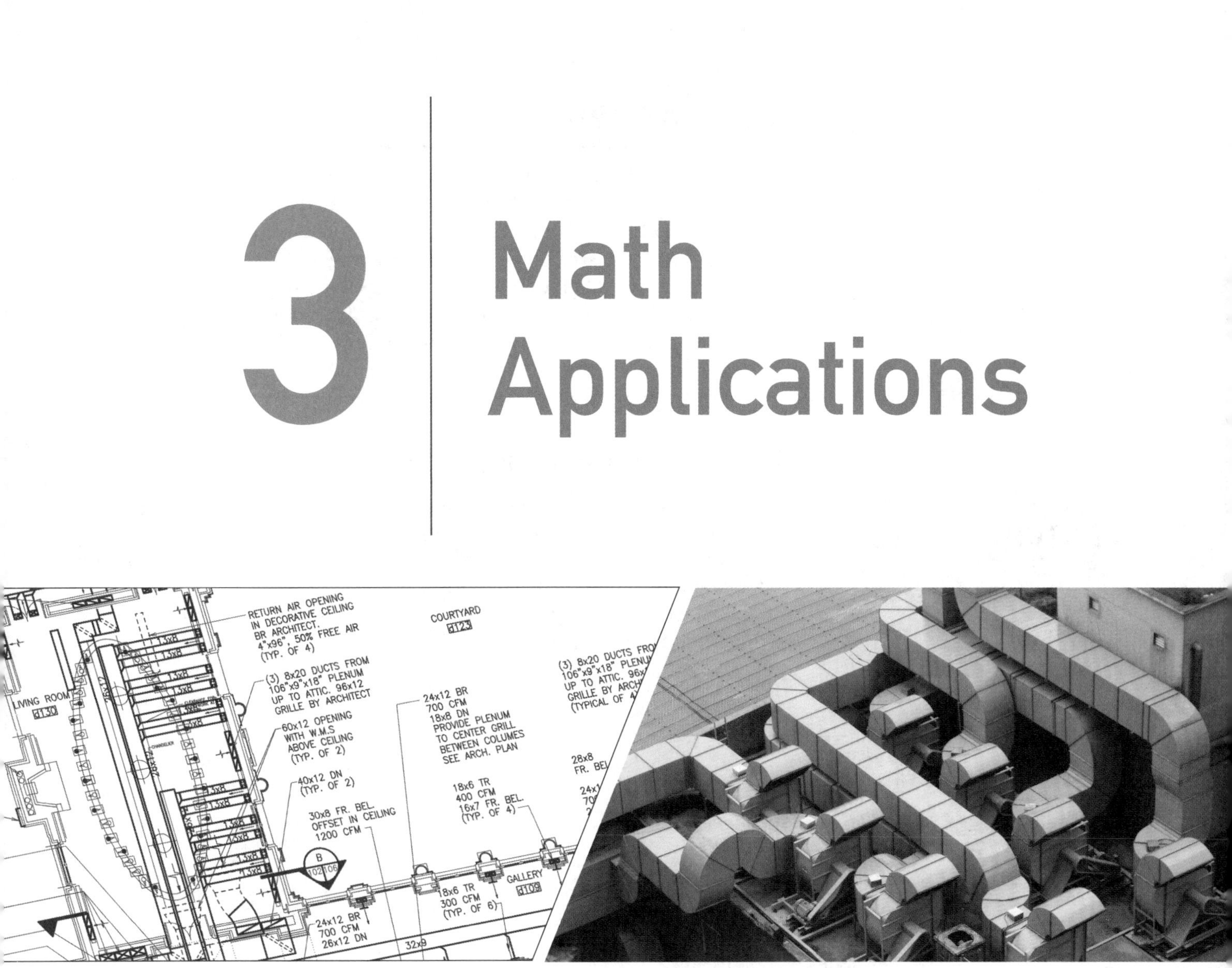

Chapter Outline

3.1 Perimeter and Circumference

3.2 Area

3.3 Volume

3.4 Average

3.5 The Metric System

3.6 Word Problems

3.6.1 How Long Will It Take to Pipe in a Split Air-Conditioning System?
3.6.2 How to Estimate the Cost of a Small Job
3.6.3 Estimating the Amount of Sheet Metal Needed
3.6.4 Sneezing and Driving (A Fun Example)

Learning Objectives

After studying this chapter, you will be able to:

- Calculate the perimeter, area, and volume of various geometric shapes.
- Calculate averages and weighted averages.
- Demonstrate a basic understanding of the metric system.
- Solve word problems by creating and solving algebraic expressions.

Technical Terms

area

arithmetic mean

centi

circumference

kilo

metric system

milli

perimeter

pi

radius

US Customary system

volume

weighted average

Introduction

This chapter will provide an overview of a wide range of mathematical concepts, along with some of their industry-related applications, to help you work effectively with numbers and prints. These concepts include the perimeter, area, and volume of two-dimensional and three-dimensional objects. Later, two common measurement systems—the metric system and US Customary system—will be examined. Practice with word problems is provided at the end of the chapter to apply some of the math concepts and learn how to best approach such problems.

3.1 Perimeter and Circumference

The ***perimeter*** is the distance around the outside of a shape or space. It is a linear measurement calculated by adding together the lengths of the sides of a shape. For example, perimeter is used to determine the amount of fencing that would be needed for a project.

Consider the drawing of a room. This drawing represents a room with actual dimensions of $16' \times 12'$.

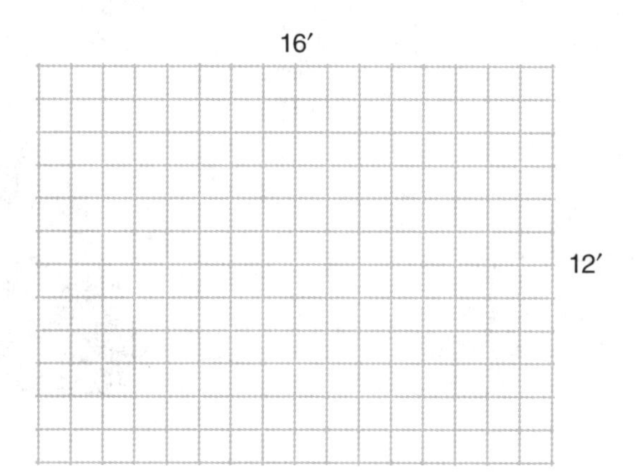

To determine how many feet would be traveled if we walked around the outer edge of the room, the room's length and width are considered.

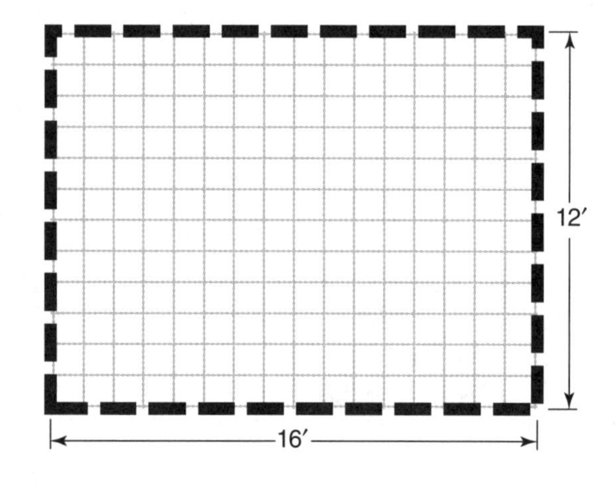

$$16' + 16' + 12' + 12' = 56'$$

The perimeter of the room is 56′, which is the number of feet around the edge of the room.

The same method is used to determine the amount of fencing material needed to enclose one's property or the amount of electrical conduit needed to encircle a building. Knowing the perimeter also aids in other calculations, such as determining how much acoustical lining material is needed for a particular section of ductwork.

The perimeter of a circle is called the ***circumference***. The circumference uses a different formula to calculate the distance around a circle:

$$\text{circumference} = 2\pi r$$

where

r = radius of the circle
π = pi (3.14159)

The constant ***pi***, π, closely estimated to be 3.14, is a relationship between the circumference and diameter and does not change. The ***radius*** of a circle is any line drawn from the center of the circle to its outer edge.

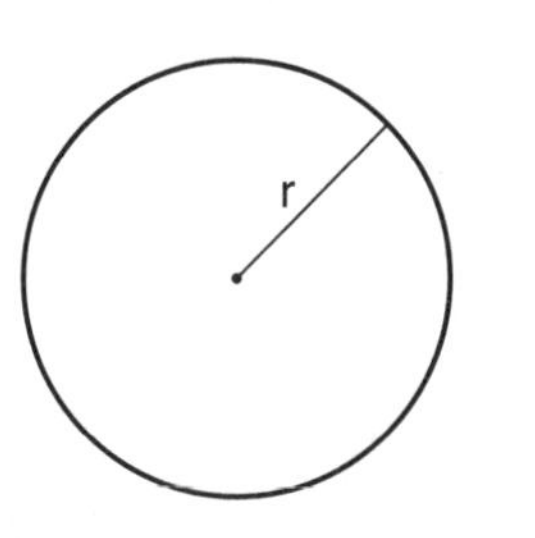

For a circle with a radius of 5″, the circumference of the circle is $2\pi r = 2\pi\,(5) = 10\pi = 10 \times 3.14 = 31.4\ \text{in}^2$. If the diameter of the circle is known, the formula for the circumference of the circle becomes πd, since the diameter of the circle is twice the radius. The diameter, d, of a circle is defined as a straight line drawn from one side of a circle to the other that passes through the center.

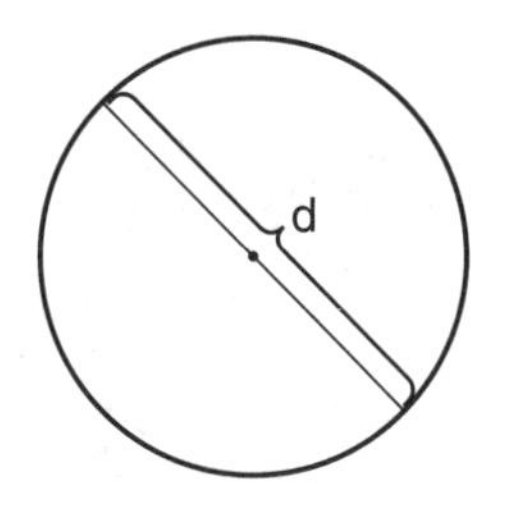

3.2 Area

The ***area*** is the amount of space within a boundary. It is measured in square units, such as square inches or square feet. Being able to determine the area of a given space can help calculate the amount of flooring material, such as carpeting or ceramic tile, needed to complete the job. In the HVACR field, the area of a room or space is needed to perform the calculations required to properly size and design radiant heating systems. The area of square and rectangular shapes is found by multiplying the length by the width of the space.

$$\text{Area of rectangle} = l \times w$$

where
l = length
w = width

For example, the area of a 12′ × 16′ room can be found by multiplying 16 × 12.

$$16' \times 12' = 192 \text{ ft}^2$$

The 192 ft^2 value can be used to calculate the heat loss (per square foot) of a room. The area of a room is an extremely important part of the calculations to ensure the radiant heating system does not overheat, or underheat, the space.

To determine the area of other shapes, the formula used for a rectangle, length × width, is not followed. For instance, when calculating the area of a triangle, the following formula is used:

$$\text{Area of triangle} = 1/2\ b \times h$$

where
b = base
h = height

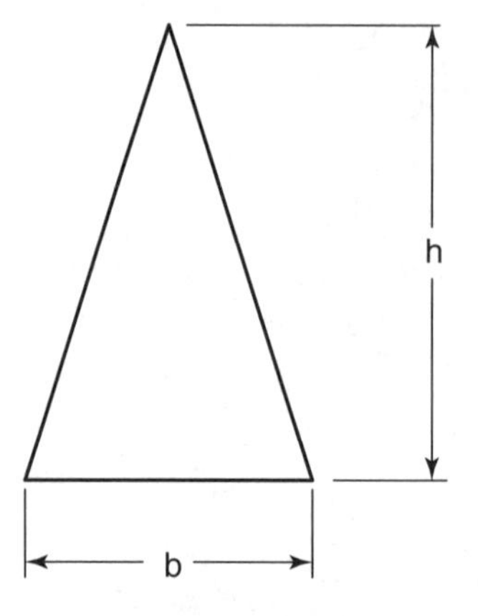

Consider a triangle with a base of 8″ and a height of 10″.

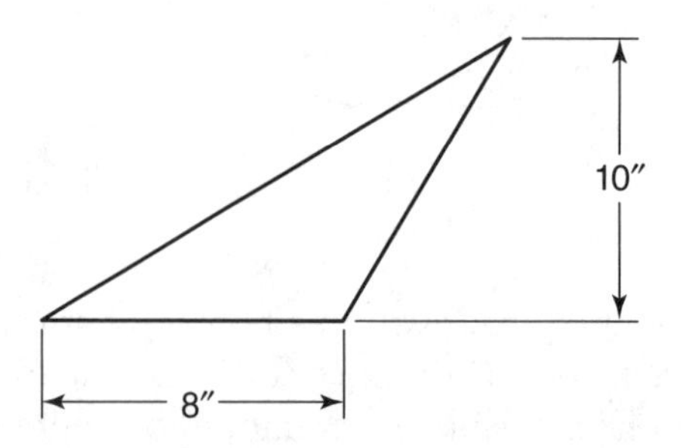

The area can be calculated as:

$$1/2 \times 8'' \times 10''$$
$$1/2 \times 80 \text{ in}^2 = 40 \text{ in}^2$$

The 10″ height measurement is not the length of a side of the triangle. Instead, it is the length of the top point of the triangle to the base itself. When this perpendicular line falls outside the boundaries of the triangle, a line is drawn to extend the base.

Other formulas to calculate the area of various shapes include the following:

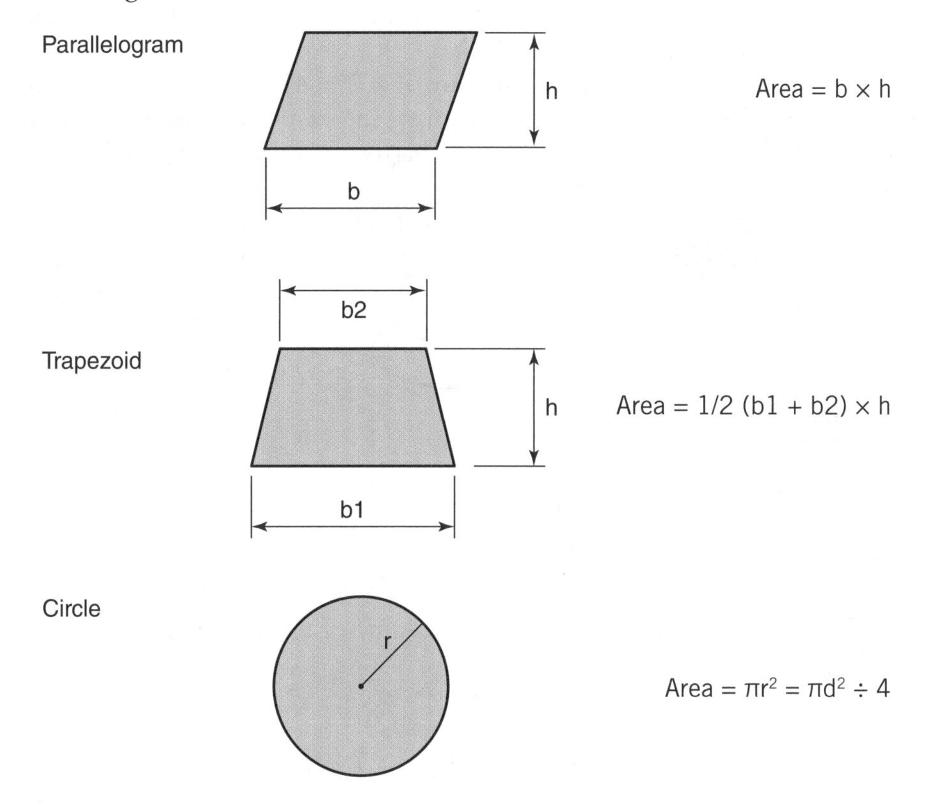

3.3 Volume

Volume describes how much space a particular substance occupies in a three-dimensional object. Volume is used to calculate structural leakage rates in the HVACR industry. It is also used when calculating ventilation rates. In this case, the term "air changes per hour" is used. Volume is expressed in cubic units, such as cubic feet or cubic inches. For example, to calculate the volume of this cube, we use the following formula:

$$\text{Volume} = l \times w \times h$$

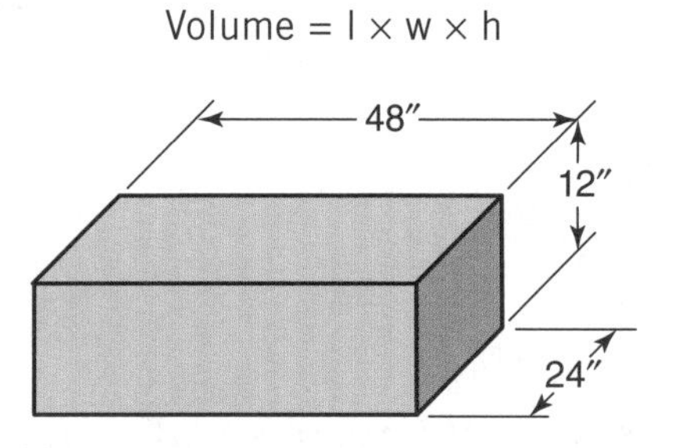

Consider a water tank that has interior measurements of 12″ × 24″ × 48″. To calculate this area, we multiply:

$$12'' \times 24'' \times 48'' = 13{,}824 \text{ in}^3$$

It is best to express volume in cubic feet. To do this, the conversion of 1,728 in^3 = 1 ft^3 is used. Convert the answer into cubic feet by dividing by 1,728 in^3/ft^3.

$$13{,}824 \text{ in}^3 \div 1{,}728 \text{ in}^3/\text{ft}^3 = 8 \text{ ft}^3$$

There is a much easier way to look at this example. Examine the original problem carefully. Notice that, because 12″ = 1′, 24″ = 2′, and 48″ = 4′, the dimensions of the tank are 1′ × 2′ × 4′.

Because the measurements are in feet, the result is in cubic feet. To avoid having to work with numbers larger than necessary, reduce or change the units to suit the needs of the problem being addressed.

Calculating the volume of a cylinder involves a similar formula, where the area of the circle base or footprint, is multiplied by its height:

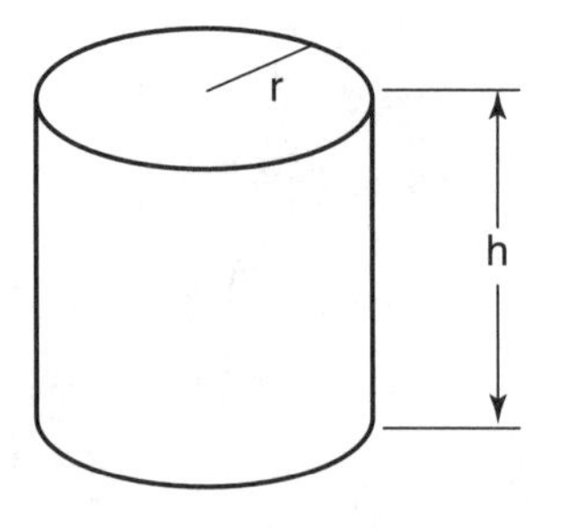

The formula used to calculate the volume of a cylinder is:

$$V = \pi r^2 h$$

Consider a cylinder that has a radius of 3″ and a height of 8″.

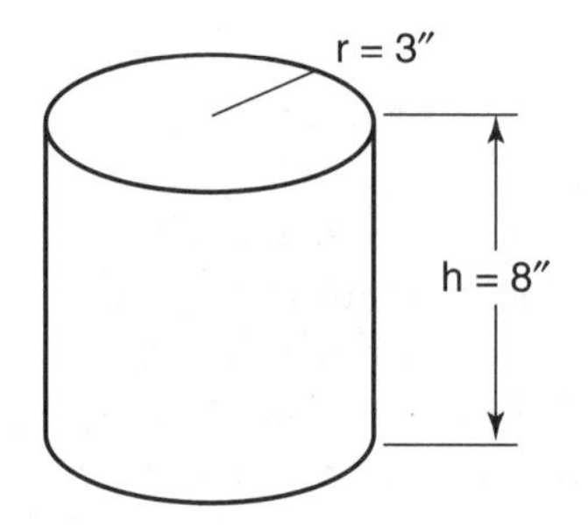

The volume of the cylinder is then calculated as the following:

$$V = \pi(3)^2(8)$$
$$\pi(9)(8) = 72 \times \pi$$
$$72 \times \pi = 226.1 \text{ in}^3$$

3.4 Average

The ***arithmetic mean*** is the average of a particular set of numbers. The average is not often equal to any numbers in the set, and will be higher than the lowest number in the set, and lower than the highest number. The average is calculated by adding up all of the numbers in the set and then dividing the sum by how many numbers there are.

In the HVACR field, the process of calculating the average is especially useful when determining the average velocity of an airstream as it flows through a duct. To accomplish this, air velocity readings are taken at various points in the duct.

For example, the following six readings were measured in a duct:

$VELOCITY_1$ = 700 feet per minute (fpm)
$VELOCITY_2$ = 650 feet per minute (fpm)
$VELOCITY_3$ = 625 feet per minute (fpm)
$VELOCITY_4$ = 725 feet per minute (fpm)
$VELOCITY_5$ = 675 feet per minute (fpm)
$VELOCITY_6$ = 750 feet per minute (fpm)

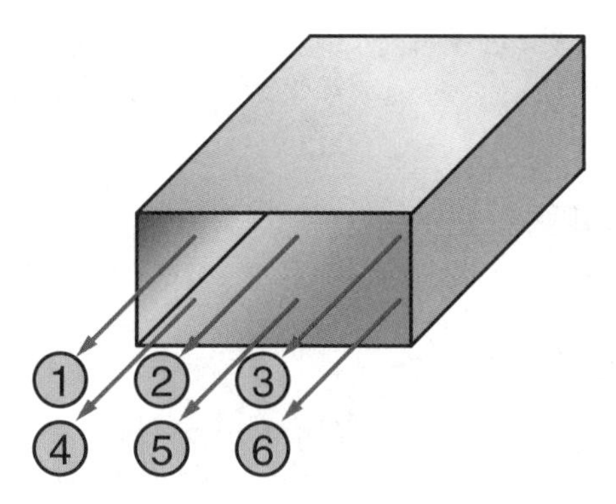

To determine the average air velocity, the six readings are first added together.

$$700 + 650 + 625 + 725 + 675 + 750 = 4125 \text{ fpm}$$

This total is then divided by 6, which is the number of readings taken.

$$4125 \div 6 = 687.5 \text{ fpm}$$

Thus, the air is flowing through the duct at an average velocity of approximately 688 feet per minute.

Sometimes, when calculating the average, a weighted average is needed for a more in-depth average calculation. A ***weighted average*** is similar to a basic average calculation, but each value is multiplied by a factor that is proportional. The weighted average is used when:

- Any or all values do not hold the same weight.
- One of the values is more or less important than the other values.
- All the values have a different effect on the outcome.

For example, if 1 cup of water at 70°F was mixed with 1 cup of water at 90°F, the resulting water temperature would be 80°F. This is because the quantities of the water are the same, so the resulting water temperature will simply be the average, which is (70°F + 90°F) ÷ 2, of the two temperatures. But what would happen if there were 2 cups of 70°F water mixed with 1 cup of 90°F water? The 70°F of water will have more impact on the final temperature of the water mixture because there is simply more of it. So, because there is twice as much 70°F water as there is 90°F water, the 70°F water will be counted twice, giving 2 parts of 70°F water to 1 part of 90°F water.

$$\text{weighted average} = 2 \times (70°F) + 1 \times (90°F) = 230$$

Since there are now three numbers being added together, the total will be divided by 3 instead of 2:

$$230 \div 3 = 76.7°F$$

This problem can be looked at a little differently. By seeing that 2/3 of the final water mixture is made up of 70°F water and 1/3 of the final water mixture is made up of 90°F water, the expression can be rewritten using decimals.

$$\begin{aligned}\text{weighted average} &= 2/3\ (70°F) + 1/3\ (90°F)\\ &= 0.67\ (70°F) + 0.33\ (90°F)\\ &= 76.7°F\end{aligned}$$

The result is still the same.

3.5 The Metric System

The ***metric system***, or *International System of Units (SI)*, is a measurement system used worldwide, but not in the United States. When equipment, materials, system components or other devices manufactured overseas are used in the United States, it is not uncommon for the technical data and information to be expressed in metric units. For example, instead of using the foot as a standard for measuring length, the metric system uses units such as meters. The metric system is relatively easy to work with and is based on multiples of 10 to determine lengths shorter or longer than a meter, **Figure 3-1**.

Since the standard unit of measurement is the meter, we use prefixes to describe multiples of the meter. For example, the prefix ***kilo*** means 1000, so 1 kilometer equals 1000 meters. Similarly, the prefix ***centi*** means one hundredth, so there are 100 centimeters in one meter. The prefix ***milli*** indicates a factor of one thousandth. Thus, there are 1000 millimeters in a meter and 10 millimeters in a centimeter.

It may be necessary to convert between the ***US Customary system***, or *inch-pound (IP) system*, to the metric system. Some linear-measurement conversions between the IP and SI system are:

1 inch = 2.54 centimeters = 25.4 millimeters
1 foot = 12 inches = 0.305 meters = 30.5 centimeters = 305 millimeters
1 mile = 5280 feet = 1.61 kilometers = 1609 meters
1 kilometer = 1000 meters = 0.62 miles = 3274 feet
1 centimeter = 0.39 inches

Common Metric Multiples for 1 Meter (m)							
Multiple	**1/1000 meter**	**1/100 meter**	**1/10 meter**	**1 meter**	**10 meter**	**100 meter**	**1000 meter**
Name	millimeter	centimeter	decimeter	meter	dekameter	hectometer	kilometer
Symbol	mm	cm	dm	m	dam	hm	km

Goodheart-Willcox Publisher

Figure 3-1. This table illustrates common metric multiples of 1 meter.

Given all of these conversion factors, converting between the IP and SI system may appear confusing. However, as far as distance goes, only one conversion rate must be known to determine all the others. Assume that only the conversion factor between centimeters and inches (2.54 cm = 1 inch) is known. We already know that there are 12 inches in a foot, so we can multiply the 2.54 by 12 to get 30.5. Because we are multiplying and dividing by the units of inches, they cancel each other out and the remaining units are cm/foot.

$$\frac{2.54\text{ cm}}{\cancel{\text{inch}}} \times \frac{12\ \cancel{\text{inches}}}{\text{foot}} = \frac{30.5\text{ cm}}{\text{foot}}$$

Because we now know there are 30.5 centimeters in one foot, we can determine the conversion factor that exists between feet and meters. We know that 100 centimeters are in 1 meter, so we can multiply by this conversion factor to obtain the result of 0.305 meters = 1 foot. Notice, in this example, we are multiplying and dividing by centimeters, so those units cancel each other out, leaving the units of meters per foot.

$$\frac{30.5\ \cancel{\text{cm}}}{\text{foot}} \times \frac{1\text{ meter}}{100\ \cancel{\text{cm}}} = \frac{0.305\text{ meter}}{\text{foot}}$$

Next, let us determine how many kilometers are in 5 miles. Given the chapter's earlier conversions, you know that 1 kilometer = 0.62 miles. It might be tempting to simply multiply 0.62 × 5 to arrive at 3.1 as an answer, but this is incorrect. Always make certain you can cancel units out. Once properly set up, the problem should look like the following:

$$\frac{5\ \cancel{\text{miles}}}{1} \times \frac{1\text{ km}}{0.62\ \cancel{\text{miles}}} = 8.06\text{ km}$$

To solve this problem correctly, we divide 5 by 0.62. Notice how the mile units cancel each other out and you are left with the units of kilometers. Dividing 5 by 0.62 yields a result of 8.06 kilometers.

BETWEEN THE LINES

Estimating Measurement

Remember, as you work through any math problem, try to estimate your result prior to performing any calculations. In this case, because 1 kilometer = 0.62 miles, we know a mile is longer than a kilometer. So, if we want to determine how many kilometers are in a mile, the number will be greater than 1. Since we are looking for the kilometer equivalent of 5 miles, the result must be greater than 5. So, if the initial calculation yielded a result such as 3.1, we know a calculation error was made.

3.6 Word Problems

Converting the written or spoken word into mathematical formulas may sound complicated, but it is a valuable tool that must be learned in the building trades. With practice, you will be able to solve these problems accurately and effectively. This section introduces several different word problems, and then discusses how to set each up and solve them correctly.

3.6.1 How Long Will It Take to Pipe in a Split Air-Conditioning System?

The Problem

It takes John 5 hours to pipe in a split air-conditioning system. It takes Michael 7 hours to pipe in the identical split system. How long will it take John and Michael to pipe in the system if they work together?

The Solution

Some people might immediately jump to the conclusion that it will take 6 hours for John and Michael to pipe in the system. This would be the correct answer if the question asked were, "If John and Michael take 5 hours and 7 hours, respectively, to pipe in a split air-conditioning system, what is the average amount of time it took for them to pipe in the systems?" This, however, is not the question being asked.

From the information provided, we know John will take 5 hours to complete the piping job, while Michael will take 7 hours to complete the job. They will work together to pipe in another system and we must determine how long it will take them to accomplish this task together. The easiest way to solve this is to look at the first hour of the project. After the first hour, John will have completed 1/5 of the job, while Michael will have completed 1/7 of the job. We know this because John needs 5 hours to complete the job, so after 1 hour, 1/5 of the job would be finished.

Michael needs 7 hours to complete the job, so after 1 hour, 1/7 of the job would be finished. So, after 1 hour, 1/5 + 1/7 of the job has been finished because the two men are working together.

$$\frac{1}{5} + \frac{1}{7} = \frac{1 \times 7}{5 \times 7} + \frac{1 \times 5}{7 \times 5} = \frac{7}{35} + \frac{5}{35} = \frac{12}{35}$$

Thus, 12/35 of the job has been finished after the first hour. We can then conclude that, if 12/35 of the job is completed after the first hour, another 12/35 of the job, for a total of 24/35 of the job, will be completed after the second hour. The question then becomes, "How many times will 12/35 of the job need to be completed?" Three times will be too many, because this gives us 36/35, which is greater than 1, which represents the entire job. To convert this to another mathematical expression, we can ask, "How many times will 12/35 fit into 1?" This then becomes a simple division problem and is written in the following manner:

$$1 \div \frac{12}{35} = 1 \times \frac{35}{12} = \frac{35}{12} = 2.92 \text{ hours}$$

The general formula for this type of question is the following:

time to complete the joint activity = (time "A" × time "B") ÷ (time "A"+ time "B")

Now let us look at another question of this type. It takes Tony 3 hours to wash his service truck, whereas it takes Gary 4 hours to wash his service truck. Assuming the trucks are the same, how long will it take the two men to wash a truck together? Using the formula presented above, we get the following:

$$\begin{aligned}\text{time to wash the truck} &= (3 \text{ hours} \times 4 \text{ hours}) \div (3 \text{ hours} + 4 \text{ hours})\\ &= (3 \times 4) \div (3 + 4)\\ &= 12 \div 7\\ &= 1.7 \text{ hours}\end{aligned}$$

For those of you with a background in basic electricity, the same formula is used to calculate the total equivalent resistance of two resistors wired in parallel with each other. Consider an electric circuit with two resistors (5Ω and 20Ω) wired in parallel with each other.

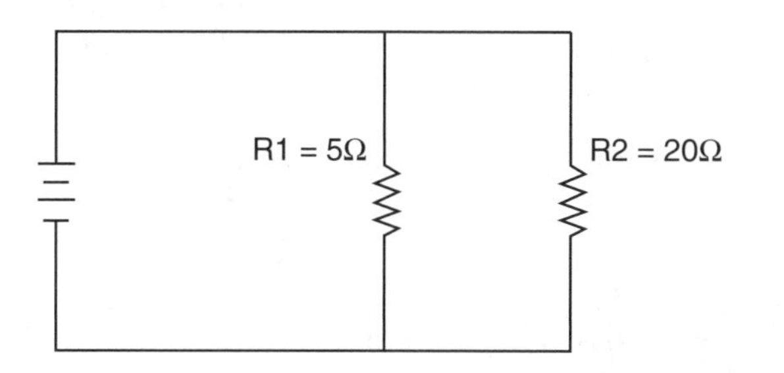

The total resistance of this circuit can be calculated as the following:

$$\begin{aligned}\text{total resistance} &= (R1 \times R2) \div (R1 + R2)\\ &= (5\Omega \times 20\Omega) \div (5\Omega + 20\Omega)\\ &= 100 \div 25\\ &= 4\Omega\end{aligned}$$

3.6.2 How to Estimate the Cost of a Small Job

The Problem

You are asked to price up a portion of a job that involves the installation of 25 pieces of equipment. Your cost of the equipment is $150 per unit, and your company will mark up the equipment by 40% to make a profit on the sale. It will take 2 hours of labor to prepare and install each piece of equipment. The billable labor rate for this project is 5 times the hourly rate paid to the worker. How much will your company bill the customer if the hourly rate for the worker is $20?

The Solution

There are a number of elements to this calculation, and each will be discussed separately. First, the cost of the equipment can be calculated by multiplying together the cost of one unit of the product and the number of units being purchased. The cost for the equipment is therefore determined as the following:

$$\begin{aligned}\text{cost of equipment} &= \text{unit cost} \times \text{number of units}\\ &= \$150/\text{unit} \times 25 \text{ units}\\ &= \$3750\end{aligned}$$

Notice that the unit cost is expressed in units of $/unit. When $/unit is multiplied by the number of units, the result is expressed in units of $.

The next element of this calculation is the markup on the equipment, which was previously established to be 40%. The markup of 40% can be expressed as 40/100 or 0.40. Because the cost of the equipment is $3750, the markup can be calculated by multiplying together the cost of the equipment and the markup percentage.

$$\begin{aligned}\text{markup} &= \text{cost of equipment} \times \text{markup percentage}\\ &= \$3750 \times 0.40\\ &= \$1500\end{aligned}$$

The customer will be billed a total of $5250 for the equipment. This amount is the sum of the cost of the equipment ($3750) and the markup ($1500).

This portion of the problem can be looked at another way. By putting the two elements of the customer cost of the equipment to one side of the equations, we get the following:

$$\begin{aligned}\text{customer cost} &= (\text{cost of equipment}) + (\text{markup})\\ &= (\text{unit cost} \times \text{number of units}) + 0.4\\ &\quad\ (\text{unit cost} \times \text{number of units})\end{aligned}$$

By factoring out the (unit cost × number of units) factor, this expression can also be presented as

$$\begin{aligned}\text{customer cost} &= (\text{unit cost} \times \text{number of units}) \times (1 + 0.40)\\ &= (\text{unit cost} \times \text{number of units}) \times (1.40)\end{aligned}$$

The manipulation of the previous expression adheres to the following distributive property format:

$$A + AB = A\,(1 + B)$$

When the actual numbers are put into this expression, we get:

$$\begin{aligned}\text{customer cost} &= (\text{unit cost} \times \text{number of units}) \times (1.40)\\ &= (\$150 \times 25) \times (1.40)\\ &= (\$3750) \times (1.40)\\ &= \$5250\end{aligned}$$

We now move on to the labor portion of the estimate. Here is what we know about the labor involved in this project:

- The worker earns $20/hour.
- The billing rate for the worker is 5 times the worker's hourly rate.
- Two hours of labor are required for the installation of each unit.
- A total of 25 units must be installed.

The total number of hours required to complete this project can be determined by multiplying together the number of units being installed and the number of hours required for each unit. This is expressed as:

$$\text{total hours} = (\text{hours/unit}) \times (\text{number of units})$$

Next, the cost per hour needs to be calculated. Because the billing rate for the worker is 5 times the hourly rate of the worker, which is \$20/hour, the billable rate for this job will be 5 × \$20/hour, or \$100/hour. The total for the labor portion of this job is calculated as

total labor charges = total hours × billable labor rate
= (hours/unit) × (number of units) × (billable labor rate)
= (2 hours/unit) × (25 units) × (\$100/hour)
= \$5000

The total cost for this project, not including any taxes or other fees, is the sum of the equipment and labor costs as shown next:

project cost = cost of equipment + equipment markup + labor charges
= cost of equipment + equipment markup + (total hours × billable labor rate)
= (\$150/unit × 25 units) + 0.4 (\$150/unit × 25 units) + (total hours × billable labor rate)
= \$3750 + \$1500 + \$5000
= \$10,250

3.6.3 Estimating the Amount of Sheet Metal Needed

The Problem

You have been asked to determine how many square feet of sheet metal will be needed to construct an 8′ section of ductwork that has cross-sectional measurements of 12″ × 24″. Assume there is no waste, and ignore any additional material needed to join or fasten the duct section.

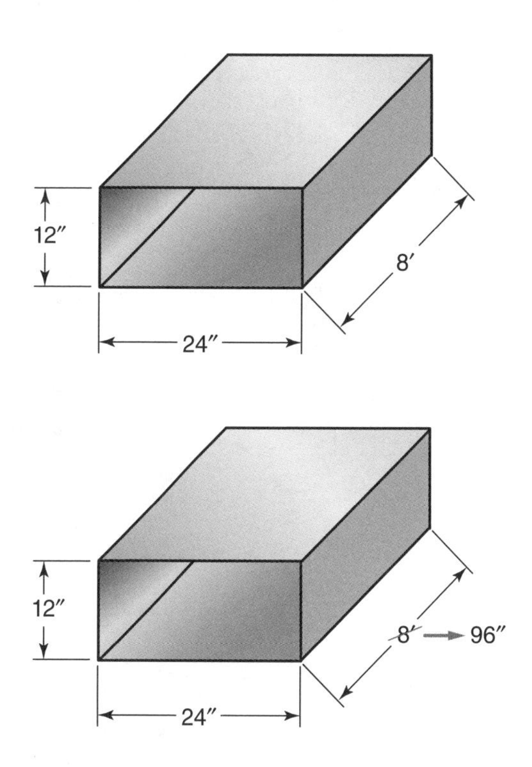

The Solution

The first step in this problem is to make certain all the dimensions are expressed in the same units. For example, the length of the duct section is expressed in feet, while the cross-sectional measurements of the duct section are expressed in inches. To ensure accurate results, the units should be expressed in either inches or feet, but not a combination of both. For this example, we will convert all measurements to inches. The length of the duct section will therefore be expressed as 96″, because 8′ × (12 in/ft) = 96″.

We can then look at each of the four sides of the duct section. Two sides have measurements of 12″ × 96″, and two sides have measurements of 24″ × 96″. We then calculate the area of each of the four sides of the duct section to determine how many square feet of sheet metal are needed to fabricate this duct section.

The surface area of the four sides of the duct section can be calculated as follows:

- Side 1: 12″ × 96″ = 1152 in^2
- Side 2: 12″ × 96″ = 1152 in^2
- Side 3: 24″ × 96″ = 2304 in^2
- Side 4: 24″ × 96″ = 2304 in^2

The total required amount of sheet metal is the sum of the areas of the four sides of the duct, or 1152 in^2 + 1152 in^2 + 2304 in^2 + 2304 in^2, for a total of 6912 in^2. The problem, however, is not completed yet. The question asked to calculate the number of square feet of sheet metal needed to fabricate this section of ductwork. The answer obtained is in units of square inches. We now need to know how many square inches are in one square foot. As discussed earlier in the chapter, 144 in^2 are in 1 ft^2, so the result obtained earlier can be converted to square feet.

$$6912\ in^2 \times (1\ ft^2/144\ in^2) = 6912\ in^2 \times 1\ ft^2/144\ in^2 = 48\ ft^2$$

Thus, 48 square feet of sheet metal are required to fabricate this section of ductwork.

Alternate Solution

In the previous solution to this problem, the measurements were all expressed in inches. However, the calculation would have been much easier if the measurements for this problem were expressed in feet:

- The length of the duct section is 8′.
- The height of the duct section is 12″, or 1′.
- The width of the duct section is 24″, or 2′.

The surface areas of the four sides of the duct can then be expressed as:

- Side 1: 1′ × 8′ = 8 ft^2
- Side 2: 1′ × 8′ = 8 ft^2
- Side 3: 2′ × 8′ = 16 ft^2
- Side 4: 2′ × 8′ = 16 ft^2

As before, the total amount of sheet metal required will be the sum of the areas of the four sides of the duct, or 8 ft^2 + 8 ft^2 + 16 ft^2 + 16 ft^2, for a total of 48 ft^2. One final way of looking at this problem would be to visually “open” or unfold the duct section. By unfolding the duct section, one large rectangle is created.

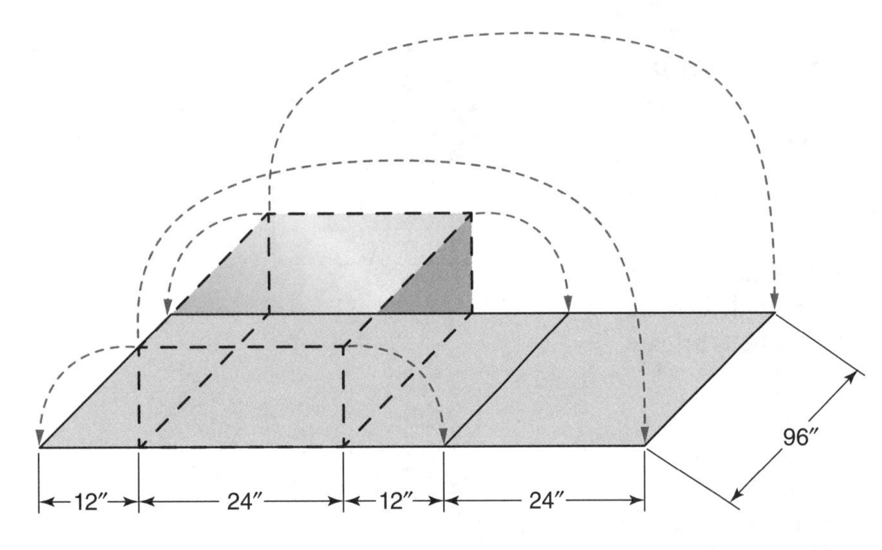

The measurements of this rectangle are 72″ × 96″.

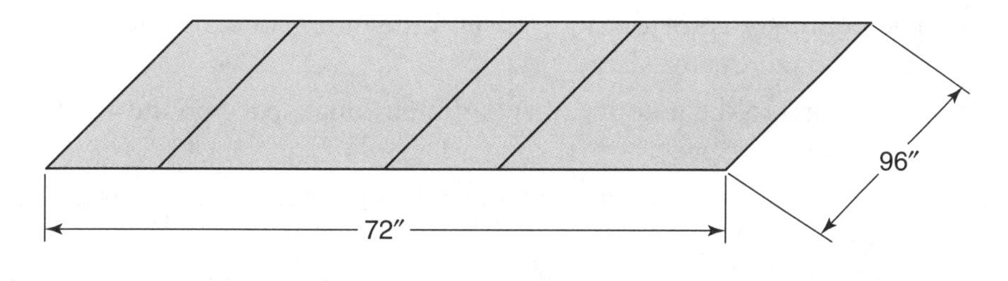

By calculating the area of this rectangle, the same result is found. As a final thought, if it was noticed that 72″ = 6′ and 96″ = 8′, the area of the rectangle could have been found by simply multiplying 6′ × 8′ to arrive at the 48 ft^2 solution.

3.6.4 Sneezing and Driving (A Fun Example)

The Problem

When you sneeze, your eyes are closed for approximately 0.25 seconds. If you are driving a car at 50 miles per hour when you sneeze, how many feet have you driven with your eyes closed?

The Solution

To solve this problem, determine what is being asked and what you have been given to work with. The important factors here are time, speed, and distance. We are looking for an answer expressed in terms of feet, but the information given us is in seconds, hours, and miles.

The best way to solve this problem is to work with the units. For starters, we know there are 5280 feet in a mile and that the driver is traveling at a speed of 50 miles per hour. Combining these terms tells us that the driver is moving 264,000 feet per hour. We need to know, however, how many feet the driver will have traveled in 0.25 seconds, so a shorter unit of time is needed.

Because there are 60 minutes in an hour, we can determine that the driver is traveling at a speed of 4400 feet per minute. We still need a shorter unit of time. Because there are 60 seconds in a minute, we can determine that the driver is traveling at a speed of 73.3 feet per second. If the driver is moving at a speed of 73.3 feet per second, in 0.25 seconds the driver has covered a distance of approximately 18.3′ (73.3 ÷ 4).

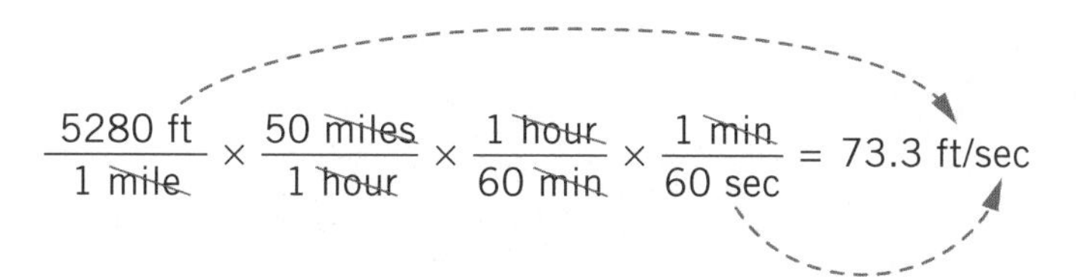

Summary

- The perimeter is the distance around the outer edge of a two-dimensional area or shape.
- Area refers to the amount of two-dimensional space bound by the perimeter of a space.
- Volume describes the three-dimensional space a particular substance occupies.
- Average is the term used to describe the central value of a particular set of numbers.
- A weighted average is used when the numbers being averaged have different impacts on the result.
- The metric system is a base-10 measurement system used throughout the world, except in the United States.
- Common units in the metric system include the meter, kilogram, and liter.
- Word problems are used to translate real-world situations into mathematical expressions that can be solved to provide answers for those problems.

Chapter Assessment

Name ______________________ Date ____________ Class ______________

Know and Understand

______ 1. The perimeter of a room that is 15′ × 10′ is _____.

A. 150 ft^2
B. 150′
C. 50 ft^2
D. 50′

______ 2. The area of a piece of property that is 100′ × 150′ is _____.

A. 500′
B. 500 ft^2
C. 15,000′
D. 15,000 ft^2

______ 3. An 80″ × 96″ area will use about _____ 8″ × 8″ floor tiles.

A. 120
B. 150
C. 180
D. 200

______ 4. What is the volume of a three-dimensional rectangular shape that has measurements of 5″ × 6″ ×6″?

A. 36″
B. 36 in^3
C. 180″
D. 180 in^3

______ 5. What is the volume of a three-dimensional rectangular shape that has measurements of 6″ × 1′ × 1′?

A. 0.5 ft^3
B. 5 ft^3
C. 8 ft^3
D. 10 ft^3

______ 6. What is the volume of a vertical cylinder that has a diameter of 6″ and a height of 1′?

A. 113 in^3
B. 226 in^3
C. 339 in^3
D. 452 in^3

______ 7. The average value of the following set of numbers (12, 14, 18, 36) is _____.

A. 12
B. 16
C. 20
D. 25

______ 8. The length of a yardstick is approximately equal to _____.

A. 1 cm
B. 50 mm
C. 1 m
D. 1 km

______ 9. Two miles is equal to approximately how many meters?

A. 804.5 m
B. 1609 m
C. 3218 m
D. 6436 m

______ 10. A particular job requires 500 ft^2 of material. If the cost of material is \$3.00 per square foot, how much will the materials for the job cost if the contractor purchases an additional 20% to account for any waste and damage?

A. \$1500
B. \$1800
C. \$2100
D. \$2400

Apply and Analyze

1. A contractor is installing a suspended ceiling in an open office area. The area measures 100′ × 50′. Assuming the ceiling tiles are 24″ × 48″ and there is no waste, how many tiles will be required to complete the ceiling?

____

2. Convert 6′ 10″ into centimeters.

____

3. If refrigerant costs a contractor $80/lb, and the contractor intends to mark the refrigerant up by 20%, how much will the contractor charge for refrigerant?

____

Critical Thinking

1. Why is area expressed in square feet, while volume is expressed in cubic feet?

____

____

____

2. Why is it a good idea to estimate the results of calculations prior to actually doing them?

____

____

____

3. What calculations are required to determine how much material is needed to wrap a section of ductwork with insulation?

____

____

____

Large Prints Activity

Name ______________________ Date ____________ Class ______________

Practice Using Large Prints

Refer to Print-21 in the Large Prints supplement to answer the following questions. Note: The large prints are created at 50% of their original size. When extracting information from the print using an architect's scale, use a scale that is one-half of the one indicated on the drawing.

1. How many feet of piping would be required to run around the entire TELECOM RM (h005b), ignoring any doors, windows, or other openings or interruptions in the walls?

 ____

2. How many feet of piping would be required to run around the entire ELECTRICAL RM (h005), ignoring any doors, windows, or other openings or interruptions in the walls?

 ____

3. How many cubic feet of air will need to be added to the PLUMBING RM (h005a) each minute if the air in this room needs to be "changed" 5 times each hour? Assume a ceiling height of 10′.

 ____

Refer to Print-26 in the Large Prints supplement to answer the following questions.

4. How many gallons of water will be in each foot of the cold water make-up line? Assume 231 in^3 = 1 gal, the pipe is completely full of water, and the pipe measurement provided is the inside diameter of the pipe.

____

5. If the water in the pipe from Question 4 flows at a speed of 20 ft/min, how many gallons per minute are flowing through the pipe?

____

4 Measurements and Scales

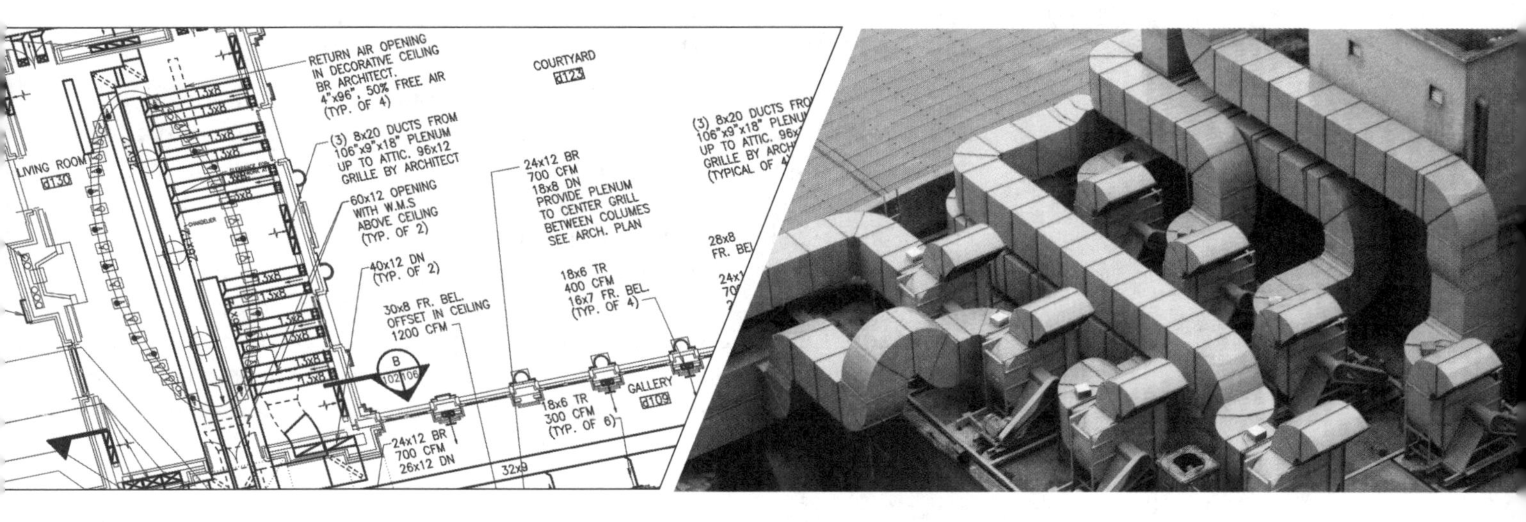

Chapter Outline

4.1 Tape Measure

4.2 Drawing Scales

4.3 Scaled Drawings

4.3.1 The Architect's Scale
4.3.2 The Engineer's Scale

Learning Objectives

After studying this chapter, you will be able to:

- Describe the use of a ruler or a tape measure.
- Discuss the types of measurements that can be performed with a tape measure.
- Define a scale and how it is used in prints.
- Create a scaled drawing using a 1/4″ = 1′ scale.
- Perform measurements using the architect's scale and engineer's scale.

Introduction

Taking and interpreting measurements accurately are two important skills for any HVACR system installer, technician, or print reader. An improper measurement can lead to errors in selecting or cutting materials, sizing HVACR equipment, and estimating projects. Most project-related drawings are not life-size renderings. Instead, scaled drawings are used to show the proper proportions of a project. A tradesperson applies their understanding of measurements to read scaled drawings and thus successfully carry out a project.

Understanding all aspects of the print reading process helps facilitate successful completion of construction and other building trade projects. Print scales and measurements are explored in this chapter to help ensure you become proficient in accurately and completely reading prints and project plans.

Technical Terms

architect's scale

engineer's scale

scale

tape measure

4.1 Tape Measure

One of the most important tools used by a tradesperson on a jobsite is a tape measure, **Figure 4-1**. A ***tape measure*** is marked with units used for measuring, and can be set at a specific length to transfer measurements from one place to another. For example, an HVACR system installer may need a piece of hard-drawn copper piping to offset two 90° elbows to complete a portion of a refrigerant line. First, the installer must measure the distance between the two fittings and then cut the pipe to the correct length. Once the measurement is obtained between the fittings, the measurement is transferred to the pipe itself to ensure it is cut to the correct length. An error in the initial measuring or in transferring this

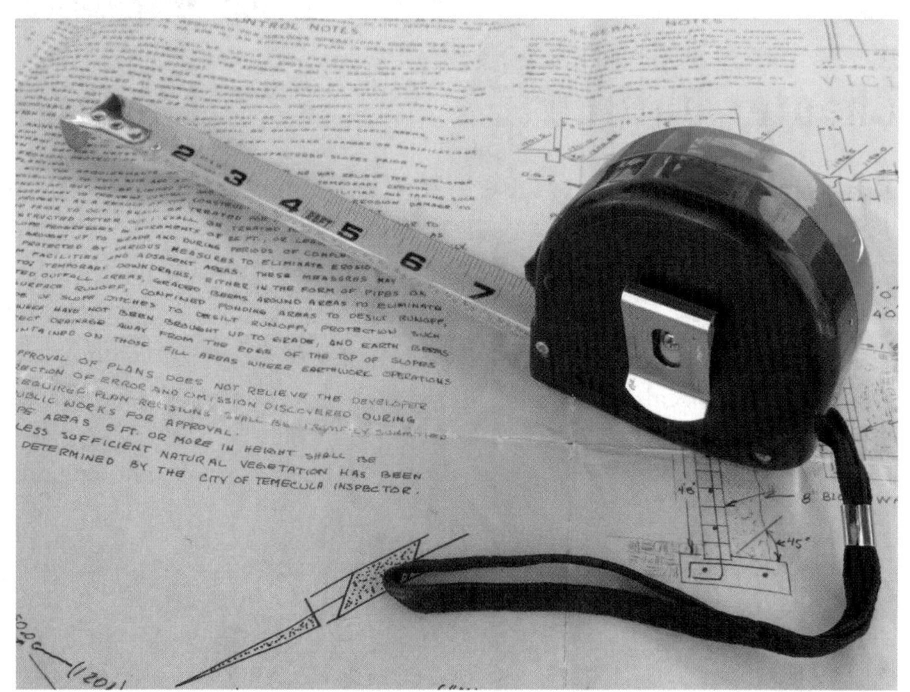

Goodheart-Willcox Publisher

Figure 4-1. A tape measure.

measurement to the pipe results in the pipe cut to the wrong length. Cutting a pipe the wrong length causes wasted material and wasted time on the jobsite.

The tape measure measures not only feet and inches, but also fractions of inches. Fractions found on rulers include:

- Halves: 1/2
- Quarters: 1/4, 3/4
- Eighths: 1/8, 3/8, 5/8, and 7/8
- Sixteenths: 1/16, 3/16, 5/16, 7/16, 9/16, 11/16, 13/16, and 15/16

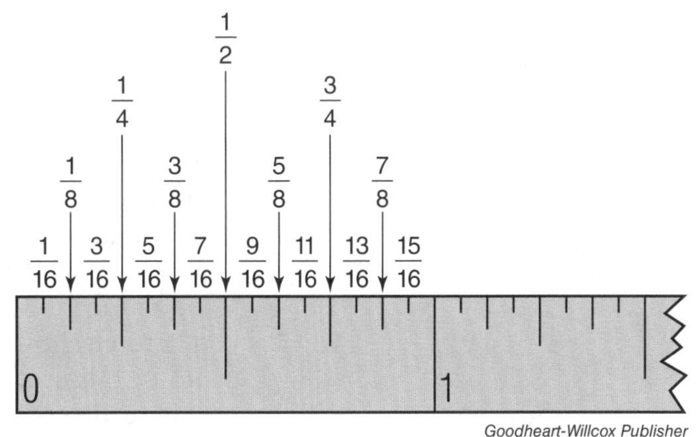

Goodheart-Willcox Publisher

Figure 4-2. Progression of the fractions of an inch.

Fractions within a one-inch segment on a tape measure are shown in **Figure 4-2**. The half-inch line divides the inch into two equal parts. Using basic addition, two half-inch sections result in a final measurement of one full inch.

The 1/4 line divides the inch into four equal sections, or quarters. Two quarters equal 1/2, so a measurement of two quarters is represented by the one-half inch line. The fraction 3/4 is represented by the quarter-inch line, which is midway between the half-inch line and the full-inch line.

The first eighth-inch line represents 1/8″. Because 2/8 equals 1/4, a measurement of 2/8 is represented by the first 1/4″ line. A measurement of 3/8″ is located midway between the 1/4″ line and the 1/2″ line, and so forth.

The smallest increment on the ruler divides the inch into sixteenths. The first sixteenth-inch line represents one sixteenth, 1/16″. Two sixteenths equal 1/8, so a measurement of 2/16 is represented on the ruler by the first 1/8″ line. Similarly, a measurement of 4/16 equals 1/4, so a measurement of 4/16 is shown on the ruler by the first 1/4″ line. Some tape measures provide even smaller increments, namely 1/32″, **Figure 4-3**.

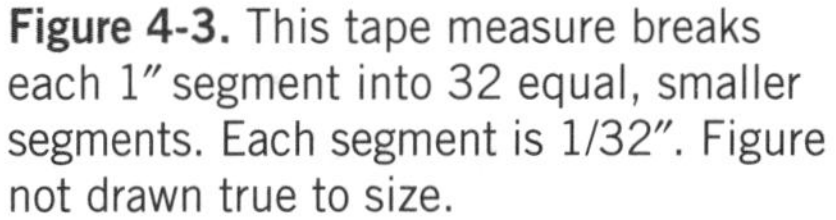

Goodheart-Willcox Publisher

Figure 4-3. This tape measure breaks each 1″ segment into 32 equal, smaller segments. Each segment is 1/32″. Figure not drawn true to size.

4.2 Drawing Scales

A print is the main form of communication between the design engineer and those who execute the plan. Thus, a print must accurately represent the structure and all that is contained within it, or how the structure should look once completed. For this reason, all prints are drawn to scale unless otherwise noted. When a figure or print is drawn to scale, the sizes and dimensions of objects in the drawing are relative to each other. For example, if a wall is drawn twice as long as another wall on a print or drawing, it is twice as long as the other when constructed. Various scales are used to create prints.

Given the size of various construction projects, along with the costs involved, it would be physically impossible to create sets of drawings that were the actual size of the project. Prints are drawn as scaled-down renditions of the final project. Each object drawn on a particular print is reduced by the same proportion, or ***scale***, giving an accurate representation of the desired project outcome.

For example, the first line in **Figure 4-4** is 2 units in length, while the second line is 4 units in length. The four-unit line is twice as long as the two-unit line.

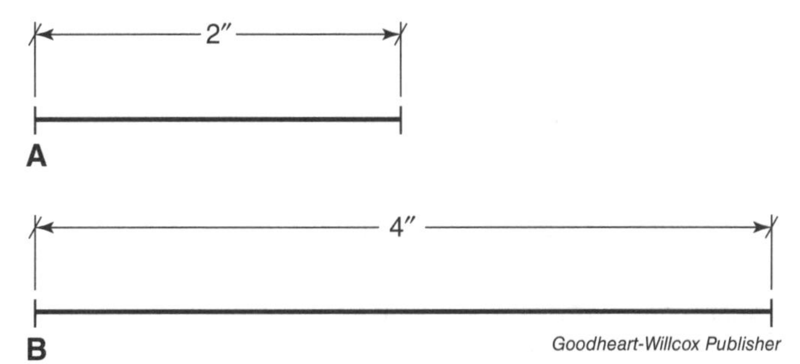

Goodheart-Willcox Publisher

Figure 4-4. The line in A represents an object that is one-half as long as the object represented by B.

If these two lines were found on the same drawing, the actual object represented by the four-unit line would be twice as long as the object represented by the two-unit line.

The size of the project and the size of the final print determine the proportion by which the original measurements are reduced. The final drawing should be large enough that construction details can easily be seen, and small enough to permit an entire view to be shown on a single print. An example of creating a scaled drawing is to let 1/4″ on the drawing represent 1′ of linear measurement on the actual project. The sketch of a room shown in **Figure 4-5** is drawn as a 4″ × 3″ rectangle. If the scale 1/4″ = 1′ is used, there would be 16 1/4″ segments across, and 12 quarter-inch segments from top to bottom.

Combining these two figures, we create a final grid with both segments represented in the rectangle. Using this scale, this figure then represents a room 16′ long by 12′ wide. To calculate the area of this room, simply multiply the room's length by its width. In this example, the area of the room is 192 ft^2 (16′ × 12′ = 192 ft^2).

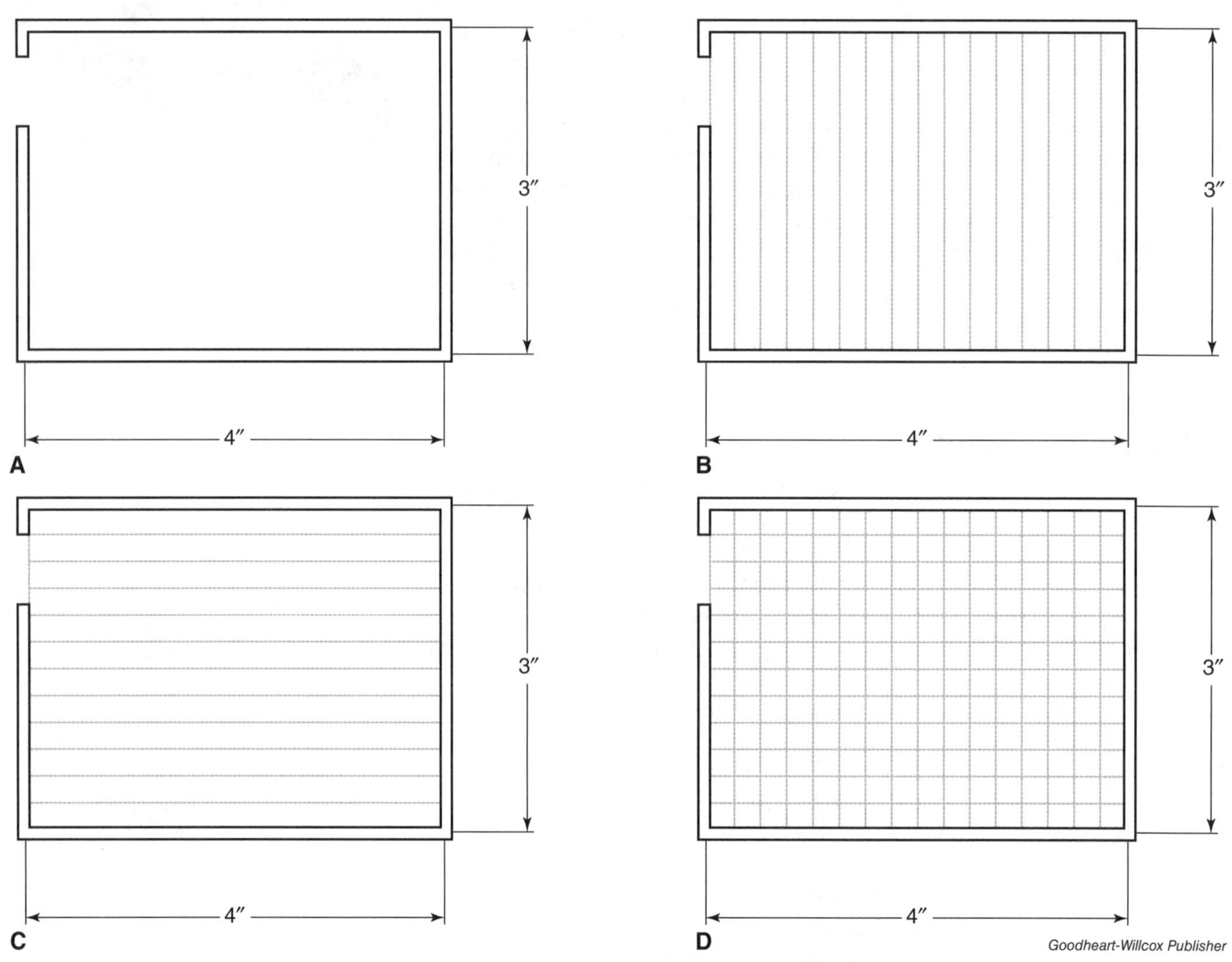

Figure 4-5. A—This 3″ × 4″ drawing represents a room that is actually 12′ × 16′. The drawing was created with a scale of 1/4″ = 1′. Figure not drawn true to size. B—There are sixteen 1/4″ sections from right to left, representing the 16′ length of the room. Figure not drawn true to size. C—There are twelve 1/4″ sections from top to bottom, representing the 12′ width of the room. Figure not drawn true to size. D—Each square in this grid represents a 1′ × 1′ square in the room.

4.3 Scaled Drawings

To create these scaled drawings and prints, specially designed scales are used. These instruments enable the drafter of the print to easily obtain the correct scale of a line without having to perform numerous calculations.

Procedure — Creating Scaled Drawings

If an individual is creating a 1/4″ = 1′ scaled drawing of a 6′6″ wide wall using a standard ruler, the following steps must be taken:

1. State the actual dimension of the wall in units. Thus, the 6′6″ wall is expressed as 6.5 units. (The 0.5 comes from the 6″ portion of the wall, which is one-half of a foot, expressed as 0.5.)
2. Multiply the number of units by the scale being used. In this case, the scale is 1/4″ = 1 foot, so we need to multiply 6.5 units by 1/4. The result is 1.625″.
3. Convert 1.625″ into fractions commonly used on a standard ruler. This measurement can be converted to 1 5/8″.
4. Determine the scale. The line that is 1 5/8″ long represents the wall in our example.

Imagine the time it would take if the architect or engineer had to perform the above calculations for each line drawn on a print. The possibility of making errors while performing these calculations increases also. Luckily, there is an alternative. Two commonly used scales eliminate the need for such calculations: the architect's scale and the engineer's scale.

4.3.1 The Architect's Scale

The ***architect's scale***, **Figure 4-6**, is a tool most often used in the HVACR field for drafting and producing scaled drawings. A typical 12″ architect's scale is a six-sided, triangular-shaped instrument with a number of different scales printed on it. Commonly used scales include the following:

- 1/8″ = 1′0″
- 1/4″ = 1′0″
- 3/8″ = 1′0″
- 1/2″ = 1′0″
- 3/4″ = 1′0″
- 1″ = 1′0″
- 1 1/2″ = 1′0″

Goodheart-Willcox Publisher

Figure 4-6. An architect's scale.

The size of each scale is labeled on each rule. The typical architect's scale has at least 10 different scales from which to choose. Some scales are read left to right, while others are read right to left. Although we will not examine all of the scales in this text, examples using the 1/8″ = 1′0″ and 1/4″ = 1′0″ scales will be given.

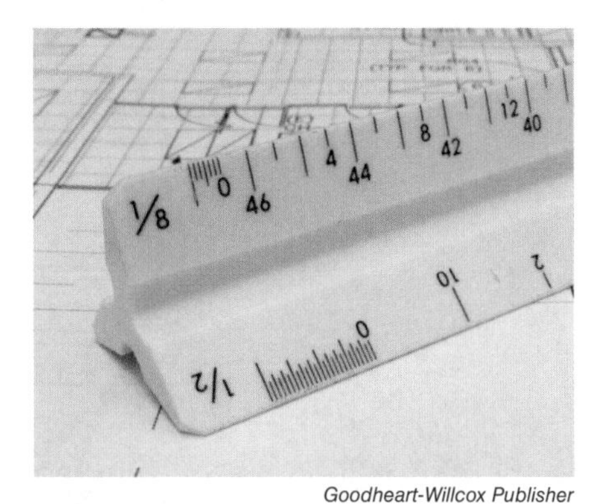

Goodheart-Willcox Publisher

Figure 4-7. The 1/8″ scale on the architect's scale.

The 1/8″ Scale

On the 1/8″ scale, **Figure 4-7**, notice that the numbers 0, 4, 8, 12, and 16 increase as we go from left to right. The larger numbers 34, 36, 38, 40, 42, 44, and 46 increase from right to left and are part of the 1/4″ scale that starts on the right-hand side of the scale. When using the 1/8″ scale, these numbers can be disregarded, but the markings will be used for both the 1/4″ and 1/8″ scales.

The numbers to the right of the zero point on the scale indicate the number of feet of a particular line. For example, a line drawn from the zero point to the number 4 on the scale represents a wall or other component that is 4′ in length, **Figure 4-8**. Similarly, a line drawn from the zero point to the number 24 on the scale represents a wall or other component that is 24′ in length.

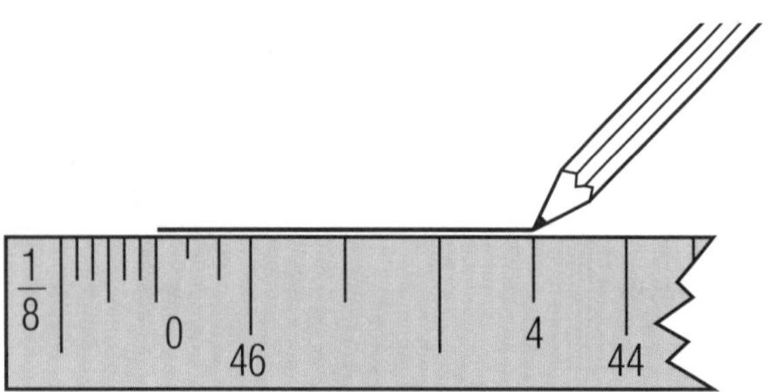

Goodheart-Willcox Publisher

Figure 4-8. The line drawn in this figure represents an actual 4′ line. The line segment is actually 1/2″ long. Figure not drawn true to size.

To the left of the zero point on the scale is a 1/8″ section divided into smaller increments, which represent fractions of a foot, **Figure 4-9**. Each small increment on this scale represents 2″ of actual length. There are six 2″ increments on the scale, the largest of which represents 6″, or one-half of a foot. Thus, to represent a construction element that is 4′4″ in length, we would draw a line from the number 4 on the scale to the second small increment in the divided section, **Figure 4-10**.

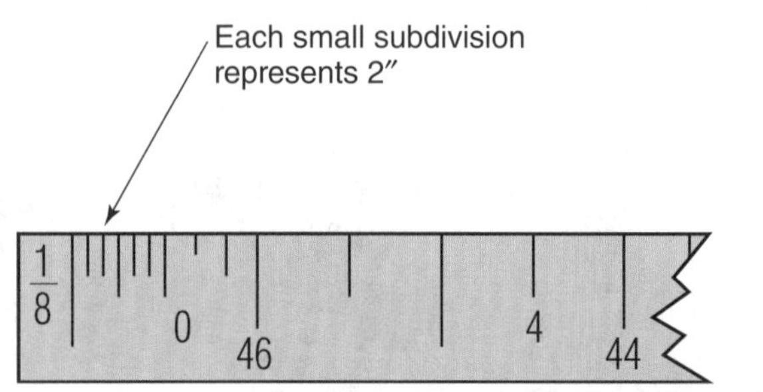

Goodheart-Willcox Publisher

Figure 4-9. The space to the left of the zero mark represents 1′, where the space is broken down into six 2″ segments. Figure not drawn true to size.

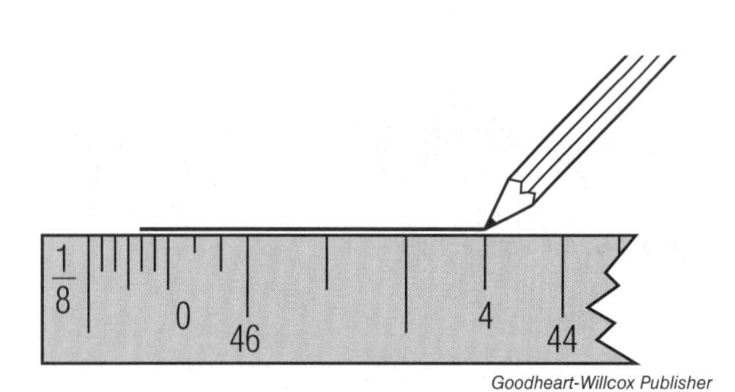

Goodheart-Willcox Publisher

Figure 4-10. The line drawn in this figure represents an actual 4′4″ line. This line extends left of the zero point by 4″. Figure not drawn true to size.

The 1/4″ Scale

On the 1/4″ scale, **Figure 4-11**, numbers, such as 0, 2, 4, 6, 8, and 10, increase as we go from right to left. The larger numbers, such as 92, 88, 84, 80, and 76, decrease from right to left. The numbers that decrease from right to left are part of the 1/8″ scale that starts on the left-hand side of the scale, as discussed in the previous section. When using the 1/4″ scale, these numbers can be disregarded.

The numbers that increase to the left of the zero point indicate the number of feet the line represents. For example, a line drawn from the zero point to the number 6 on the scale represents an actual measurement

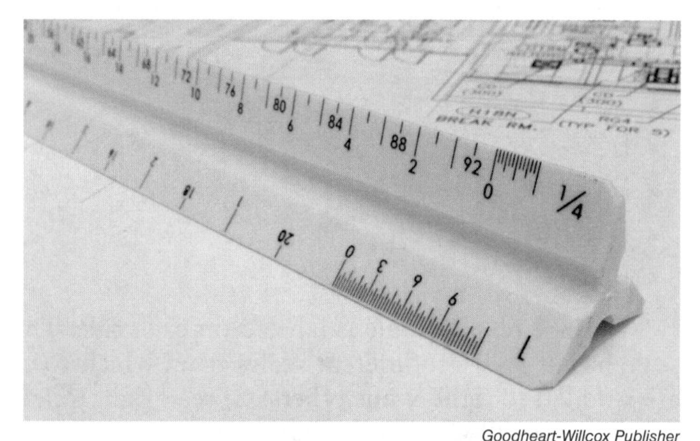

Goodheart-Willcox Publisher

Figure 4-11. The 1/4″ scale on the architect's scale.

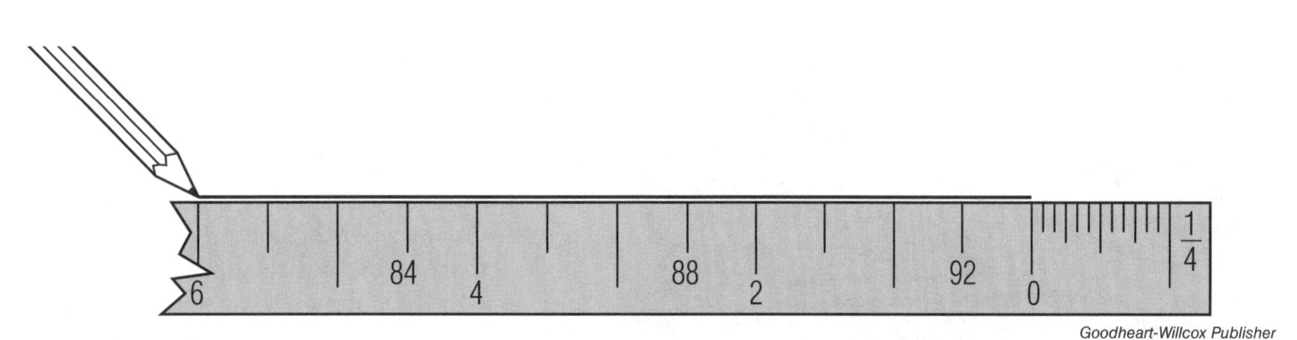

Goodheart-Willcox Publisher

Figure 4-12. The line drawn in this figure represents an actual 6′ line. In actuality, the line segment is 1 1/2″ long. Figure not drawn true to size.

of 6′, **Figure 4-12**. Similarly, a line drawn from the zero point to the number 10 on the scale represents a wall or other component 10′ in length.

To the right of the zero point on the scale is a 1/4″ section that is divided into 12 fractions of a foot, **Figure 4-13**, each representing 1′ of actual length. There are 12 one-inch increments on the scale, the largest, center increment which represents 6″, or 1/2′. The two medium-size lines represent 3″ increments. For example, to represent a wall that will be 5′7″ in length, we would draw a line from the line representing 5′, which is midway between the labels for 4′ and 6′, on the scale to the seventh line, which represents 7″, in the subdivided section, **Figure 4-14**.

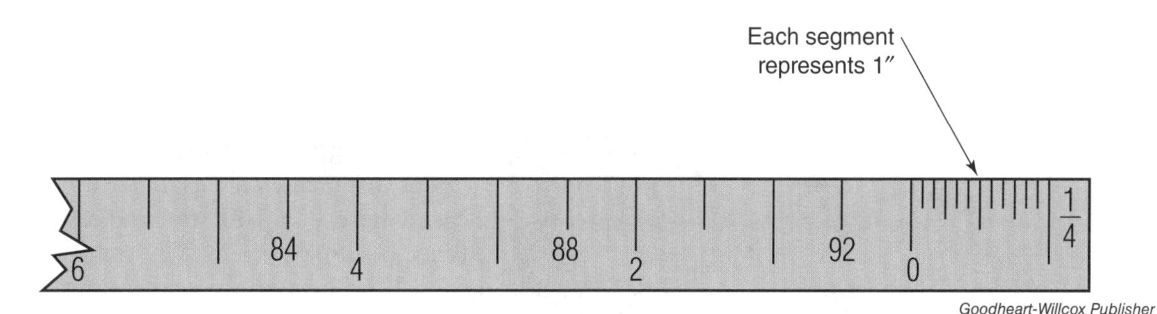

Goodheart-Willcox Publisher

Figure 4-13. Each of the segments to the right of the zero on the 1/4″ = 1′ scale represents 1″. Figure not drawn true to size.

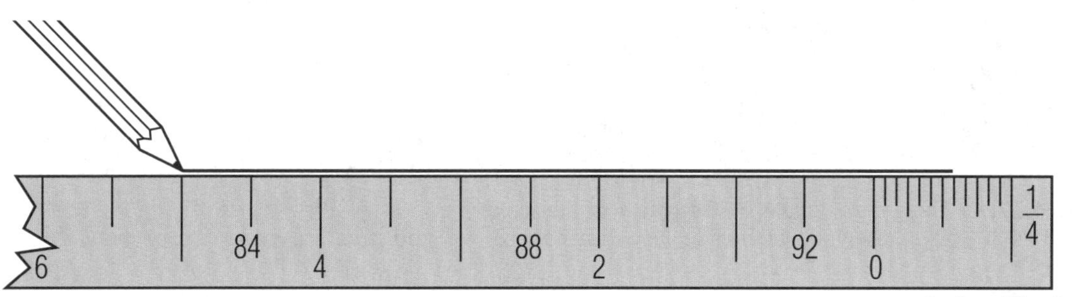

Goodheart-Willcox Publisher

Figure 4-14. This line represents 5′ 7″ on the 1/4″ = 1′ scale. Figure not drawn true to size.

4.3.2 The Engineer's Scale

The ***engineer's scale***, **Figure 4-15**, can be used to reduce larger dimensions to a smaller scale. It is similar to the architect's scale in that it enables individuals to construct scaled drawings. While the architect's scale often contains scales that range from 1/8″ = 1′0″ to 3″ = 1′0″, the engineer's scale contains scales that often range from 1″ = 10′0″ to 1″ = 100′0″. The engineer's scale is frequently used to render drawings of

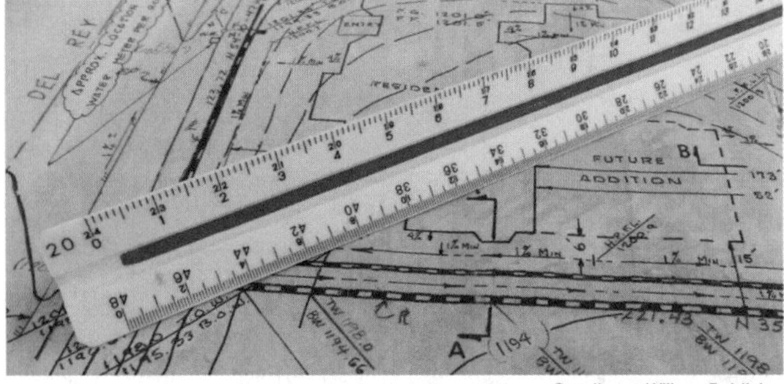
Goodheart-Willcox Publisher

Figure 4-15. An engineer's scale.

construction sites that include the land on which the building or structure is to be built. These construction sites are often hundreds, if not thousands, of feet in length and width. Without the use of the architect's scale, the drawings would be too large and impractical to meet the needs of the construction trades.

In **Figure 4-16**, notice that the 1″ = 10′ scale is labeled with the number 10 at the left-hand side of the scale. Each 1″ increment on this scale represents 10′ of actual length or distance on the project. On this scale, each 1″ section is divided into 10 subdivisions. Each of these subdivisions represents 1′. A piece of property that is 85′ in width is therefore represented by a line that is 8.5″ in length. The same scale can be used to represent larger distances, such as 1″ = 100′ or 1″ = 1000′. In the case where 1″ = 100′, the line represents an actual distance of 850′, not 85′. The small subdivisions between each 1″ section on the scale represents 10′.

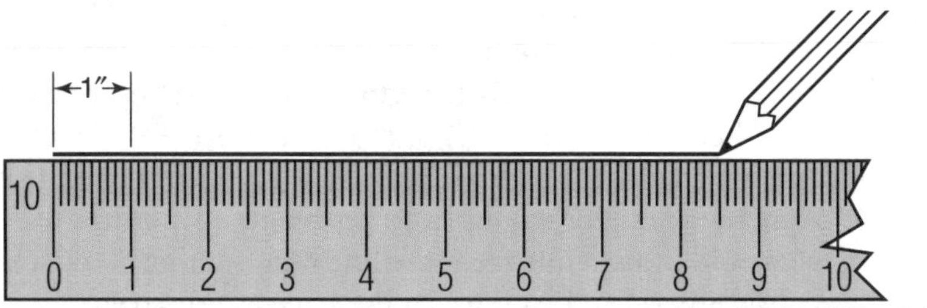

Goodheart-Willcox Publisher

Figure 4-16. Each 1″ segment on the 1″ = 10′ scale represents 10′. The line drawn here represents 85′. Figure not drawn true to size.

Other scales on the engineer's scale include 1″ = 20′ and 1″ = 30′. On the 1″ = 20′ scale, each 1″ segment is divided into 20 smaller sections, each of which represents 1′. On the 1″ = 30′ scale, **Figure 4-17**, each 1″ segment is divided into 30 smaller sections, each of which represents 1′. In a similar manner to the 1″ = 10′ scale, these two scales can be used to create scales by powers of 10, such as 1″ = 2000′ or 1″ = 300′.

Goodheart-Willcox Publisher

Figure 4-17. On the 1″ = 30′ scale, a 1″ segment on the scale represents 30′, as represented by the midpoint between the 2 and the 4.

Taking a look at the 1″ = 30′ scale in **Figure 4-18**, the line drawn can represent 75′, 750′, 7500′, or even 75,000′.

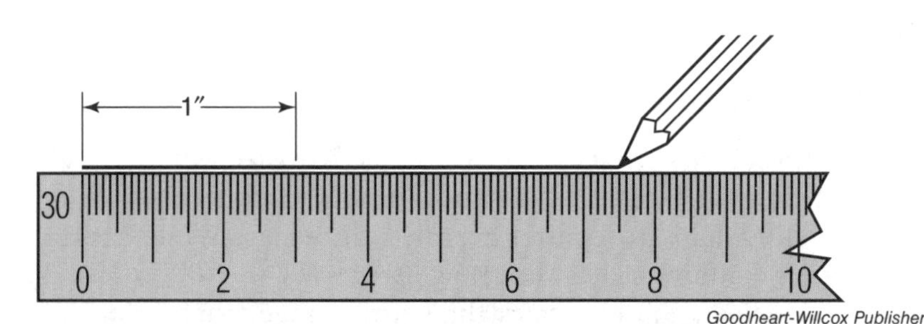

Goodheart-Willcox Publisher

Figure 4-18. The line drawn here represents 75′. Figure not drawn true to size.

Procedure — Determining a Print Scale Using 1″ = 20′

A plan of a house property has been created using the 1″ = 20′ scale, **Figure 4-19**. Use the following steps to calculate the scale accurately and read the plan correctly.

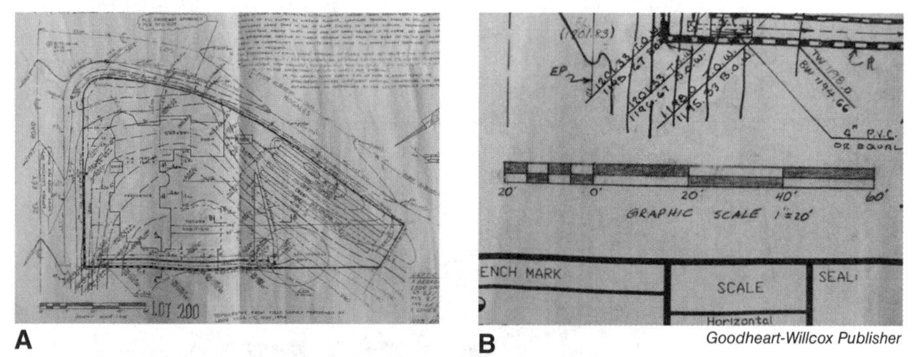

A B

Goodheart-Willcox Publisher

Figure 4-19. A—A site or plot plan. B—This notation on the plan indicates it was created using a 1″ = 20′ scale.

1. Line up the 0 on the engineer's scale with the 0 on the print. The 2 on the engineer's scale should line up perfectly with the 20′ mark on the print, the "4" with the 40′ mark on the print, and the "6" with the 60′ mark on the print, **Figure 4-20**.

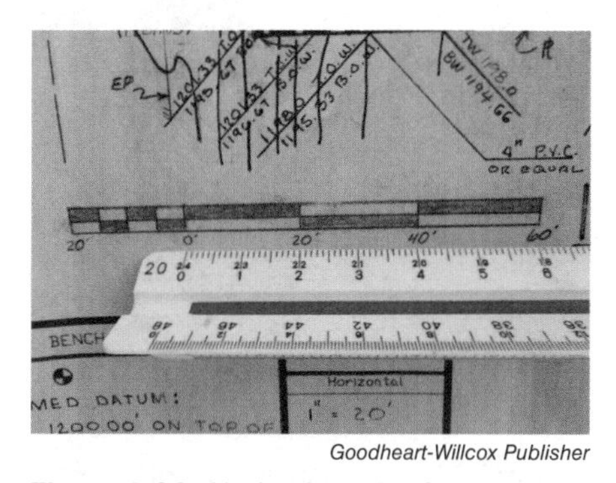

Goodheart-Willcox Publisher

Figure 4-20. Notice how the 2, 4, and 6 on the 1″ = 20′ scale line up perfectly with the 20′, 40′, and 60′ markings on the plan.

2. Determine the scale. The actual distance between the 0 and the 2 on the engineer's scale is 1″, which represents 20′, **Figure 4-21**.

Goodheart-Willcox Publisher

Figure 4-21. On the 1″ = 20′ scale, a 1″ segment on the scale represents 20′.

3. Use the 1″ = 20′ scale to determine the dimensions of the shop area in this residence.
4. Line up the 0 on the scale to one end of the line representing one of the walls in the shop. It can be determined that the shop is 25′ wide, **Figure 4-22**. The line representing the shop wall extends from the "0" on the scale to a point halfway between the 2 and the 3. Because the 2 represents 20′ and the 3 represents 30′, the line in the plan indicates a dimension of 25′.

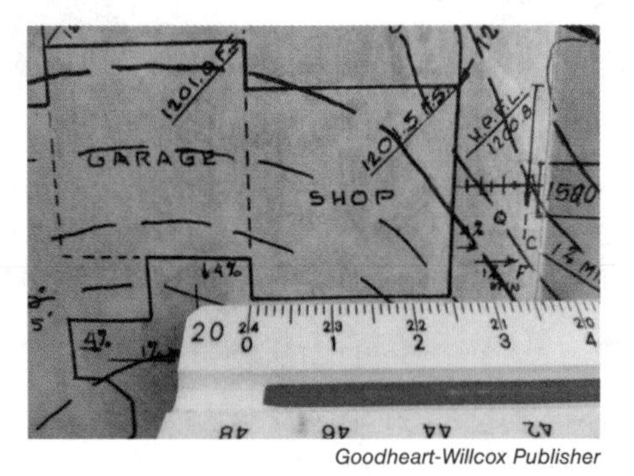

Goodheart-Willcox Publisher

Figure 4-22. The shop area is 25′ wide.

5. Use the same method to determine the length of the shop. This shows the shop is 24′ long, **Figure 4-23**.

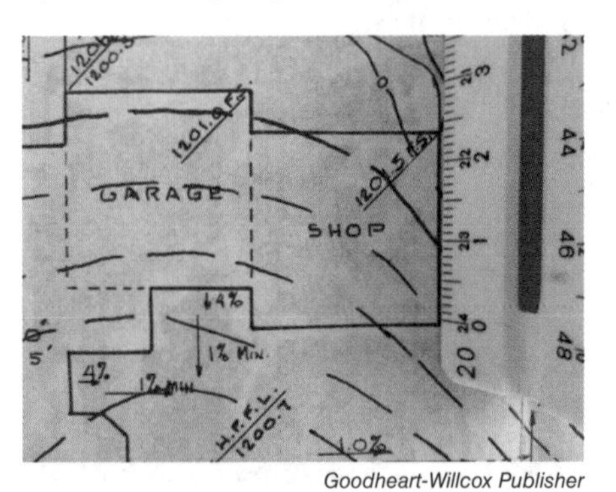

Goodheart-Willcox Publisher

Figure 4-23. The shop area is 24′ deep.

6. Determine the size of the property by using the same scale, **Figure 4-24**. By examining the length of the line, the property is calculated to be 223′ deep.

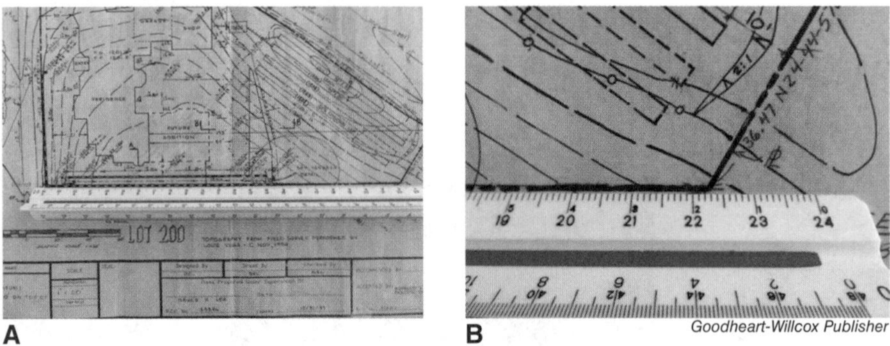

Goodheart-Willcox Publisher

Figure 4-24. A—Using the 1″ = 20′ scale to determine the length of the property. B—The length of the property is 223′.

Summary

- A tape measure is marked with units used for measuring. It can be set at a specific length to transfer a measurement from one object to another.
- A tape measure can measure feet, inches, and fractions of an inch.
- Objects on a print are reduced by the same proportion or scale to provide an accurate representation of the project's actual size.
- Scales can be created by the drafter using simple multiplication and decimal conversion.
- Prints are often drawn using the architect's scale or the engineer's scale.
- The architect's scale is a common tool in the HVACR field used for drafting and producing scaled drawings. It utilizes smaller scales, such as 1/8″ = 1′.
- The engineer's scale is used to reduce much larger dimensions to a smaller scale. It utilizes larger scales, such as 1″ = 20′, and is often used to create site plans.

Chapter Assessment

Name ______________________ Date ______________ Class ______________

Know and Understand

_______ 1. The device that has scales such as 1/4″ = 1′ and 1″ = 1′ is the _____.

A. tape measure
B. ruler
C. architect's scale
D. engineer's scale

_______ 2. The tool typically used to create scaled site plans is the _____.

A. tape measure
B. ruler
C. architect's scale
D. engineer's scale

_______ 3. Using the 1/4″ = 1′0″ scale, a line that is 4″ long represents an object that is _____ in length.

A. 4′
B. 1′
C. 16′
D. Cannot be determined from the information provided.

_______ 4. Using the 1/2″ = 1′0″ scale, a line that is 2″ in length represents an object that is _____ in length.

A. 4′
B. 1′
C. 16′
D. Cannot be determined from the information provided.

_______ 5. Using the 1″ = 10′ scale, a line that is 3 1/2″ in length represents an object that is _____ in length.

A. 10′
B. 15′
C. 30′
D. 35′

Apply and Analyze

1. A 5″ line denoting the length of a wall section appears on a project drawing drawn at a 1/2″ = 1′ scale. How long will the completed wall be when the job is completed?

____

____

2. Consider a site plan where a rectangular piece of commercial property is represented by a 5″ × 6″ rectangle that was drawn at a 1″ = 150′ scale. The parking lot on the property will take up exactly 10% of the property area. The parking lot will be a poured concrete slab that is 6″ thick. Assuming no material waste or shrinkage, and a uniform pour throughout the pad, how many cubic feet of concrete will be required to complete the parking area?

____

____

3. An R-410A, split air-conditioning system is shown on a project plan using a 1/8″ = 1′ scale. The refrigerant line set is represented on the drawing by a line 14.5″ in length. If the system is installed exactly as indicated on the drawing, determine the actual length of the line set.

____

____

Name ______________________ Date ____________ Class ______________

4. On a separate sheet of paper, draw a line that is 5.75″ in length. Determine the actual length this line would represent if it was drawn at each of the following scales:

   A. 3/32″ = 1′

   ____

   B. 1/8″ = 1′

   ____

   C. 1/4″ = 1′

   ____

   D. 1/2″ = 1′

   ____

   E. 1″ = 10′

   ____

   F. 1″ = 20′

   ____

5. If a floor plan of a house that is 25′ wide and 60′ long is drawn on a single sheet of 8.5″ × 11″ paper, what is the largest scale on a traditional architect's scale that can be used to create the complete drawing?

   ____

   ____

6. A cylindrical tank is represented as a rectangle on an elevation drawing using a scale of 1/4″ = 1′. The rectangle is 1.5″ wide and 2.5″ high. If completely full, how many gallons of water can the tank hold?

____

____

7. A pipe is represented on a scaled drawing by a 16″ line. If the scale used to draw this pipe is 1/8″ = 1′, how many pipe hangers are required to support this piping run if one hanger is required every 8′? In your calculations, account for the fact that pipe hangers will be installed at each end of the piping run.

____

____

8. A cross-sectional view of a duct section is represented by a rectangle that is 1″ high and 2″ wide. If a canvas connector is needed for this duct section, how many feet of canvas connector material are required if the drawing uses a 1/2″ = 1′ scale? Ignore any overlapping material when performing the calculation.

____

____

Name ______________________ Date ____________ Class ______________

9. A length of straight, round ductwork is represented by a line 5″ in length. If the duct needs supports every 6′, how many hangers are required to properly support it if the duct run was drawn at a 1/8″ = 1′ scale?

____

____

10. A plan view of an attic-mounted air handler is represented by a rectangle that measures 1.25″ × 2″. An auxiliary drain pan is to be mounted underneath the unit. If the drain pan extends beyond the unit in both directions by at least 4″ on each side, what are the minimum measurements of the drain pan if the drawing uses a 1/2″ = 1′ scale?

____

____

Critical Thinking

1. Why should you know the scale used to create a particular project drawing?

____

____

____

____

2. How can an engineer's scale be used to create drawings with unique scales such as 1″ = 150′?

____

____

____

3. Why are different scales often used to create the different drawings in a set of project plans?

____

____

____

____

Large Prints Activity

Name ______________________ Date ____________ Class ______________

Practice Using Large Prints

Refer to Print-25 in the Large Prints supplement to answer the following question. Note: The large prints are created at 50% of their original size. When extracting information from the print using an architect's scale, use a scale that is one-half of the one indicated on the drawing.

1. If the Security Office was drawn at a scale of 1/8″ = 1′, but was measured using a 1/4″ = 1′ scale, how will the measured area of the room compare to the actual measurements of the room?

   ____

Refer to Print-26 in the Large Prints supplement to answer the following question. Note: The large prints are created at 50% of their original size. When extracting information from the print using an architect's scale, use a scale that is one-half of the one indicated on the drawing.

2. The water lines in the garage ceiling are to be wrapped with heat trace to prevent freezing. If it will take 20″ of heat trace to wrap each foot of pipe in the garage, how many feet of heat trace will be required to wrap the lines?

   ____

Basic Duct Symbols and Drawings

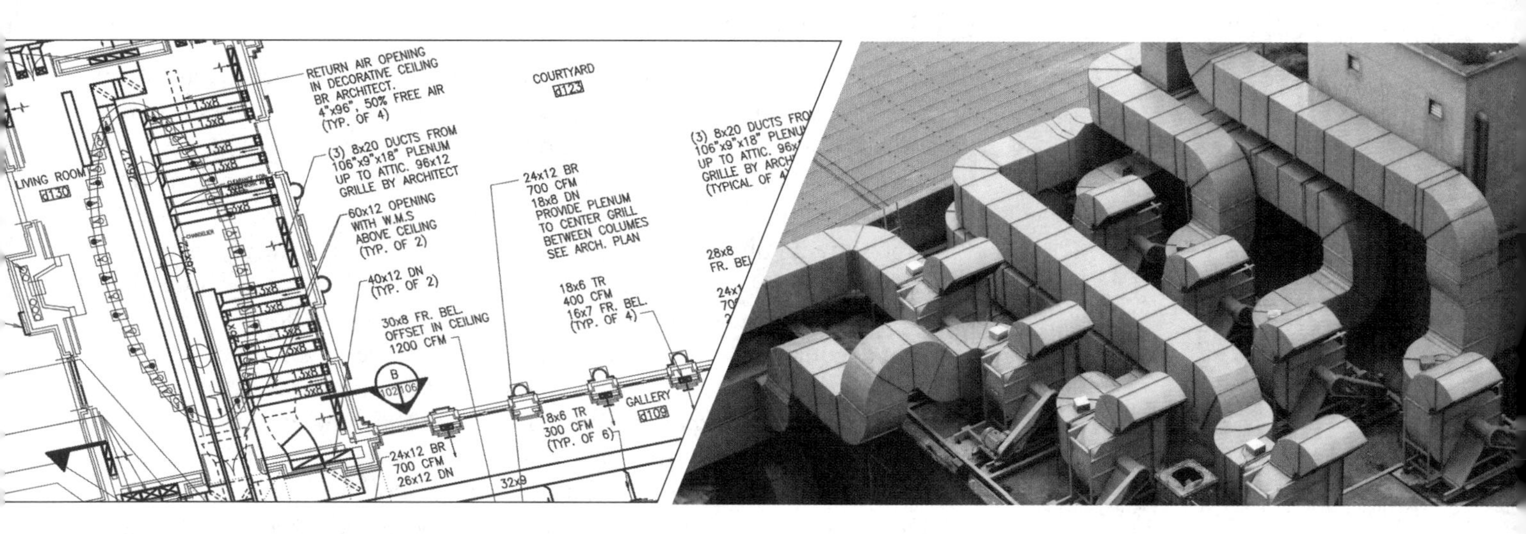

Chapter Outline

5.1 Duct Systems
- 5.1.1 Anatomy of a Duct System

5.2 Straight Duct Sections
- 5.2.1 Duct Section Symbols

5.3 Fittings
- 5.3.1 Supply Air Fittings
- 5.3.2 Return Air Fittings
- 5.3.3 Boots
- 5.3.4 Transitions
- 5.3.5 Offsets
- 5.3.6 Elbows
- 5.3.7 Tees

5.4 Duct System Accessories
- 5.4.1 Registers, Grilles, Louvers, and Diffusers
- 5.4.2 Ceiling Supply Registers
- 5.4.3 Dampers
- 5.4.4 Additional Components and Accessories

Learning Objectives

After studying this chapter, you will be able to:

- Explain the purpose and anatomy of a duct system.
- Discuss the difference between a plan view and an elevation view in a duct drawing.
- Identify various duct fittings on a print.
- Describe the function of duct fittings.
- Identify the direction of intended airflow through a duct section.
- Identify various air distribution system components based on their symbols.
- Determine the intended airflow through a duct run based on print notes.
- List common duct system accessories.
- Differentiate between louvers, grilles, and registers.
- Determine the location of ceiling supply registers, return grilles, and other system components from prints.

Introduction

An air distribution, or duct, system is one of the most common items found on an air-conditioning print. A properly designed and installed duct system helps ensure the continued satisfactory operation of the air-conditioning system for years to come.

It is the responsibility of the print reader to perform proper installation using duct drawings as a guide. Duct drawings vary based on the methods and symbols used by the individual creating them. For instance, elevation and plan views have the same concept but the notes and symbols used on them likely differ. This chapter introduces you to a variety of duct symbols and skills that allow you to extract information from duct drawings and use them for installation. These skills are used to estimate, fabricate, and install air distribution systems.

Technical Terms

branch runout
construction number
converging transition
damper
diffuser
diverging transition
duct
elbow
fitting
grille
louver
offset
plenum
primary return trunk
primary supply trunk
register
return air duct
secondary supply trunk
supply air duct
takeoff fitting
tee fitting
transfer grille
transition
turning vane

5.1 Duct Systems

All HVACR systems should provide thermal comfort to occupants of the space. Thermal comfort includes temperature control, air cleanliness, humidity, and air movement. When an air-conditioning system does its job properly, the occupants should barely be aware the air-conditioning system is running. An air distribution system is responsible for helping meet this goal. The air distribution system pulls air from the conditioned space, treats the airstream, and then returns the treated air back to the conditioned space. A properly designed and installed air distribution system helps ensure there is no excessive noise and that adequate quantities of air are delivered to areas within the structure.

The air distribution system is made up of a series of ducts connected to the air handler. A ***duct*** is a round or rectangular pipe used to carry conditioned air. Poorly designed or installed ductwork will almost guarantee that the occupants of the structure will not be comfortable.

Duct systems are designed, fabricated, and installed in strict compliance with industry guidelines set forth by several organizations:

- **Air Conditioning Contractors of America (ACCA).** ACCA is an industry association mainly comprised of contractors that provide indoor environment and energy services. ACCA writes standards for the design, maintenance, installation, testing, and performance of indoor environmental systems.
- **American Society of Heating, Refrigerating and Air Conditioning Engineers (ASHRAE).** ASHRAE is a global society that focuses on building systems, energy efficiency, indoor air quality, refrigeration, and sustainability within the industry. ASHRAE concentrates on research, standards writing, and publishing and providing continuing education materials for the industry.
- **National Fire Protection Agency (NFPA).** The NFPA is a US trade association that creates and maintains standards and codes for use and adoption by local governments.
- **Sheet Metal and Air Conditioning Contractors' National Association (SMACNA).** SMACNA is an accredited, international trade association that develops technical standards widely accepted by the HVACR and construction community.

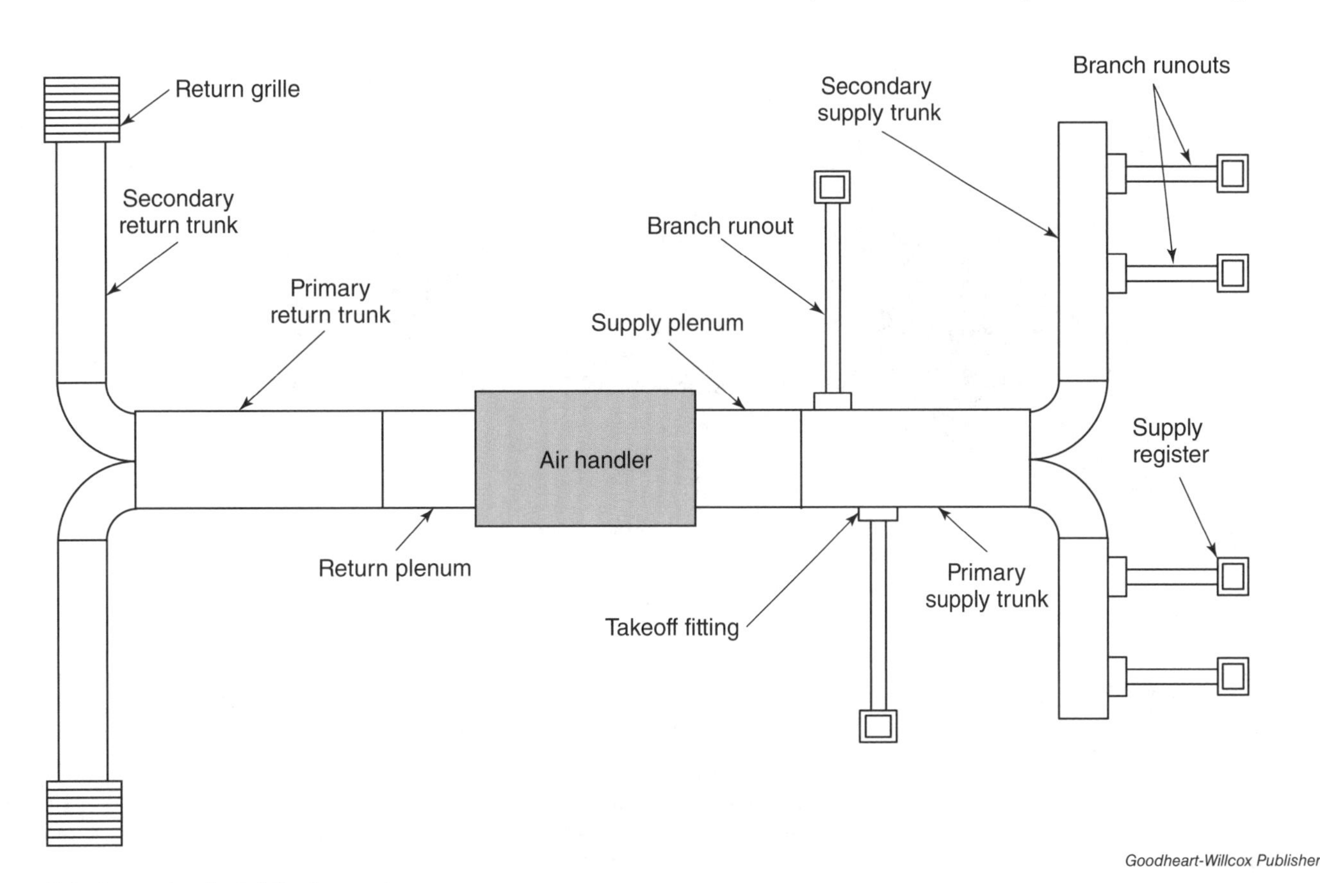

Figure 5-1. A sample air distribution system.

5.1.1 Anatomy of a Duct System

Air distribution systems vary based on the requirements of the space, but all have the same basic components and perform the same function. A system's ductwork consists of return air and supply air sections. ***Return air ducts*** carry air from the conditioned space to an air handler, while ***supply air ducts*** deliver the treated air to the conditioned space. An example of an air distribution system is shown in **Figure 5-1**. The air handler, which consists of a blower, heat transfer surfaces, and sometimes air cleaning and filtering media, is located between the return and supply ducts. Connected to each side of the air handler are ***plenums***, which are chambers for treated air leaving an air-conditioning unit that connect to primary supply and return trunks.

A ***primary supply trunk*** carries all of the air treated by the air handler. From the primary supply trunk, treated air is then distributed, via the network of supply ducts, back to the conditioned space. A ***primary return trunk*** carries all of the air from the conditioned space to the air handler. A primary supply trunk can feed air into secondary supply trunks and branch runouts, as shown in **Figure 5-2**.

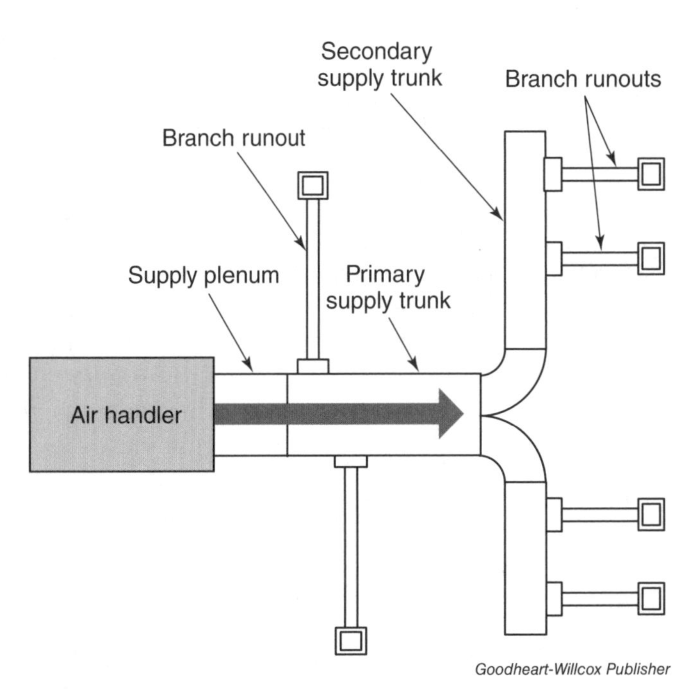

Figure 5-2. The primary supply trunk supplies air to branch runouts and/or secondary supply trunks.

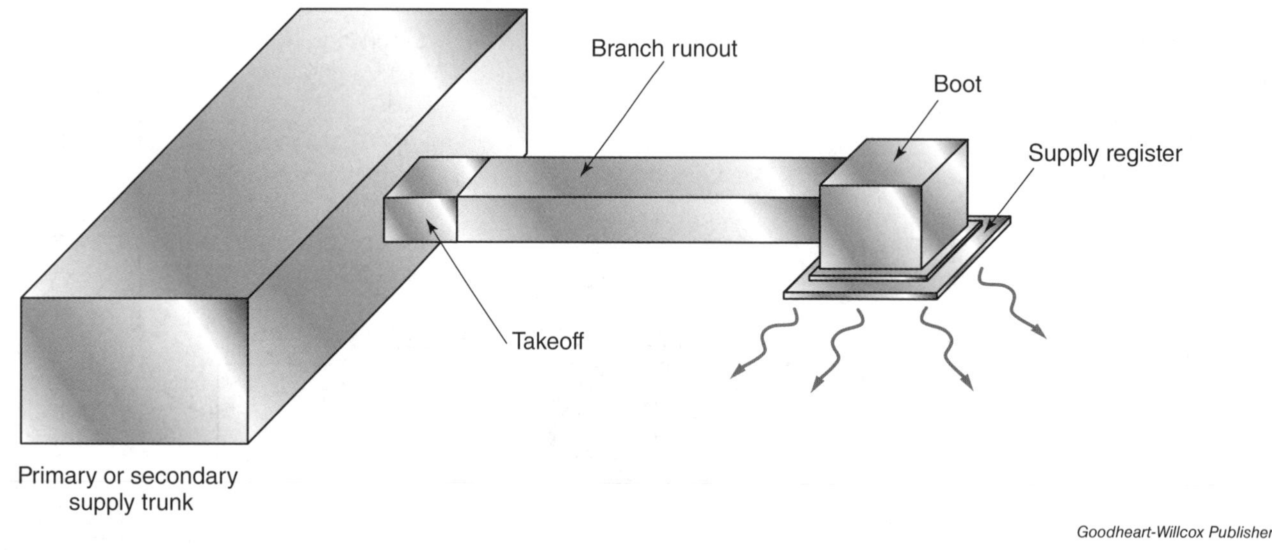

Figure 5-3. A branch runout connects the takeoff fitting to the boot.

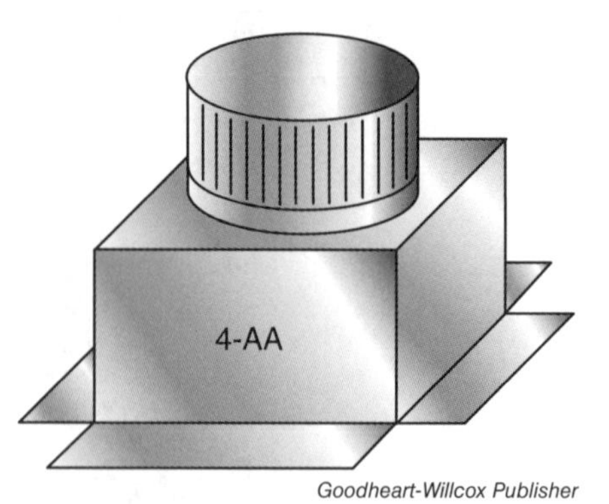

Figure 5-4. One type of boot fitting.

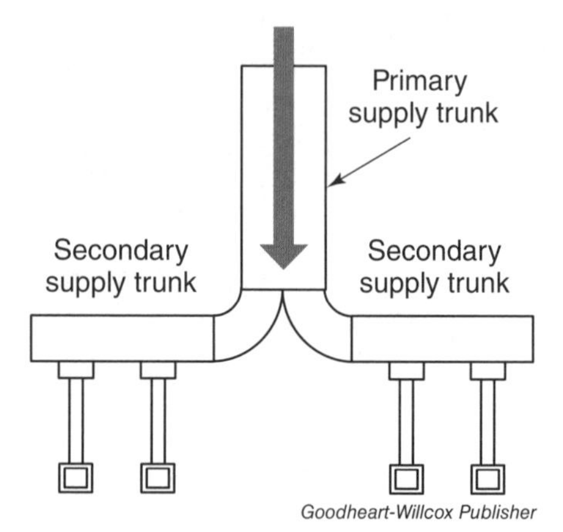

Figure 5-5. A primary supply trunk supplies air to the secondary supply trunks.

A ***secondary supply trunk*** supplies air to branch runouts or other secondary supply trunks. ***Branch runouts*** are duct runs that carry air from a takeoff fitting to a boot, as shown in **Figure 5-3**. A boot is the duct fitting that connects the runout duct to the supply register, **Figure 5-4**. In some installations, if local codes permit, flexible duct can be used for the runout. Otherwise, rigid duct sections must be used.

Air distribution systems on smaller heating and air-conditioning systems typically only have branch runouts, because the supply registers are typically close to each other. When the structure is larger, or when the supply register locations are far from each other, secondary trunks are often used. Secondary trunks carry a portion of the total amount of air. **Figure 5-5** shows how the primary supply trunk feeds air to the secondary trunks. The total airflow through all secondary supply trunks and branch runouts connected to the primary supply trunk should be the same as the airflow in the primary supply trunk.

5.2 Straight Duct Sections

The majority of ductwork consists of straight runs. Square ducts have equal width and depth dimensions, such as 20″ × 20″ or 18″ × 18″. These ducts are drawn the same on both plan and elevation views. When the cross-sectional dimensions are different, as in the case of rectangular duct sections, the sections are represented differently on plan and elevation views.

Consider a section of straight, horizontal ductwork with cross-sectional dimensions of 28″ × 14″, **Figure 5-6**. When a duct section is labeled on a drawing, the dimension listed first represents the side that is seen in the drawing. In **Figure 5-6B**, the 28″ side of the duct is shown in the drawing. Since the duct is installed with its longer sides parallel to the floor, this figure represents a plan view of this duct. **Figure 5-6C** represents an elevation view of the same duct section showing the 14″ side of the duct. The 28″ sides of the duct are still parallel to the floor, but are not seen in this view.

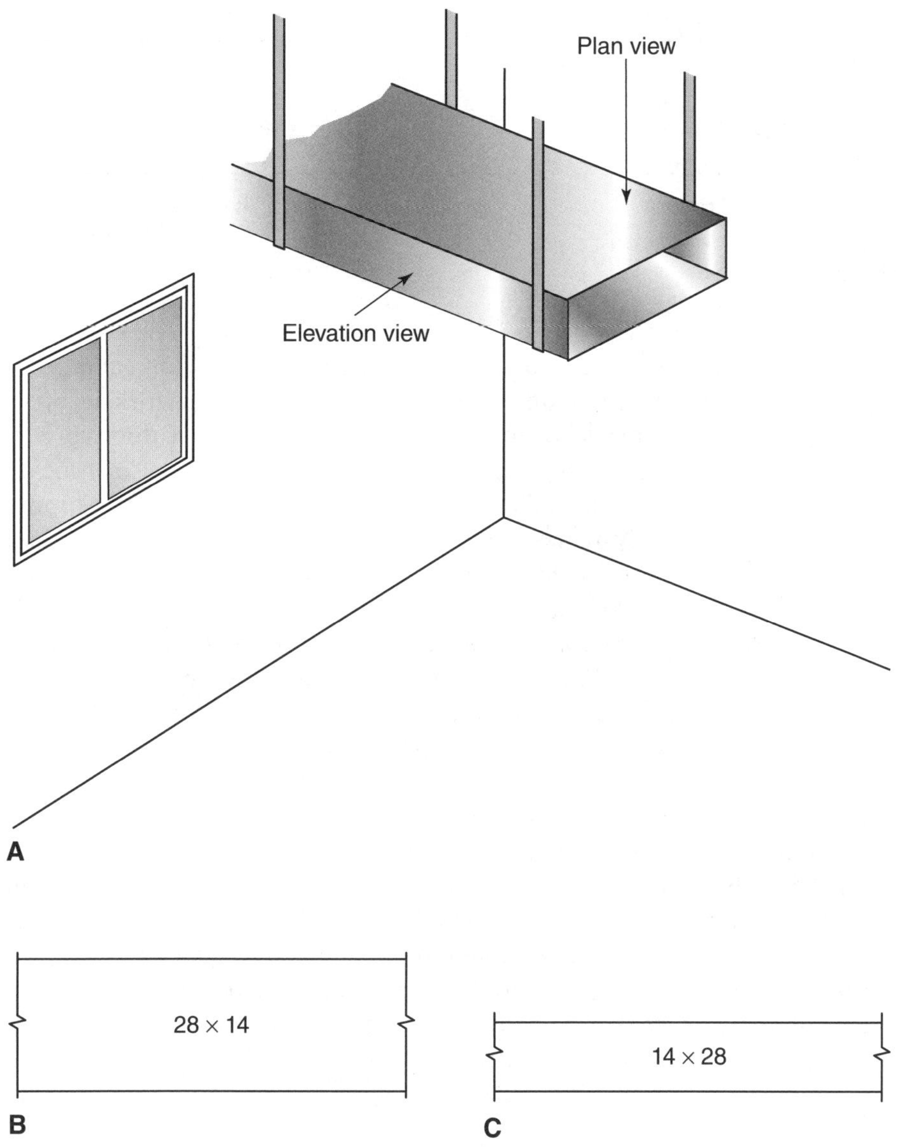

Goodheart-Willcox Publisher

Figure 5-6. A—A horizontal section of rectangular duct. B—Plan view of the duct run expressed on a print. The 28″ measurement is the side being viewed. C—Elevation view of the duct run. The 14″ measurement is the side being viewed.

5.2.1 Duct Section Symbols

A duct section symbol can provide different information based on how it is drawn. A symbol with a break line indicates that only a portion of the entire duct run is being shown, **Figure 5-7**. A section represented with a solid side shows a complete duct run or a particular section of ductwork.

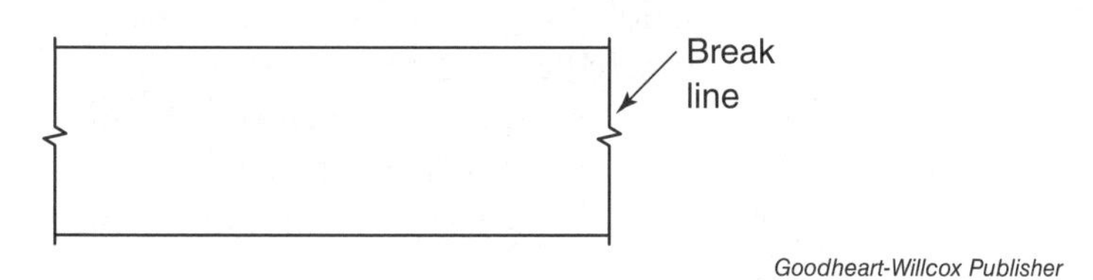

Goodheart-Willcox Publisher

Figure 5-7. A duct portion is shown as indicated by the break line.

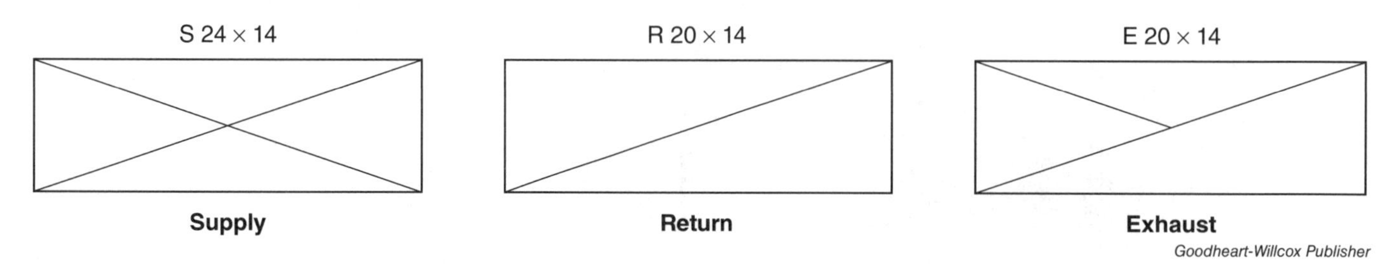

Figure 5-8. Representations of various duct systems.

Supply ducts and return ducts can be distinguished from each other by the lines within the symbol, **Figure 5-8**. Supply ductwork is often indicated by an *X* marking in the section, while return ductwork is often identified by a single diagonal line. Drawings may also indicate the cross-sectional dimensions of the duct preceded by an *S* for supply or an *R* for return. Systems with exhaust ducts use a symbol that is a cross between the supply duct and return duct symbols, and are often identified by the letter *E*.

Although air enters an air distribution system at the return grille, moves to the air handler, and then to the conditioned space, it is often difficult to identify the direction of airflow through these components in a print. The direction of airflow through a duct is usually indicated on plans using arrows, **Figure 5-9**. A straight-shafted arrow indicates the direction of supply airflow, while an arrow with a break mark on the shaft indicates the direction of return airflow.

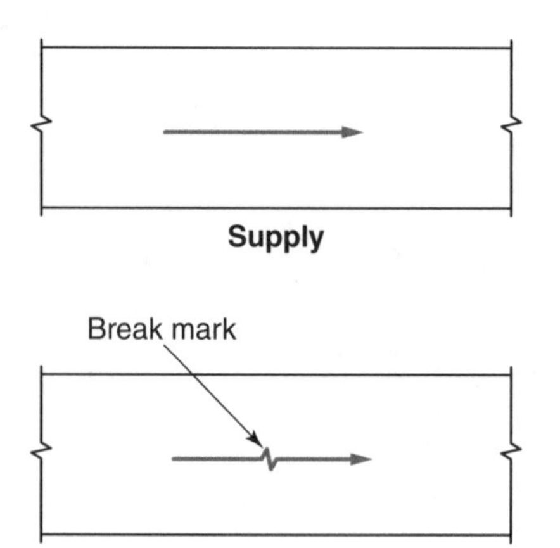

Figure 5-9. The direction of airflow through supply and return ducts.

Dashed lines and parentheses are commonly used on duct section drawings to represent hidden objects, **Figure 5-10**. When a duct is fabricated with acoustical lining for sound dampening, dashed lines are drawn within the duct drawing. When a duct run, or a portion of it, is hidden from view, the duct section is drawn with dashed lines and the duct's dimensions are drawn in parentheses.

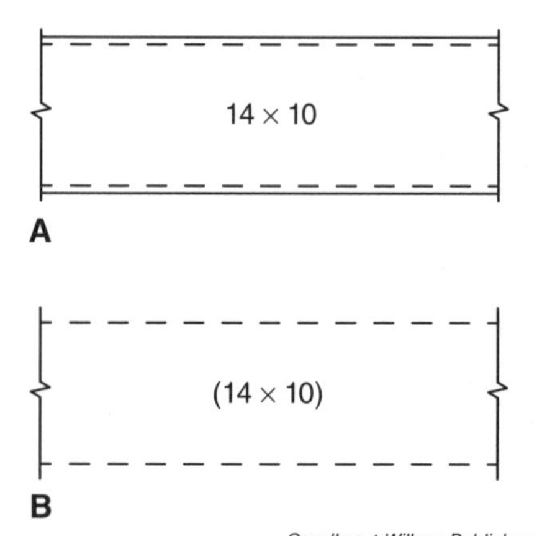

Figure 5-10. A—Dashed lines indicate that the duct section has acoustical lining. B—Dashed lines indicate that the duct section is hidden. Hidden ducts also have their measurements shown in parentheses.

5.3 Fittings

Fittings are sections of ductwork used to change the direction or size of a duct run. They can also be used to split an airstream into two or more airstreams or connect a duct to another air-side system component. A number and letter combination used to label the fitting is the ***construction number***. Each fitting has a distinct construction number.

5.3.1 Supply Air Fittings

Depending on the system configuration, the primary supply trunk may extend in a direction that is in-line with the direction of the discharge airstream from the air handler, **Figure 5-11**. Other times, the primary supply trunk needs to be redirected using fittings. Most often, this is due to space constraints, obstructions, or other design considerations. One of the most commonly used methods of redirecting the primary supply airstream is to connect a straight section of ductwork to the supply plenum, **Figure 5-12**.

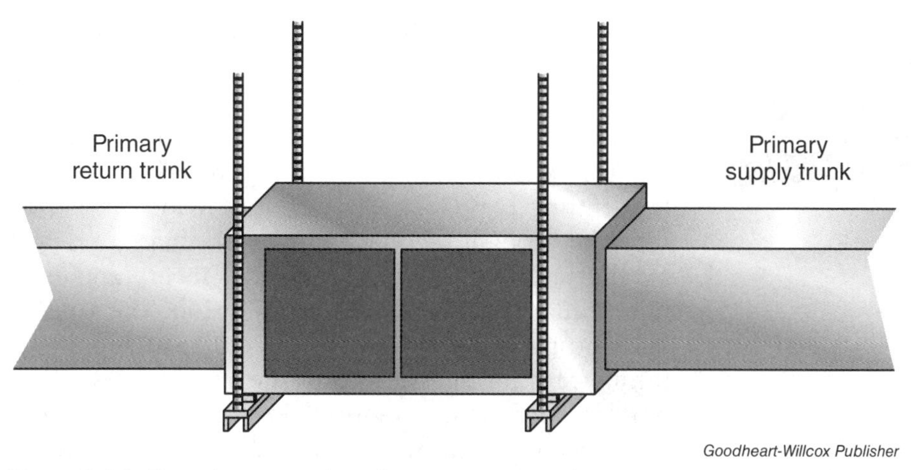

Goodheart-Willcox Publisher

Figure 5-11. The primary supply and return trunks are in-line with the direction of airflow through the air handler.

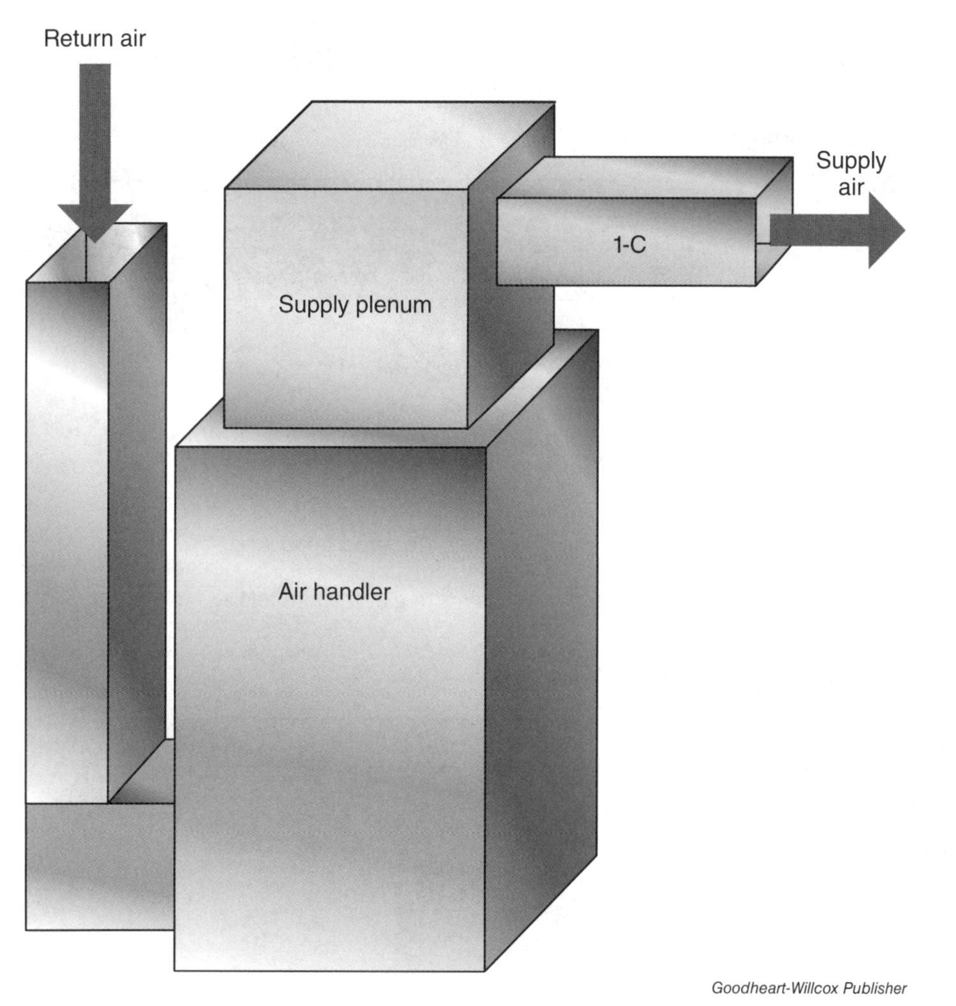

Goodheart-Willcox Publisher

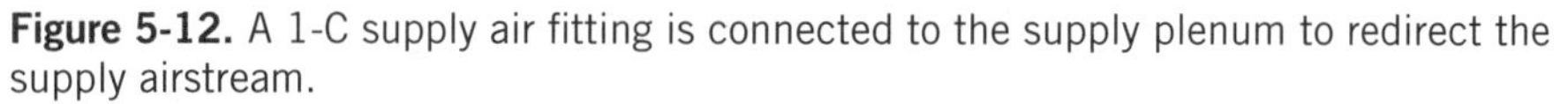

Figure 5-12. A 1-C supply air fitting is connected to the supply plenum to redirect the supply airstream.

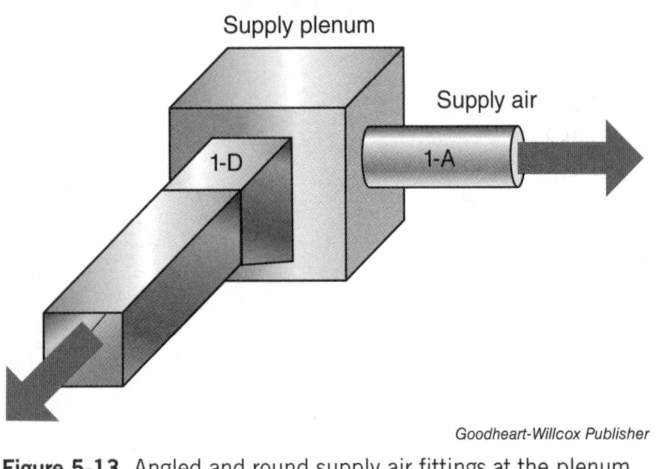

Goodheart-Willcox Publisher

Figure 5-13. Angled and round supply air fittings at the plenum.

This section can also be angled, as shown in **Figure 5-13** to help reduce air turbulence in the ductwork, or it can be round.

Two commonly used supply air fittings are the bullhead and the tapered head, **Figure 5-14**. The bullhead fitting is popular because it is made up of two separate, straight sections of ductwork that are joined by the installer. It allows for adjustments and last-minute alterations to be made on the job. The tapered head fitting is a variation of the bullhead supply air fitting. This fitting, because of the angled sides of the bottom duct section, reduces turbulence, but is more expensive to fabricate and more difficult to modify in the field.

Single-Piece Fittings

Some other supply air fittings are single-piece fittings that are rectangular in shape, **Figure 5-15**. Some have no turning vanes, as in the case of the 1-H fitting, while others, such as the 1-I fitting, have turning vanes. ***Turning vanes*** help reduce air turbulence within the duct.

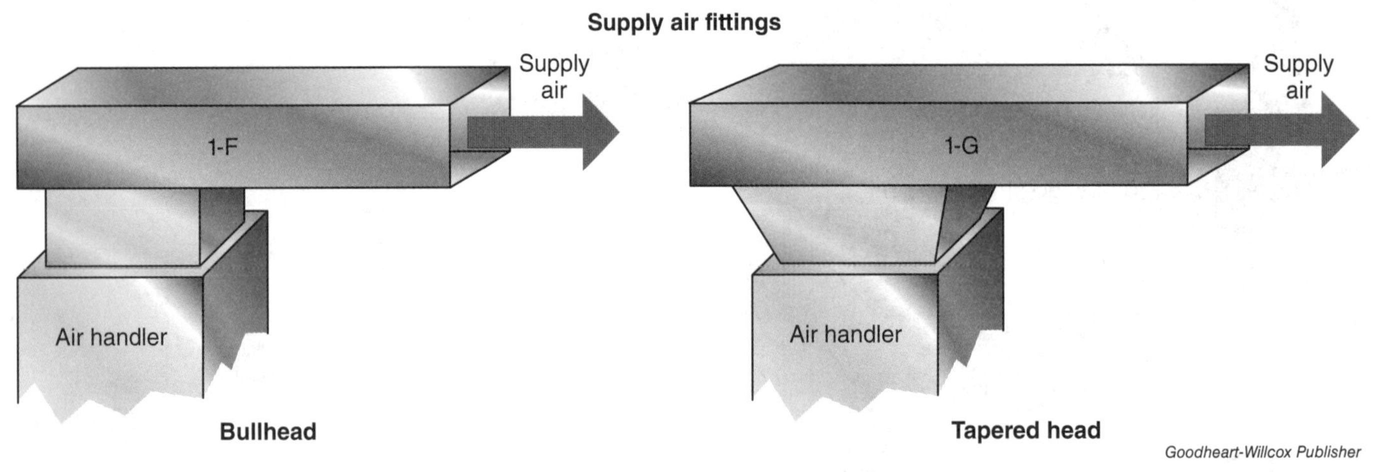

Goodheart-Willcox Publisher

Figure 5-14. A bullhead supply air fitting at the air handler and a tapered head supply air fitting at the air handler.

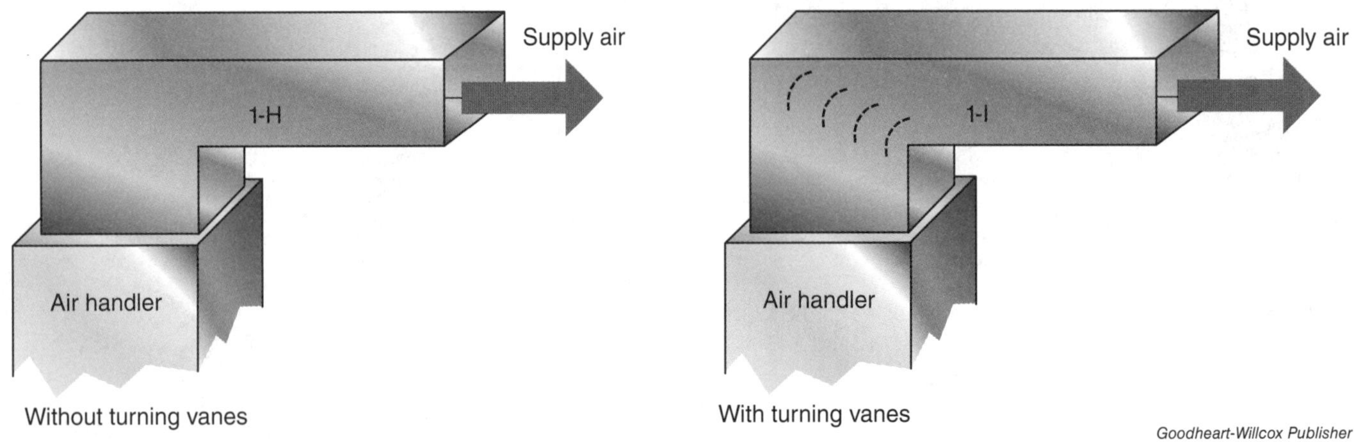

Goodheart-Willcox Publisher

Figure 5-15. Rectangular supply elbows at the air handler.

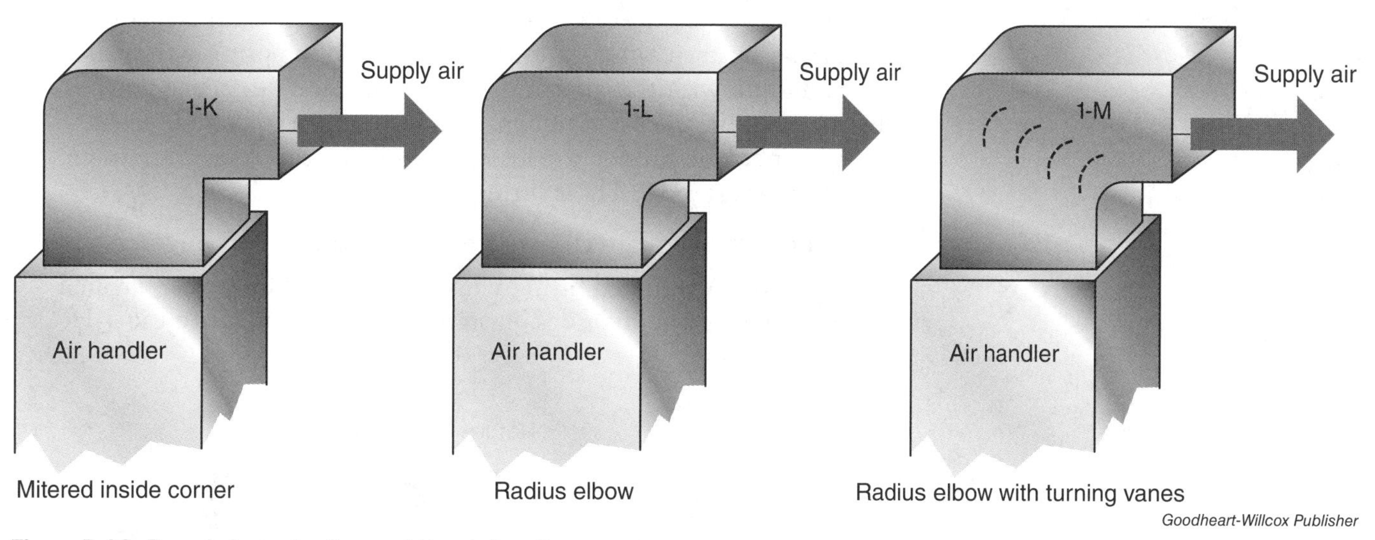

Figure 5-16. Rounded supply elbows at the air handler.

Another way to reduce turbulence is to create rounded fittings. Although rounded fittings, **Figure 5-16**, are more expensive to fabricate, they help direct air through the fitting more smoothly than rectangular-shaped fittings:

- **1-K fitting.** A rounded outer edge with a mitered inside corner.
- **1-L fitting.** A radius elbow fitting that has both rounded outer and inner edges. This fitting has better airflow than the 1-K fitting.
- **1-M fitting.** Obtained by adding turning vanes to the 1-L fitting. This is the best of the three with respect to airflow and the most expensive to fabricate.

Upon initial inspection, these three fittings might appear to be identical, but there are important differences that will affect the operation of the air distribution system. Pay close attention to the notes and references included in duct drawings. Ignoring or misinterpreting the information can lead to installation and operational problems later in the project.

Tee Fittings

In some cases, the supply air must immediately be directed in two different directions. When such is the case, tee fittings are often used, **Figure 5-17**. The most common tee is the 1-O bullhead tee, which has no turning vanes. A 1-P vaned tee is more desirable in most applications.

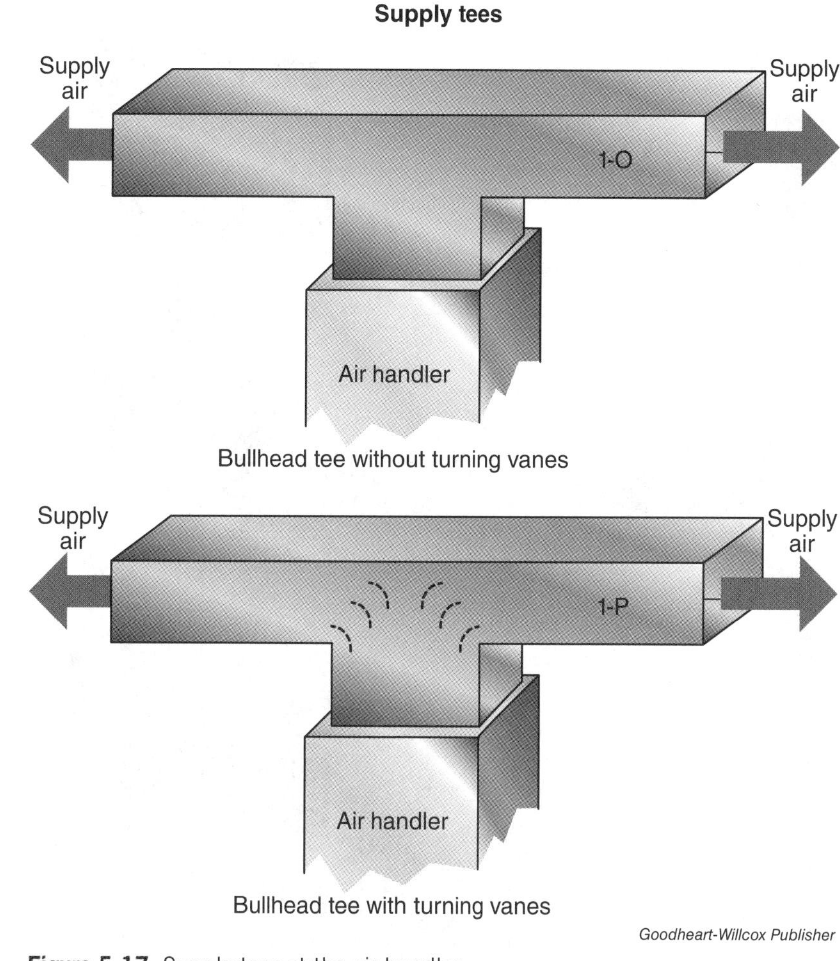

Figure 5-17. Supply tees at the air handler.

5.3.2 Return Air Fittings

Return air fittings are also connected to the air handling equipment. Depending on the configuration of the system, the primary return trunk can be in-line with the air handler, or it may, due to space constraints, obstructions, or other design considerations, have to be redirected.

A number of fittings are used to redirect the return airstream, as shown in **Figure 5-18**. These fittings include:

- **5-H fitting.** A square elbow return duct fitting.
- **5-I fitting.** A rounded fitting with a mitered inside corner. From a performance and duct design standpoint, it is similar to the 5-H return fitting.

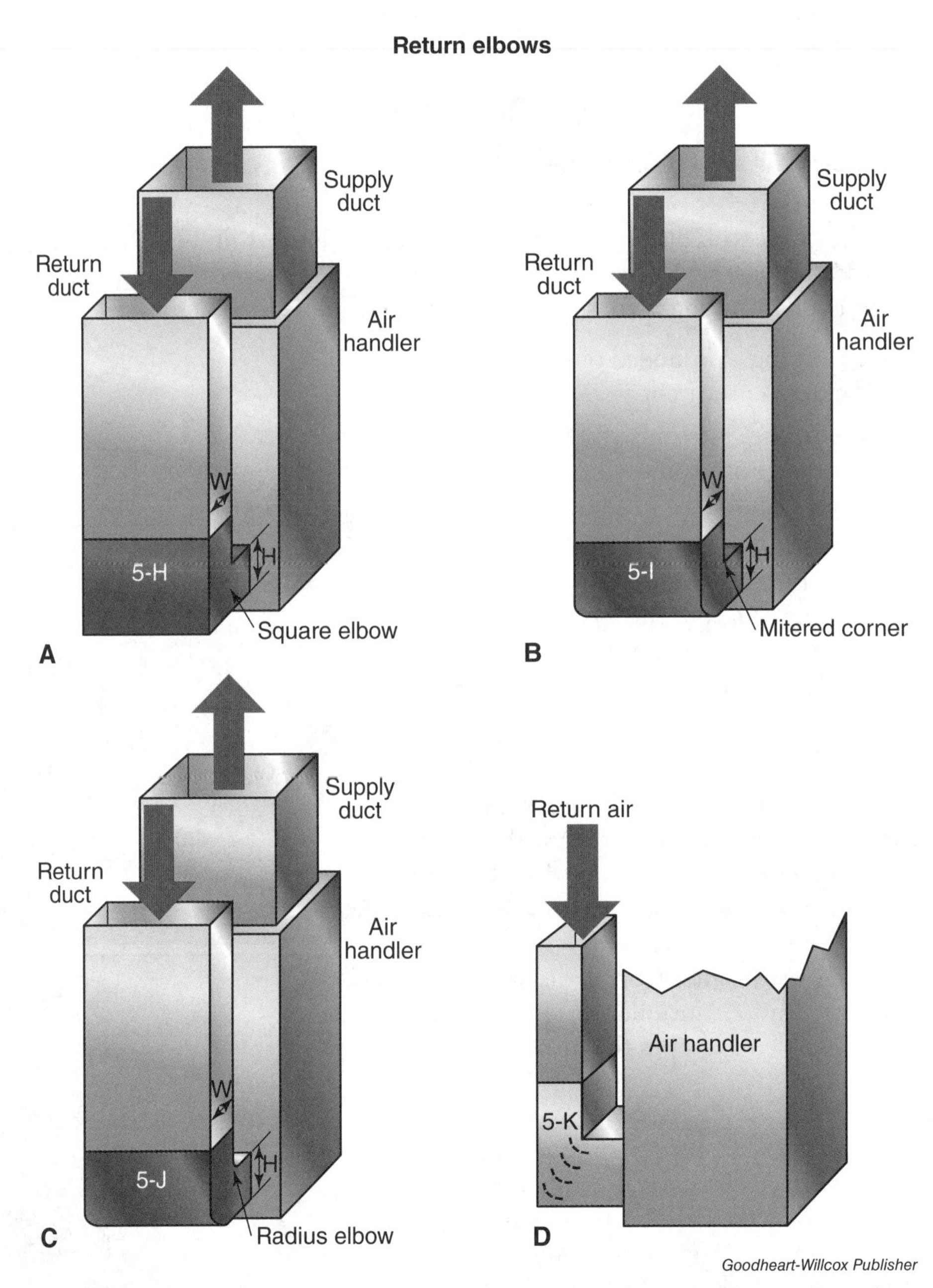

Figure 5-18. A—Rectangular/square return elbow at the air handler. B—Return elbow with a mitered inside corner. C—Return radius elbow. D—Rectangular/square return elbow with turning vanes at the air handler.

- **5-J radius elbow fitting.** Helps reduce air turbulence much better than a 5-H and 5-I fitting.
- **5-K fitting.** Created by adding turning vanes to the 5-H square elbow fitting. This fitting, even though it is a square elbow fitting, performs better than the 5-I and 5-J rounded fittings, due to the addition of turning vanes.

Bullhead Fittings

Another commonly used return air fitting is the bullhead, **Figure 5-19.** This fitting, just as with its supply-side equivalent, is popular because it is made up of two separate, straight sections of ductwork. This allows for adjustments and last-minute alterations to be made on the job. The bullhead return air fitting, with the tapered head, reduces turbulence. In some cases, the return air is returning to the air handler from two different directions. When this is the case, tee fittings are often used, **Figure 5-20.** The most common tee is the 5-L bullhead tee, which has no turning vanes. A 5-M vaned tee is more desirable in most applications.

Takeoff Fittings

On primary or secondary supply trunks, branch runouts bring air closer to the point where it will enter the occupied space. The fitting that connects the branch runout to the main supply duct is called a ***takeoff fitting***. There are many different types of takeoff fittings.

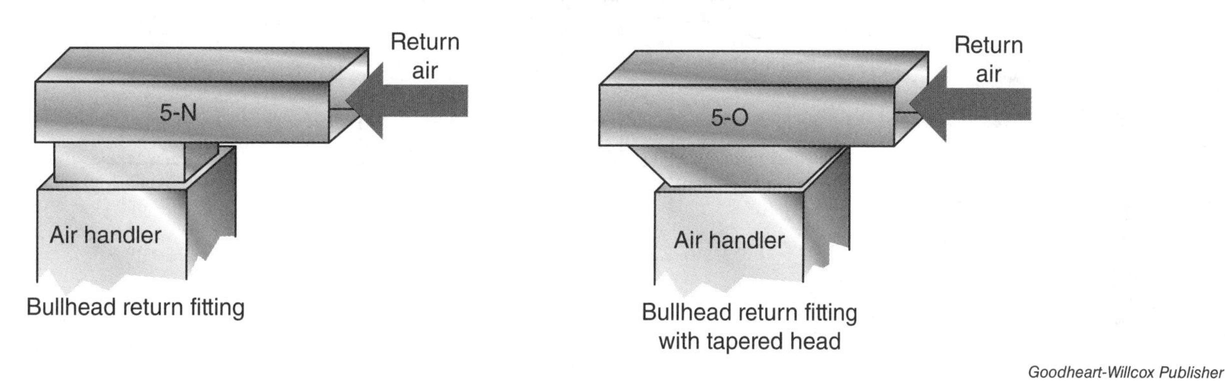

Goodheart-Willcox Publisher

Figure 5-19. Bullhead and bullhead with tapered head return air fittings at the air handler.

Return tees

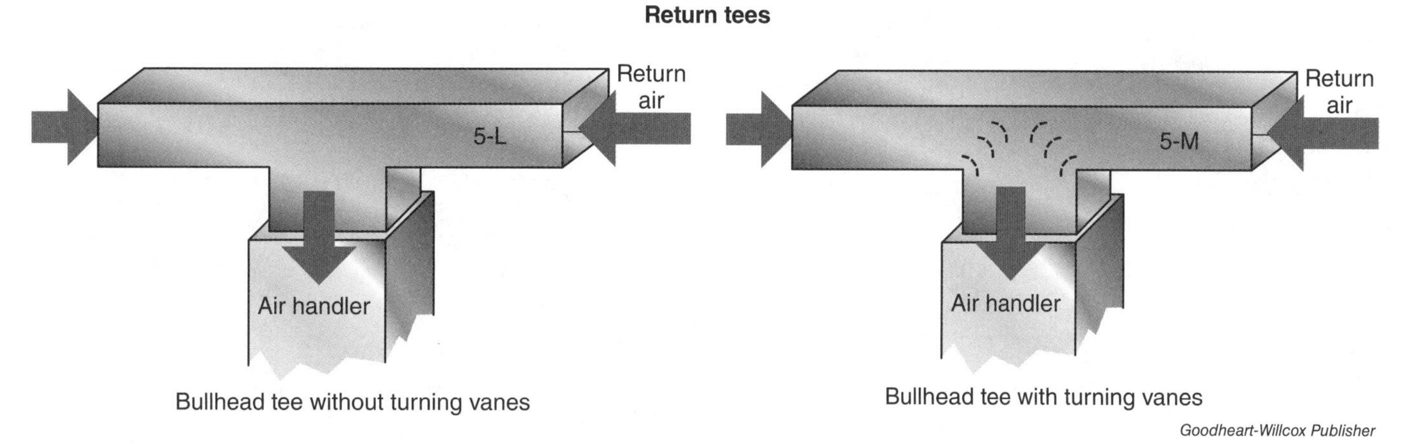

Goodheart-Willcox Publisher

Figure 5-20. Return bullhead tees with and without turning vanes at the air handler.

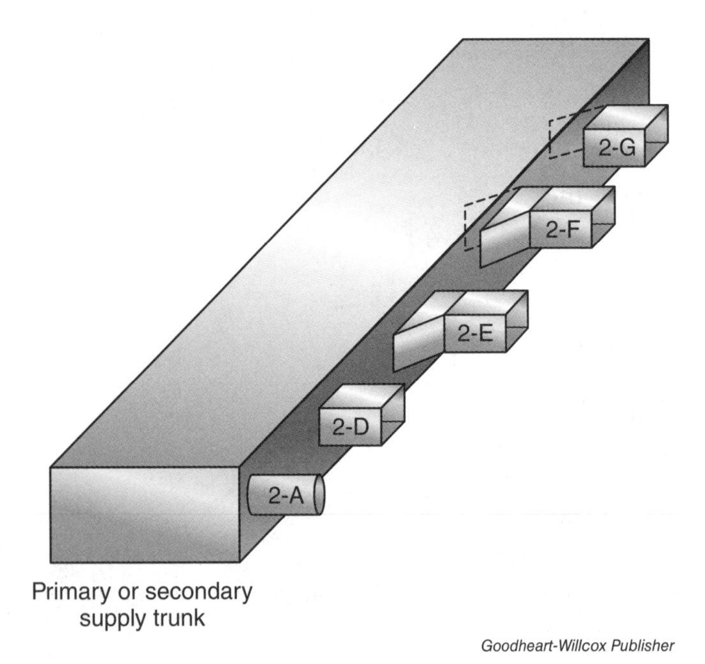

Goodheart-Willcox Publisher

Figure 5-21. Some commonly used supply takeoff fittings.

BETWEEN THE LINES

Selecting Takeoff Fittings

Deciding which takeoff fitting to use is determined by the individual who designed the system. Because different takeoff fittings affect the airflow through the system differently, the type or style of the takeoff fitting should not be changed without written approval and authorization from the system designer.

Takeoff fittings can be round, square, or rectangular sections of duct connected to the main supply duct, **Figure 5-21**. Some takeoff fittings, such as those labeled 2-F and 2-G, have scoops or dampers added to them. These scoops and dampers extend into the airstream to smoothly, and with less turbulence, direct air to the branch runouts. In duct drawings, takeoff fittings are easily identified and labeled with their construction number, **Figure 5-22**. Some branch takeoff fittings are meant to be installed on the top of the supply duct, **Figure 5-23**. Other branch takeoff fittings are installed on round supply ducts, **Figure 5-24**.

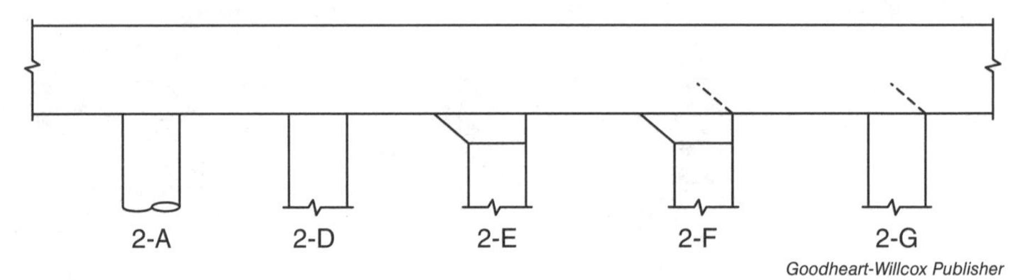

Goodheart-Willcox Publisher

Figure 5-22. How supply takeoff fittings appear in duct drawings.

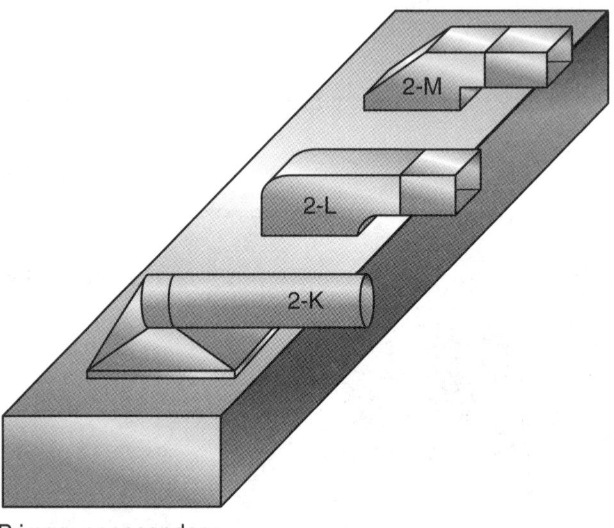

Goodheart-Willcox Publisher

Figure 5-23. Some top-of-duct mounted supply takeoff fittings.

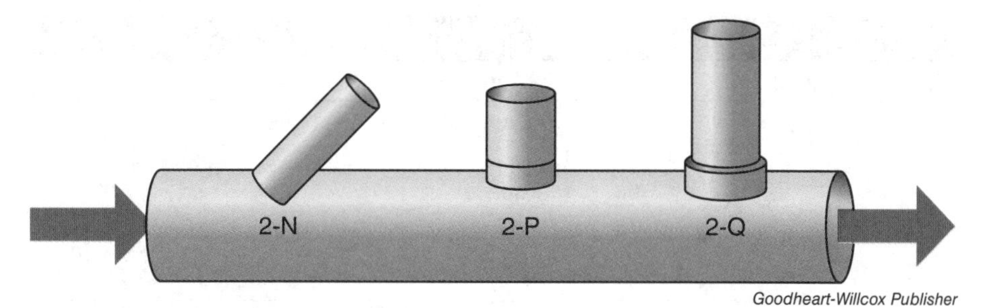

Goodheart-Willcox Publisher

Figure 5-24. Takeoff fitting for round supply ducts.

5.3.3 Boots

A branch runout starts at the main supply duct and terminates at the supply register, louver, or grille in the occupied space. The duct fitting that connects the branch runout to the supply register is called a boot. Boots, **Figure 5-25**, are designed for dozens of applications and come in multiple styles for use with wall, floor, ceiling, and other specialty supply air applications.

Boot Types and Applications

Type	Symbol	Application
4-C		Ceiling applications Floor applications
4-M		Wall applications Floor applications
4-O		Wall applications Floor applications
4-Q		Floor supply air applications
4-R		Floor supply air applications

Goodheart-Willcox Publisher

Figure 5-25. Boot fittings are used to connect the branch runout to the supply registers, grilles, or louvers. *(continued)*

Boot Types and Applications

Type	Symbol	Application
4-V		New construction projects Round duct is left exposed within occupied space
4-Z		Ceiling applications
4-AA		Ceiling applications
4-AO		Wall applications Used when occupied space on one side of wall is supplied with air from particular duct
4-AP		Wall applications Used when both areas on side of wall is supplied air from duct

Goodheart-Willcox Publisher

Figure 5-25 ***(continued)*****.**

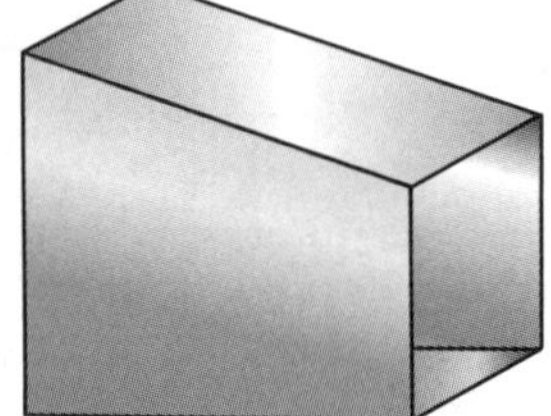

Goodheart-Willcox Publisher

Figure 5-26. A transition fitting.

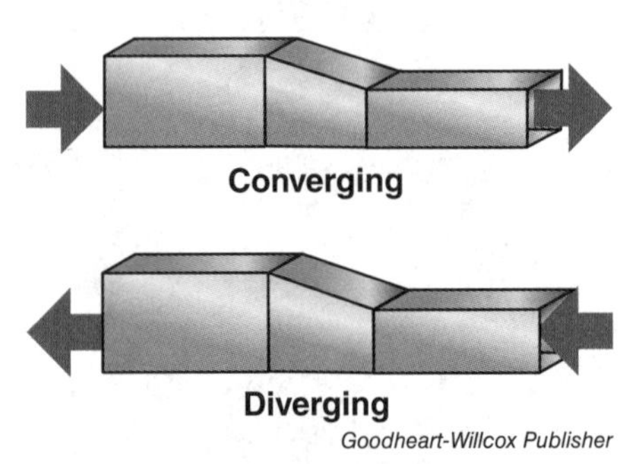

Goodheart-Willcox Publisher

Figure 5-27. A transition fitting is called converging or diverging depending on which side the air enters and travels through first.

5.3.4 Transitions

In an air distribution system, it is often necessary to join two sections of different-sized ductwork. A ***transition***, **Figure 5-26**, is a duct fitting used to connect duct sections of different sizes. Most often, three sides of the transition fitting are parallel to the two sections being connected. In some cases, especially when the duct system will be exposed, two or more sides of the transition fitting are angled, to provide a more symmetric appearance.

Transition fittings can be either diverging or converging, **Figure 5-27**. A ***converging transition*** is one that has air flowing through it from the larger size to the smaller size. A ***diverging transition*** is one that has air flowing through it from the smaller size to the larger size.

Different information can be pulled from a duct drawing based on the type of view in the drawing. **Figure 5-28** shows a plan view and an elevation view of a duct run. Upon initial inspection of the plan view, it appears to be a straight section of the same-sized ductwork. However, in reality, the actual duct run transitions from one size to another, which is determined by the different measurements on each side of the transition fitting. The left side of the fitting shows a measurement of 24″ × 20″ with the 24″, which is parallel to the floor or ceiling, listed first. Since it is listed first, this is the side of the duct that is seen in the plan view. The right side of the fitting shows a measurement of 24″ × 14″ and, since the 24″ side of the duct is listed first, it is the side of the duct that is seen in the plan view. The other measurements, 20″ and 14″ are hidden in the plan view.

An elevation view of the duct run provides more information on the transition from one duct size to the other. In this view, the different duct sizes become evident. In the elevation view, the 20″ and 14″ measurements are seen, as they are listed first, while the 24″ measurements are hidden. In this elevation view, it can be seen that the bottom of the duct run is flat, while the top of the transition section is angled to facilitate the connection of the two different size ducts.

It is helpful to identify on a plan view whether the top or bottom of the transition is flat, **Figure 5-29.** This is accomplished by adding the abbreviation BF (bottom of duct is flat), FOB (flat on bottom), TF (top of duct is flat), or FOT (flat on top) to the drawing. Once again, the measurement of the side of the duct facing the viewer is always listed first, so here, the viewer is looking at the 24″ side of the duct.

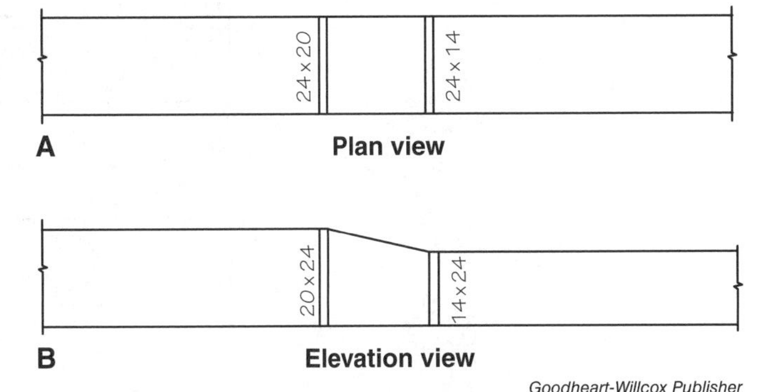

Figure 5-28. A—Plan view of a transition duct. B—Elevation view of a transition with the bottom of the duct flat.

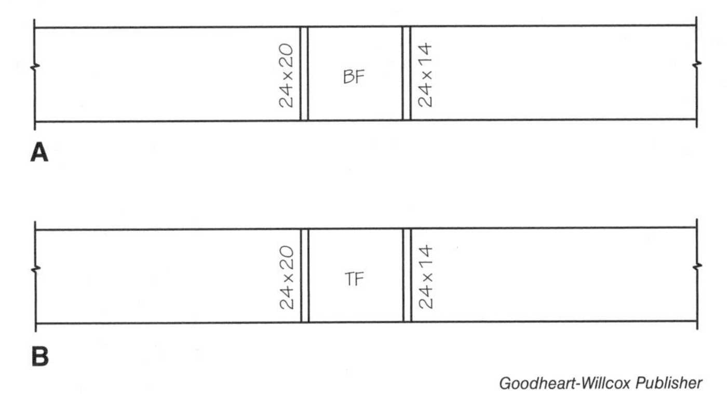

Figure 5-29. A—Notation to indicate that the bottom of the duct is flat. B—Elevation view of a transition with the top of the duct flat.

5.3.5 Offsets

An ***offset*** is a duct fitting that is used when a straight run of ductwork needs to bend around an object or obstruction. The main characteristic of an offset is that the direction of airflow flowing into the offset is parallel to the direction of the airflow leaving the offset. Typically, an offset fitting is not used to change the cross-sectional measurements of the duct. Offsets can be either vertical or horizontal.

Vertical Offset Fittings

Vertical offset fittings, **Figure 5-30**, can either raise or lower a duct run. A plan view of a vertical transition does not show if a side is higher or lower than the other, which is why an elevation view of the section is often used.

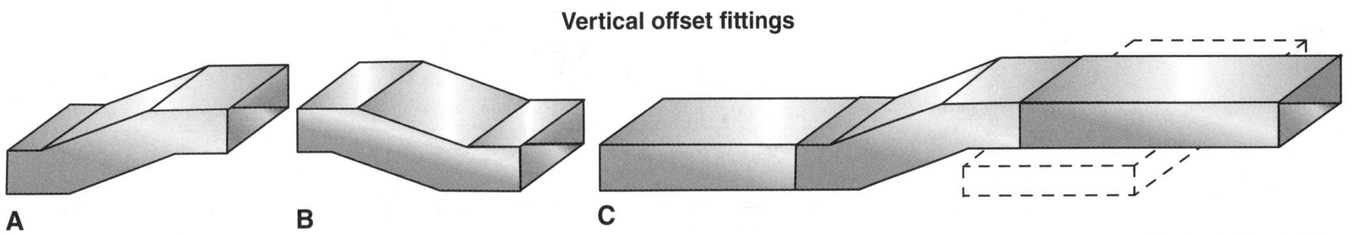

Figure 5-30. Vertical offset fittings. A—Fitting is rising from left to right. B—Fitting is dropping from left to right. C—Vertical offset fittings used to avoid horizontal obstacles.

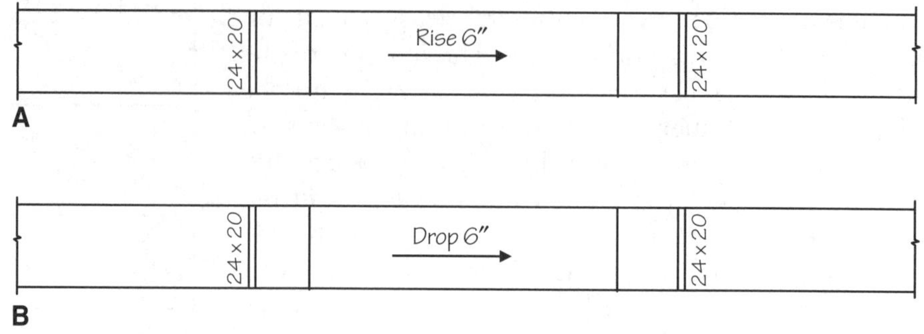

Goodheart-Willcox Publisher

Figure 5-31. Vertical offsets. A—Fitting is rising 6″ from left to right. B—Fitting is dropping 6″ from left to right.

To provide more details, the drawing of the offset fitting often includes the direction of airflow, if the fitting rises or drops in the direction of the airflow, and how much the fitting rises or drops. This is shown in **Figure 5-31.**

Another way to indicate whether the fitting is rising or dropping is to identify the distance of each side of the transition from either the floor or ceiling. The following abbreviations are often used in duct drawings:

- **TU (Top Up).** How far the top of the duct section is from the floor.
- **TD (Top Down).** How far the top of the duct section is from the ceiling.
- **BU (Bottom Up).** How far the bottom of the duct section is from the floor.
- **BD (Bottom Down).** How far the bottom of the duct section is from the ceiling.

Figure 5-32 shows the plan view of a portion of a duct run that has a transition fitting. The left side of the transition fitting is labeled TU 8′1″, which indicates the top of the duct is 8′1″ up from the floor. The right side of the transition fitting is labeled TU 8′7″, which indicates that the top of the duct is 8′7″ up from the floor. From this information, it can be determined that the fitting is rising from left to right and that the amount of rise is 6″.

Horizontal Offset Fittings

A horizontal offset fitting, **Figure 5-33**, is used to move a duct run to avoid vertical barriers. Most of the time, horizontal offset fittings are not included as part of the original ductwork plan for the project. However, this type of fitting, along with the vertical offset, are often added during the installation process when unforeseen obstacles present themselves.

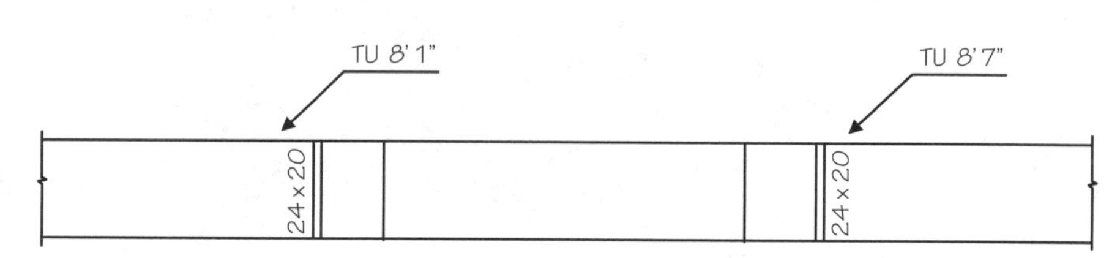

Goodheart-Willcox Publisher

Figure 5-32. A plan view of a portion of a duct run with a transition fitting.

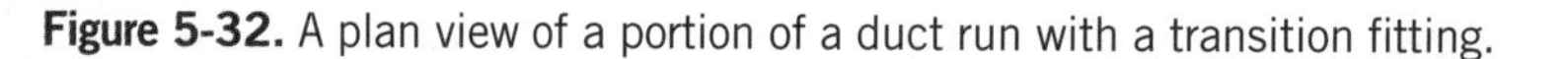

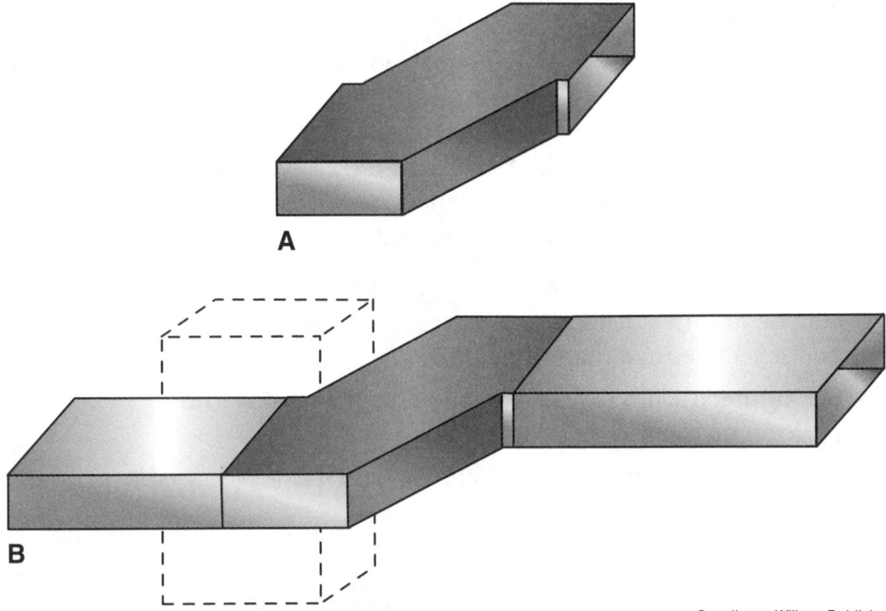

Goodheart-Willcox Publisher

Figure 5-33. A—A horizontal offset fitting. B—This fitting is used to avoid horizontal obstacles.

BETWEEN THE LINES

Rerouting Ductwork

Sometimes, a pipe or other obstruction is installed in the space where a section of duct should go. In other cases, construction elements are present that were not shown in the project drawings. If a duct run needs to be redirected, alert the general contractor who will ultimately decide whether the obstacle will be removed or the duct will be rerouted around it.

If ductwork needs to be rerouted, additional compensation to the HVACR contractor is often required, because additional materials, fittings, and labor will be needed. Any and all changes to the original plan must be documented, agreed upon, and signed off on to prevent problems later in the project.

Contrary to the vertical offset fittings, horizontal offset fittings are not readily identifiable in elevation views. They are more evident in the plan view, **Figure 5-34.**

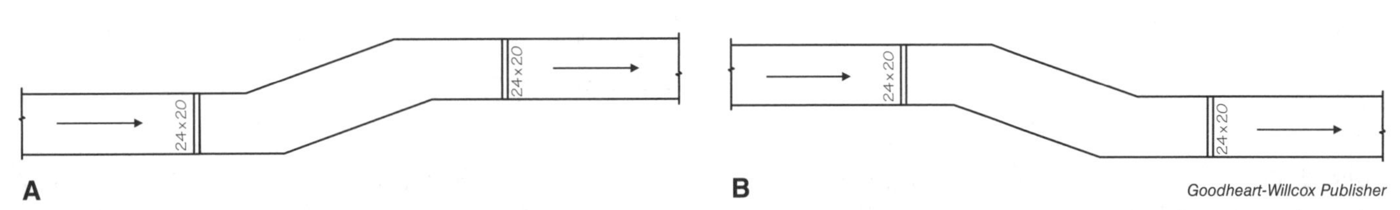

Goodheart-Willcox Publisher

Figure 5-34. A—Plan view of a horizontal offset fitting that shifts to the left in the direction of airflow. B—Plan view of a horizontal offset fitting that shifts to the right in the direction of airflow.

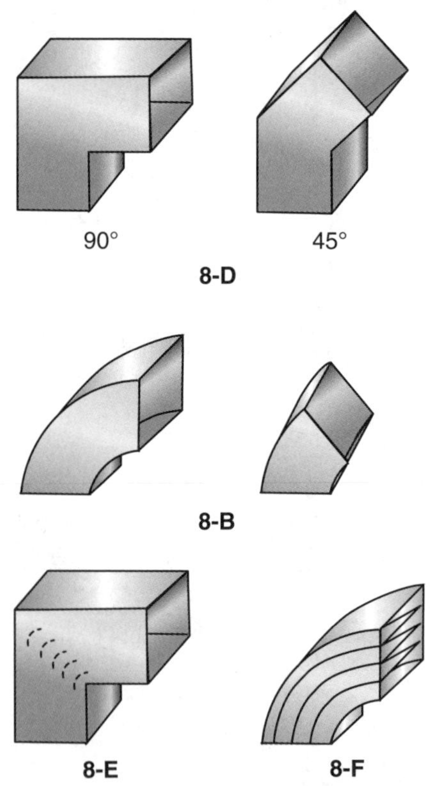

Figure 5-35. Samples of elbow fittings.

5.3.6 Elbows

Elbows are duct fittings that are connected between two sections of ductwork to facilitate a change in the direction of airflow. Elbows are commonly 90° or 45° fittings but can be created at any angle based on job specifications. Elbows can be fabricated either with or without turning vanes, as specified by the project plans.

Common elbows, **Figure 5-35,** include the following:

- **8-D fitting.** A common elbow. Often, a 90° version and a 45° version are used. This fitting is popular because it is easy to fabricate and less expensive than other styles.
- **8-E fitting.** Turning vanes can be added to improve fitting performance.
- **8-B fitting.** Another common elbow, which is a radius elbow.
- **8-F fitting.** Like square elbow fittings, turning vanes can be added to improve fitting performance.

Consider the elevation view of the duct section shown in **Figure 5-36.** Notice that this duct run has dashed lines and is using an 8-B 90° elbow to redirect the duct run to a vertical orientation. The airflow is also directed from left to right, and then upward after the 90° turn is made. It cannot be determined whether the duct is a supply duct or a return duct. If we examine the plan view of the same duct run, more information can be obtained.

Recall that supply ducts are identified by crossing diagonal lines, so from the plan view, **Figure 5-37A,** we can see that this drawing represents a portion of the supply duct system. Because these crossing diagonal lines are solid, the duct is directed upward. Also, the acoustical lining in the vertical portion of the duct run is visible and is therefore drawn with solid lines. If the crossing diagonal lines were dashed, this means the duct is being directed downward, **Figure 5-37B.**

If the portion of the system was a part of the return air ductwork, the crossing diagonal lines would be replaced by a single diagonal line, **Figure 5-38.** The upturning return ducts show airflow direction from left to right and from right to left, while the downturning return ducts show airflow direction from left to right and from right to left.

If the duct run were part of an exhaust system, the sections would appear as in **Figure 5-39.** If the duct had been round instead of rectangular, the plan view would have looked similar to that shown in **Figure 5-40.**

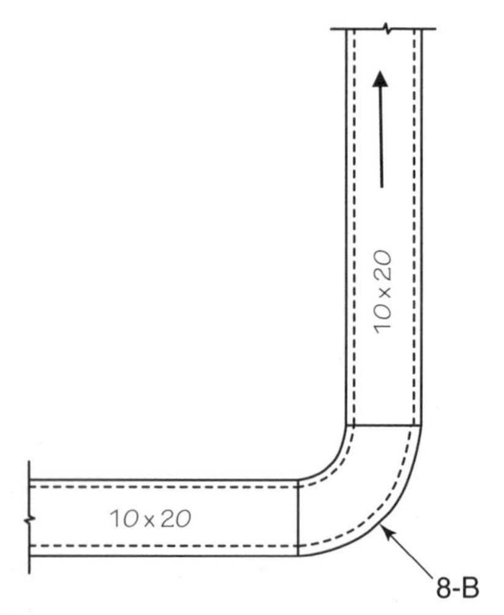

Figure 5-36. Elevation view of a duct section of an air distribution system.

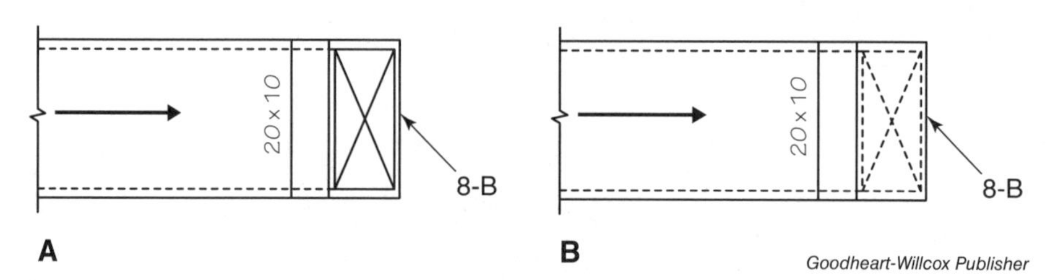

Figure 5-37. A—Plan view of an upturning section of supply ductwork. B—Plan view of a downturning section of supply ductwork.

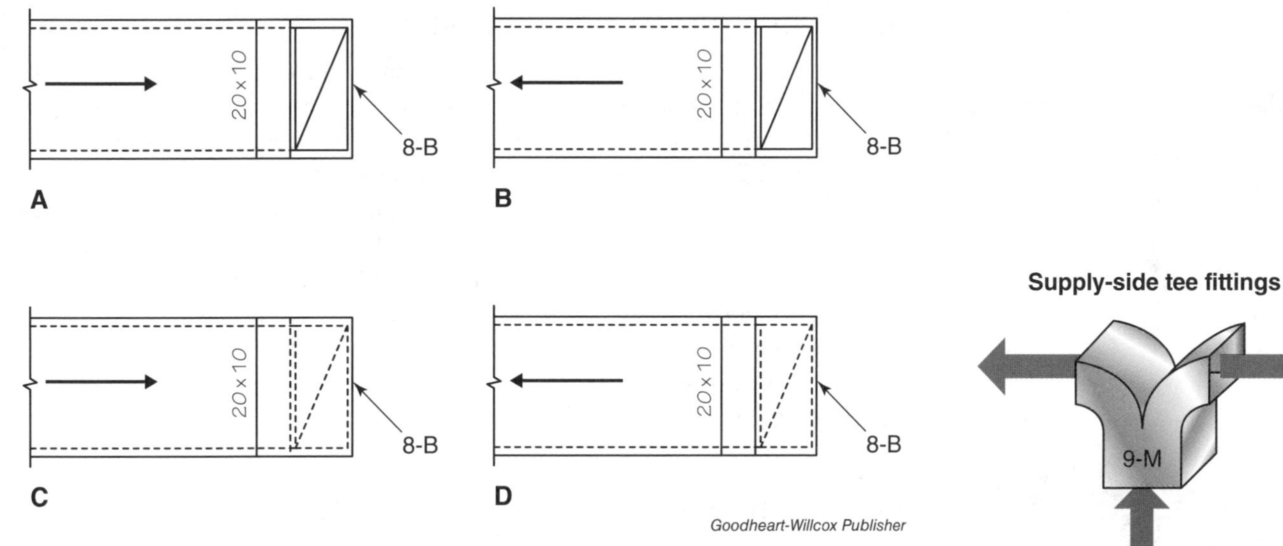

Goodheart-Willcox Publisher

Figure 5-38. A and B plan views show upturning return ducts. C and D plan views show downturning return ducts. The arrow indicates the direction of airflow.

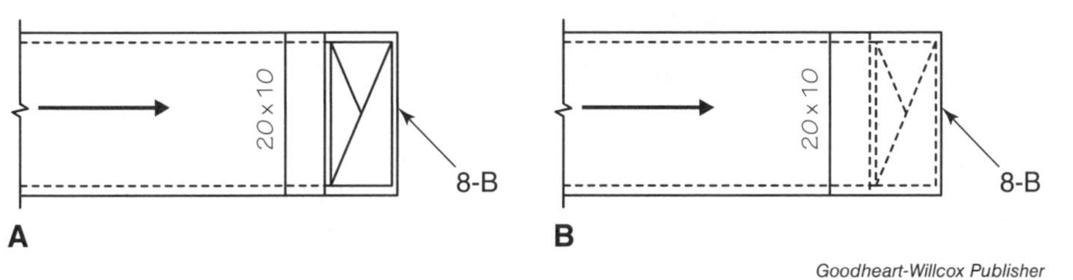

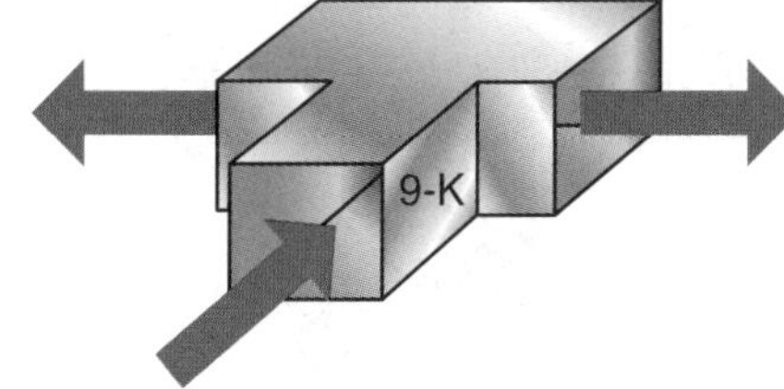

Goodheart-Willcox Publisher

Figure 5-39. A—Upturning exhaust duct with airflow from left to right. B—Downturning exhaust duct with airflow from left to right.

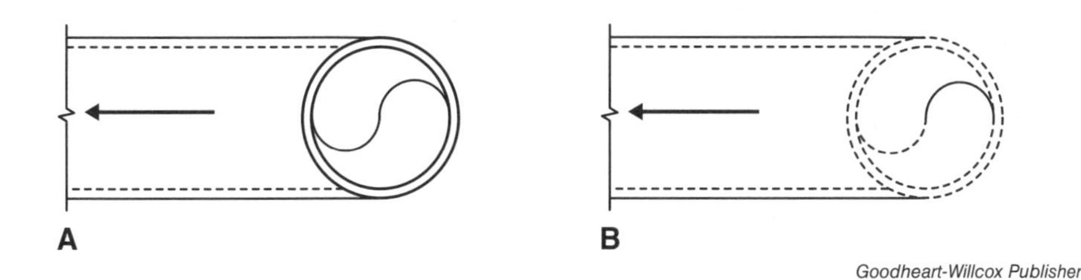

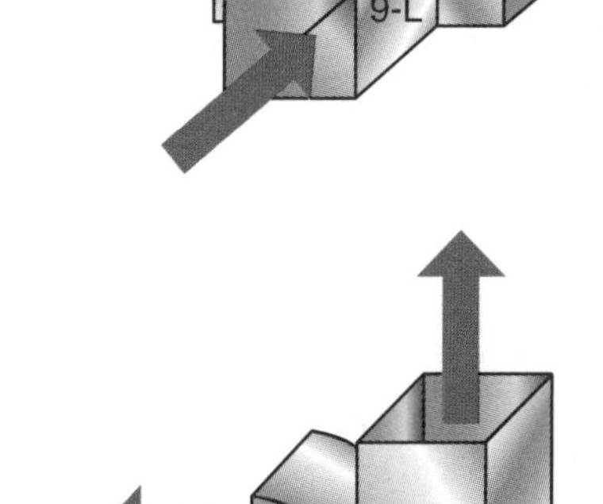

Goodheart-Willcox Publisher

Figure 5-40. A—Upturning round duct with airflow from right to left. B—Downturning round duct with airflow from right to left.

5.3.7 Tees

Tee fittings are used at various points in the air distribution system to direct the ductwork in two different directions. This can happen on both the supply and return duct systems. As with other fittings, tees can be fabricated with or without turning vanes, as specified. **Figure 5-41** shows some commonly used supply-side tee fittings, as well as return-side tee fittings. Tee fittings must be designed and sized to ensure that the total volume of air entering the tee fitting is equal to the total volume of air leaving the fitting.

9-G

Goodheart-Willcox Publisher

Figure 5-41. Various supply tee fittings and various return tee fittings.

(continued)

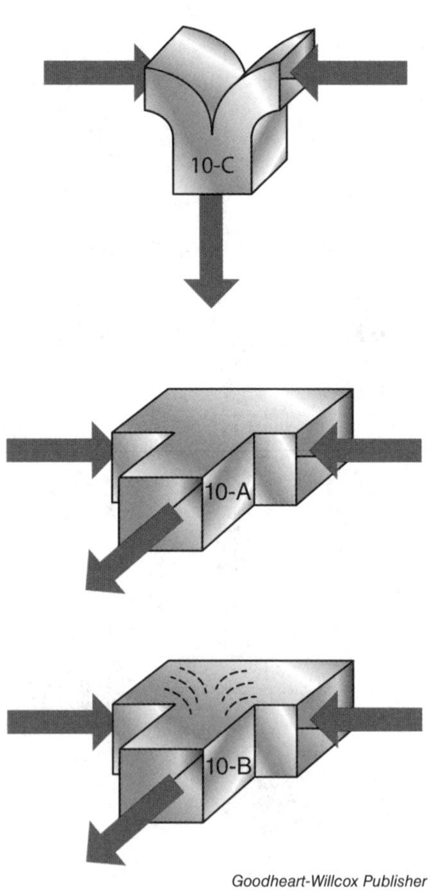

Goodheart-Willcox Publisher

Figure 5-41 (*continued*).

5.4 Duct System Accessories

Other accessories can go into a duct system to help it function in an effective and efficient manner. Some accessories help a service technician gain access to unreachable components in the duct system, while others act to ensure the safety and well-being of the occupants.

Registers, grilles, and louvers are the portions of an air distribution system that are located in the conditioned space and are visible to the occupants. These are the decorative framed panels that are the transition points between the occupied space and the duct system. Although the terms for them are sometimes used interchangeably in the HVACR industry, these accessories are not the same. In different industries and applications, these terms take on different meanings.

5.4.1 Registers, Grilles, Louvers, and Diffusers

Typically, in the HVACR industry, the term ***louver*** is used to describe a framed set of fixed slats that cover an opening in a duct, doorway, or other construction element. Air is able to pass through the louver and be directed in a particular path, while somewhat limiting the ability to see through the louver to the other side.

Grilles, like louvers, are a framed set of slats. However, some grilles can have their slats adjusted, **Figure 5-42**. Both grilles and louvers are unable to control the volume of air that passes through them, but grilles can with the use of a register. A ***register*** is the combination of an adjustable grille and a damper. It can control the direction of airflow and the volume of air passing through it.

Diffusers are designed to minimize the stratification of air and effectively mix the supply airstream with air already in the occupied space. Diffusers help ensure uniform air conditioning by reducing hot or cold spots due to ineffective or uneven airflow patterns. They can dispense air in many directions, which makes them more effective than registers in this regard. Be sure to refer to the notes that accompany duct drawings in any project. They will clearly identify the specific devices to be used, thus eliminating any confusion.

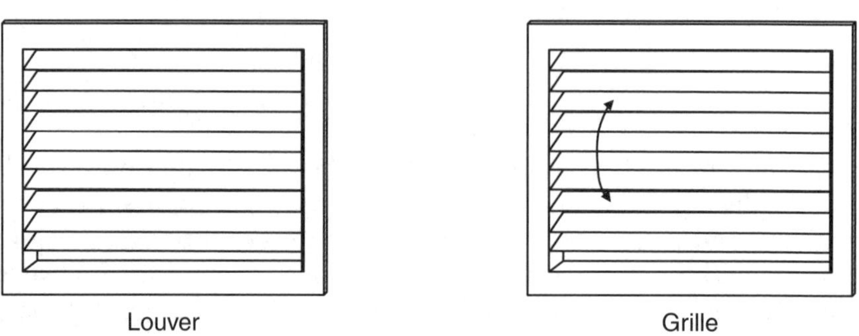

Goodheart-Willcox Publisher

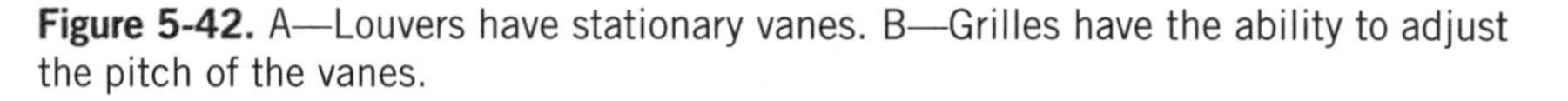

Figure 5-42. A—Louvers have stationary vanes. B—Grilles have the ability to adjust the pitch of the vanes.

5.4.2 Ceiling Supply Registers

Ceiling supply registers can be configured in a number of different ways and can take on many shapes. The most popular shape is square. Air can be directed from the register in one or more directions. The desired air patterns are indicated on duct drawings, **Figure 5-43**.

- **Four directions.** When located in the middle of a room or area, these registers often direct the supply airstream in all four directions.
- **Three directions.** When located against a wall, the registers often direct the supply airstream in three directions.
- **Two opposite directions.** When located in a long, narrow space, such as a hallway, the registers typically direct the supply airstream in two opposite directions.
- **Two directions.** For installations where the supply register is located in a corner, the register is used to direct airflow in two directions, creating a 90° angle.
- **One direction.** Although not a popular option, square registers are available that will direct the air supply in only one direction.

Return air inlets located in the ceiling are drawn similarly to supply registers, but are not directional. These inlets are simply pulling air from the conditioned space back to the air handling equipment. Some commonly used return air symbols are shown in **Figure 5-44**. The symbols are used primarily on plan views to identify the location and type of ceiling supply and return locations.

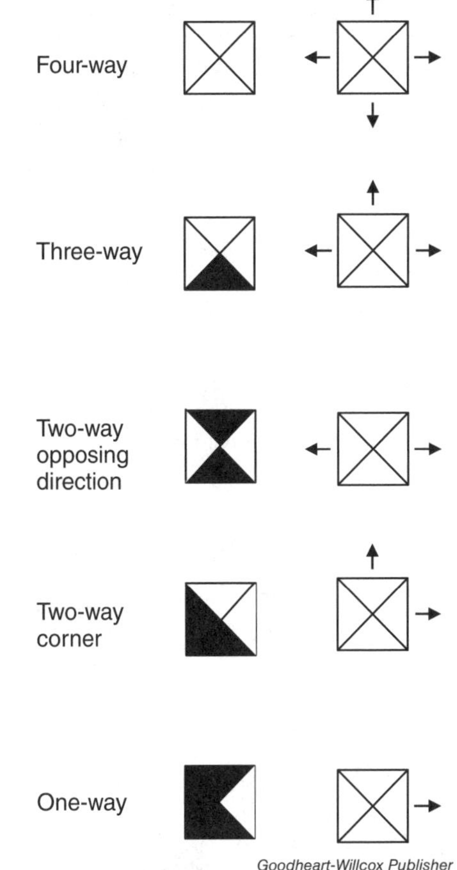

Figure 5-43. Symbols that indicate various air patterns for supply registers.

Return Grilles and Registers

A plan view indicates the location of the supply and return grilles and registers when an airstream enters or leaves the space from the wall. Drawings for wall-mounted supply grilles are typically labeled with the size of the grille and the amount of air intended to flow through the grille, **Figure 5-45**. This figure represents a 20 × 6 supply grille, *SG*, and a 20 × 6 return grille, *RG*. These are indicated by the direction of airflow and the device abbreviations. The two horizontal lines in the figure represent a wall, and the arrow points either out of the wall, as in the case of the supply grille, or into the wall, as with the return grille.

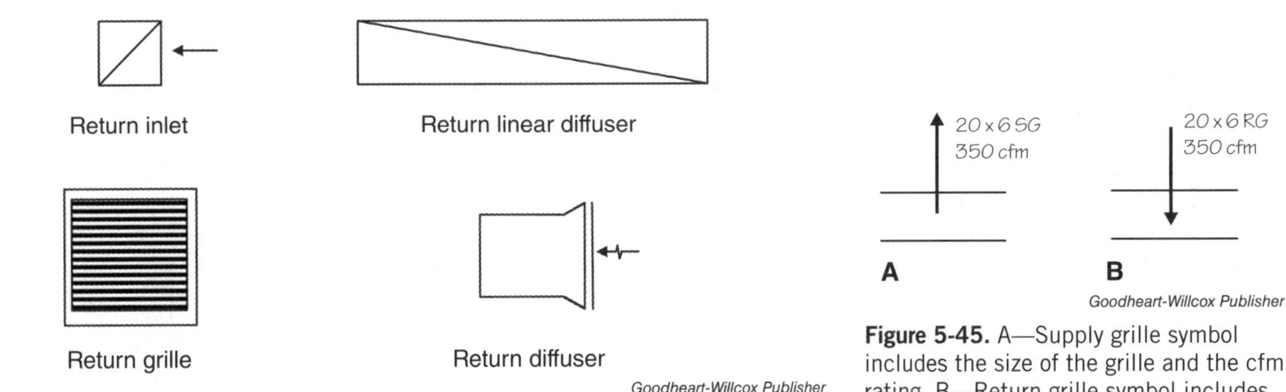

Figure 5-44. Various return air symbols.

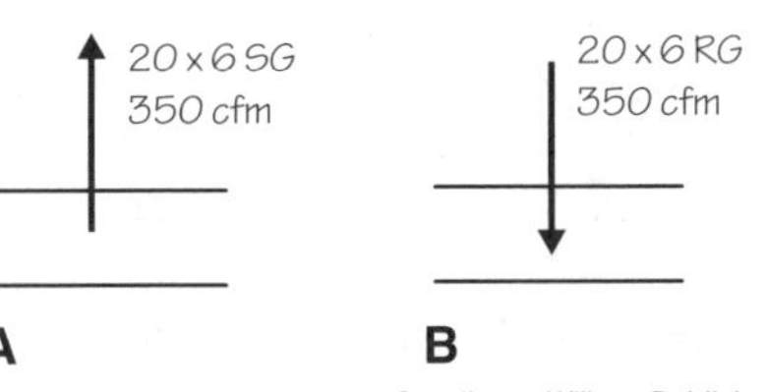

Figure 5-45. A—Supply grille symbol includes the size of the grille and the cfm rating. B—Return grille symbol includes the size of the grille and the cfm rating.

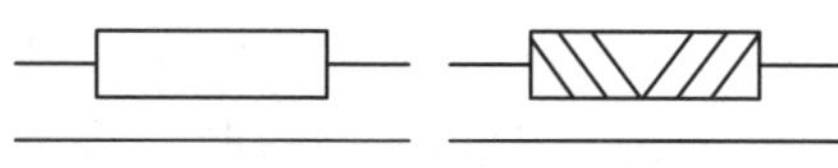

Wall supply louver Wall supply register

Goodheart-Willcox Publisher

Figure 5-46. Alternate wall supply air symbols.

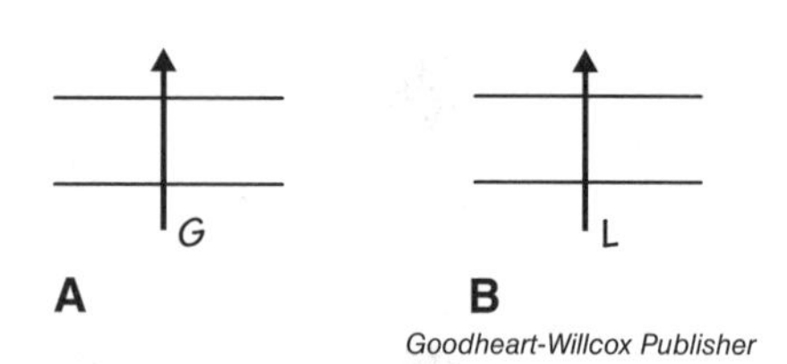

Goodheart-Willcox Publisher

Figure 5-47. A—Symbol for a transfer grille. B—Symbol for a transfer louver.

Wall louvers and registers can be drawn as shown in **Figure 5-46.** The first symbol represents a wall louver and is simply a rectangle that is positioned on the side of the wall where the louver will be installed. The second symbol is a wall register symbol. This is similar to that of the wall louver, but the V-shaped lines indicate the slats on the register can be adjusted.

Sometimes, transfer grilles are installed between rooms in the same conditioned zone. ***Transfer grilles*** allow for free airflow between the two rooms through a penetration in the wall between the two spaces. **Figure 5-47** shows two symbols commonly used to indicate transfer grilles and louvers. Notice that the arrows on the transfer grille and louver symbols extend completely into both areas, separated by the wall. This indicates that air is free to flow from one space, through the wall, and into the other. Transfer grille symbols, therefore, differ from supply register and return grille symbols, because the arrows in these do not completely pass through the wall.

On systems with a single, centrally located return grille, air must flow freely into the conditioned space and then freely back to the return grille. In applications where there are doors that can be closed, airflow back to the return grille can be impeded when the doors are closed. To facilitate sufficient airflow throughout the conditioned space, it is often desirable to install grilles or louvers in the doors themselves, **Figure 5-48.**

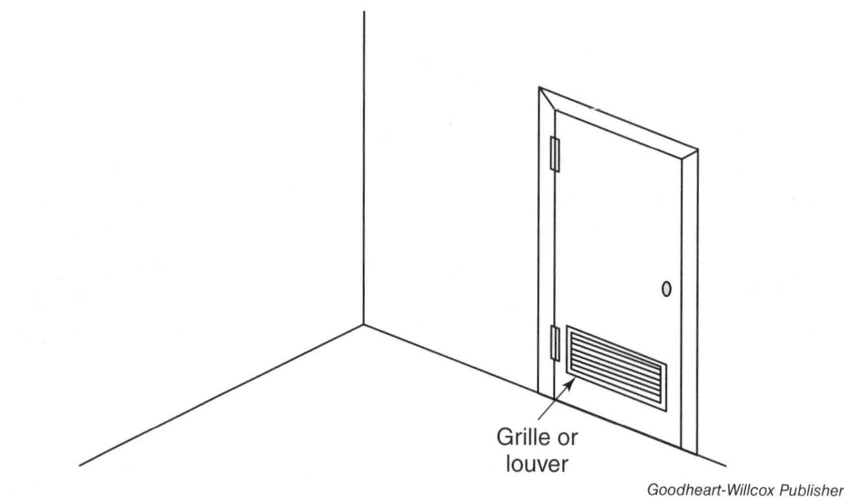

Goodheart-Willcox Publisher

Figure 5-48. Door grille or louver is installed to improve airflow in the space.

BETWEEN THE LINES

Undercutting a Door

When grilles or louvers are installed in doors, the drawing reflects this by using symbols, **Figure 5-49**. When installing a grille or louver in a door is not aesthetically pleasing, another option is to undercut the door. The "G", "L", and "U" represent grille, louver, and undercut, respectively. Undercutting a door is the process of cutting the bottom of the door to leave a gap through which air can pass, even when the door is closed, **Figure 5-50**.

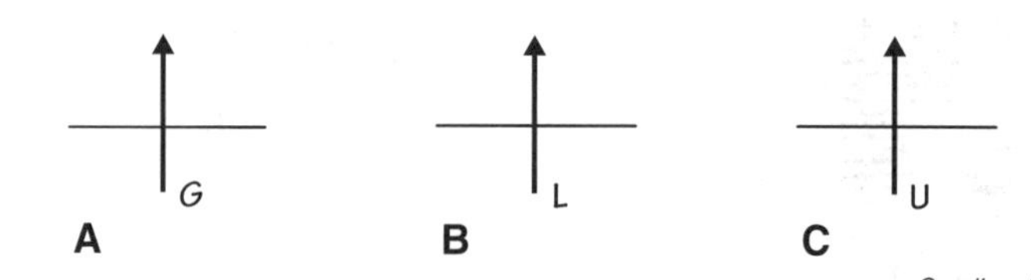

Goodheart-Willcox Publisher

Figure 5-49. A—Symbol for a door grille. B—Symbol for a door louver. C—Symbol for an undercut door.

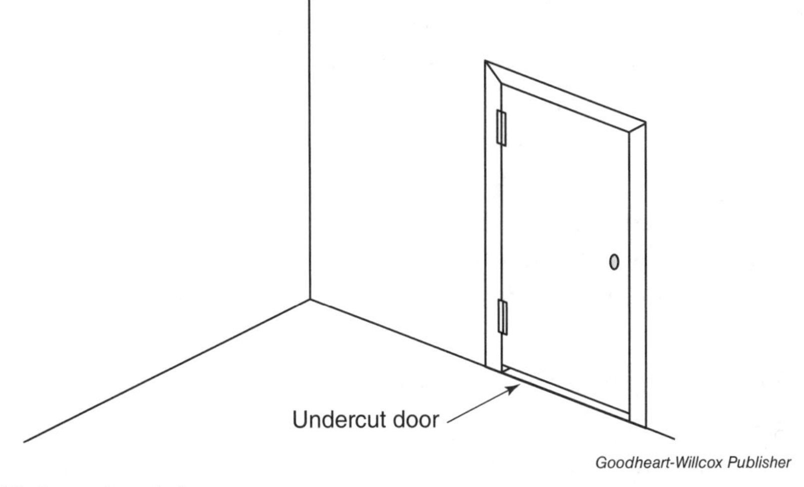

Figure 5-50. An undercut door.

5.4.3 Dampers

Even the simplest air distribution system has numerous dampers installed in it. ***Dampers*** are moveable panels that change their position, manually or automatically, to control airflow through a particular duct system. A variety of dampers and their symbols are shown in **Figure 5-51**.

Dampers perform many functions and are operated in a number of ways:

- **Manually operated dampers.** This most basic style of damper is used to balance the air distribution system to ensure the proper amount of air is delivered to each area the system serves.

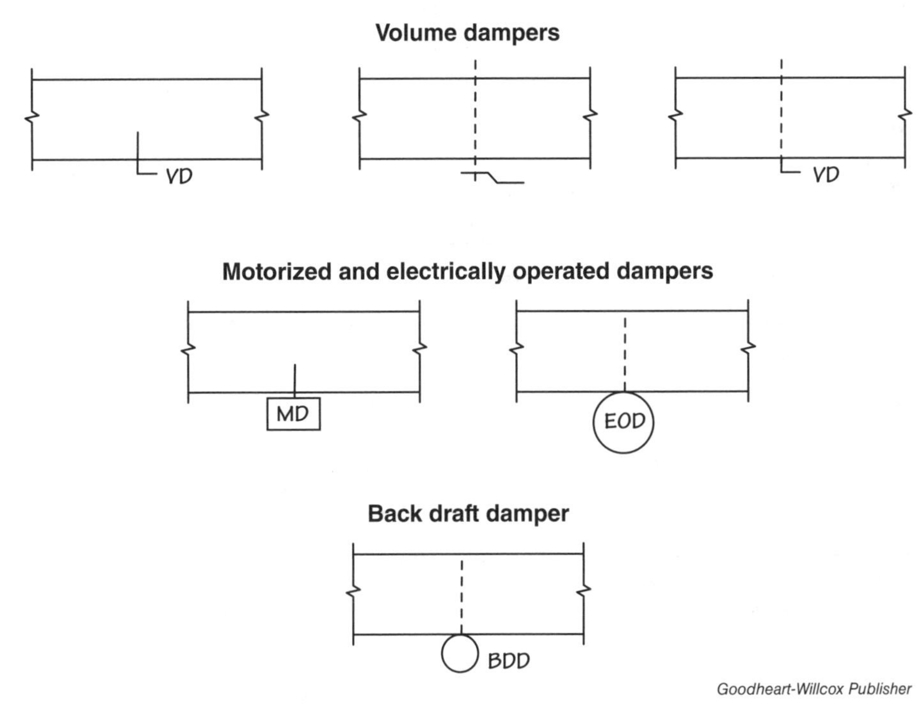

Figure 5-51. Symbols for various dampers.

- **Motorized dampers.** These open and close to increase or decrease the amount of air introduced to a particular area in response to some condition, such as occupancy.
- **Back draft dampers.** Allow air to flow in only one direction and closes if air flows in the opposite direction.

Dampers also play a large role in ensuring the safety of occupants in a conditioned space. If smoke or fire is detected in a duct, dampers close to prevent the air-conditioning system's blower from potentially feeding the fire. Symbols for smoke and fire dampers are shown in **Figure 5-52.**

5.4.4 Additional Components and Accessories

Although a legend showing the symbols used on the particular set of drawings will be provided with the plans used on the job site, **Figure 5-53** displays some additional symbols you will likely encounter when reading air-distribution prints.

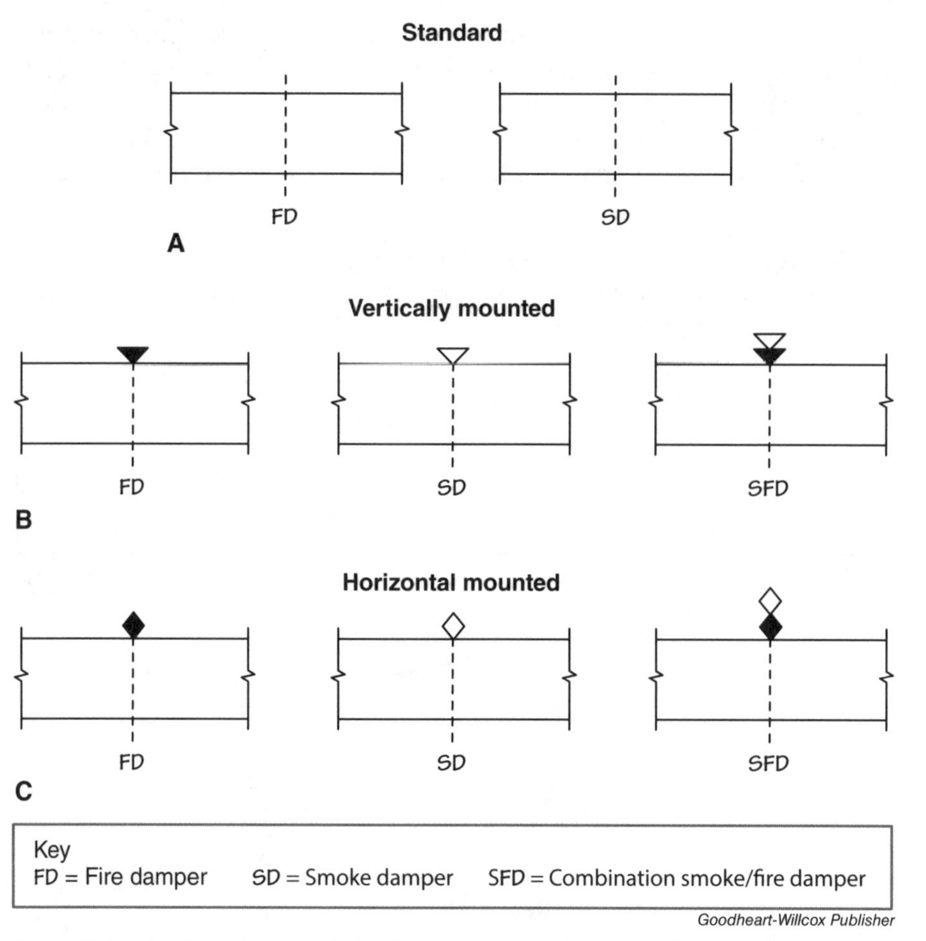

Goodheart-Willcox Publisher

Figure 5-52. A—Generic symbols for fire and some dampers. B—Symbols for vertically installed fire, smoke, and combination smoke/fire dampers. C—Symbols for horizontally installed fire, smoke, and combination smoke/fire dampers.

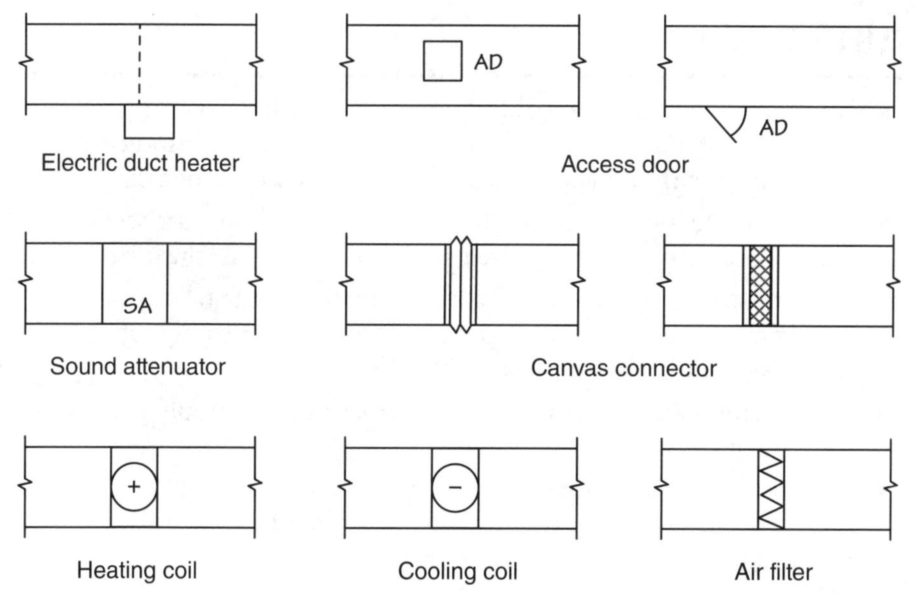

Figure 5-53. Additional air distribution symbols.

Summary

- Ducts are round or rectangular pipes used to carry conditioned air. Duct systems should be designed, fabricated, and installed in compliance with industry-accepted guidelines and standards.
- A system's ductwork consists of return air and supply air sections. Supply and return plenums are connected directly to the air handler.
- Primary supply trunks supply air to branch runouts or secondary supply trunks. Secondary supply trunks supply air to branch runouts or other secondary supply trunks.
- Branch runouts connect the primary or secondary supply trunks to supply registers.
- Plan views and elevation views for ductwork provide different information. The measurement provided first in a dimension repesents the side of duct viewed in the drawing.
- Notes and symbols on duct drawings must be read carefully to avoid missing potentially important information.
- Duct fittings are sections of ductwork that change the direction or size of a duct run. They have construction numbers assigned to them.
- Two common supply air fittings are the bullhead and tapered head.
- Common duct fittings include takeoff, transition, offset, and elbow fittings.
- Transition fittings change the cross-sectional measurements of a duct, while offset fittings route ductwork runs around vertical or horizontal obstructions.
- Takeoff fittings connect the branch runout to the main supply duct.
- Boot fittings connect the branch runout to the register or grille.
- Registers, grilles, louvers, and diffusers are duct system accessories that help a duct system function effectively and efficiently. Louvers have fixed vanes, whereas grilles can have adjustable vanes.
- Registers are grilles that have dampers to adjust the direction and volume of air passing through them. Transfer grilles allow air to flow freely between two adjacent conditioned spaces in the same zone.
- Dampers are used to increase or decrease airflow to a particular area or zone. They can be manually or automatically controlled.
- Smoke and fire dampers are used to close off sections of duct in the event of a fire.

Chapter Assessment

Name ______________________ Date ____________ Class ______________

Know and Understand

_______ 1. The duct fitting used to connect two different-sized duct sections is the _____.

A. offset
B. boot
C. takeoff
D. transition

_______ 2. The duct fitting used to direct a duct run around an obstruction is the _____.

A. offset
B. boot
C. takeoff
D. transition

_______ 3. The duct fitting that connects a duct runout to a main supply duct is the _____.

A. offset
B. boot
C. takeoff
D. transition

_______ 4. The duct fitting that connects a branch duct to a supply register is the _____.

A. offset
B. boot
C. takeoff
D. transition

_______ 5. Which of the following best describes the elbow shown in the following figure?

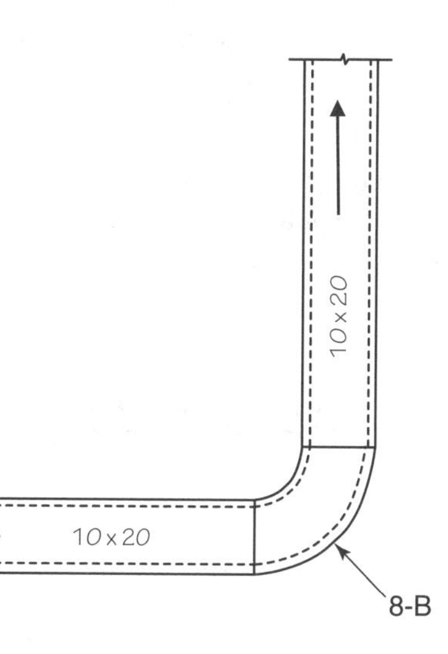

A. Easy bend radius elbow with turning vanes
B. Easy bend radius elbow without turning vanes
C. Hard bend radius elbow with turning vanes
D. Hard bend radius elbow without turning vanes

_______ 6. A main difference between a grille and a louver is that a _____.

A. louver has fixed slats, whereas a grille may have adjustable slats
B. grille has fixed slats, whereas a louver may have adjustable slats
C. grille is installed in ceilings, whereas a louver is installed in walls
D. louver is installed in ceilings, whereas a grille is installed in walls

_______ 7. Which of the following duct system elements smoothly directs air from a primary or secondary trunk line to a branch runout?

A. Elbow

B. Scoop

C. Turning vane

D. Transition

_______ 8. Dashed lines and parentheses are commonly used to identify _____.

A. duct sections to be insulated

B. existing duct sections to be removed

C. existing duct sections to remain in place

D. duct sections hidden from view

_______ 9. All of the following are industry organizations that create standards for the safe and proper installation of HVACR systems *except* _____.

A. NFPA

B. USAC

C. ASHRAE

D. ACCA

_______ 10. The two duct fittings connected to the air handler are the _____.

A. return plenum and the supply plenum

B. return transition and the supply plenum

C. return plenum and the supply transition

D. return transition and the supply transition

Apply and Analyze

Refer to the following image for questions 1 and 2.

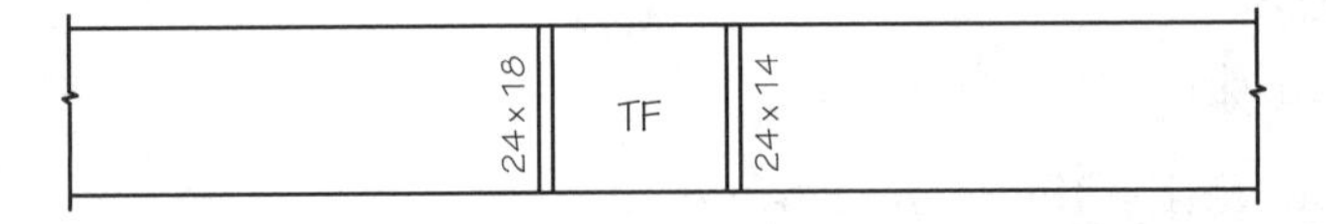

_______ 1. Which of the following drawings best represents the elevation view of the above duct run?

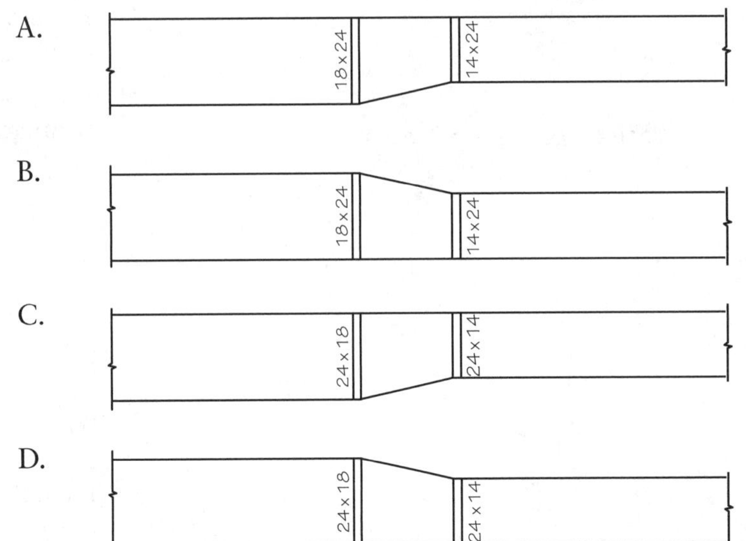

Name ______________________ Date ____________ Class ______________

_______ 2. The portion of the air distribution system shown contains which of the following?

A. Straight duct sections and an offset
B. Straight duct sections and a transition
C. Straight duct sections, an offset, and a transition
D. Offsets and one transition

_______ 3. Consider a tee fitting that has three duct sections connected to it. A 300-cfm airstream is entering the tee from one duct, while a 200-cfm airstream is leaving the tee through another duct. What is the airflow and its direction in the third duct connected to the tee fitting?

A. An airstream of 500 cfm is flowing out of the tee.
B. An airstream of 500 cfm is flowing into the tee.
C. An airstream of 100 cfm is flowing into the tee.
D. An airstream of 100 cfm is flowing out of the tee.

Critical Thinking

1. Why is it important for all of the tee "inflows" to equal all of the tee "outflows"?

____
____
____
____

2. Why do offset fittings have the same inlet and outlet cross-sectional measurements?

____
____
____
____

3. With respect to system and space pressures, what function does a transfer grille perform?

____
____
____
____

Large Prints Activity

Name ______________________ Date ____________ Class ______________

Practice Using Large Prints

Refer to Print-03 in the Large Prints supplement to answer the following question.

1. Determine the total length of 13″ × 8″ straight ductwork used to supply air to the courtyard.

   ____

Refer to Print-05 in the Large Prints supplement to answer the following question.

_______ 2. The 13″ × 8″ ducts are best described as _____.

A. runouts
B. secondary trunks
C. primary trunks
D. transitions

Refer to Print-14 in the Large Print supplement to answer the following questions.

_______ 3. Which of the following best describes the three sections (two 45° elbows and a straight section that connects them) of 46 × 18 supply duct that carry air from AH-G4?

A. They make up a horizontal transition.
B. They make up a horizontal offset.
C. They make up a vertical transition.
D. They make up a vertical offset.

_______ 4. In the mechanical plan, what are the dimensions of the horizontal straight duct run with the longest length?

A. 28 × 10
B. 32 × 8
C. 46 × 18
D. 22 × 10

_______ 5. Locate the duct run highlighted in the figure below and determine its total straight length.

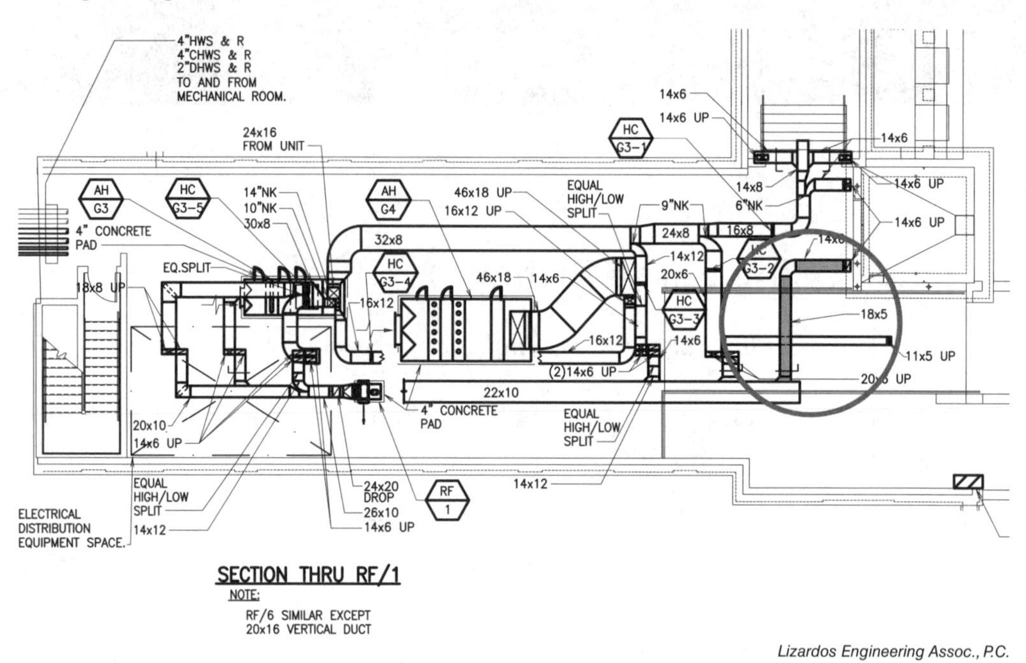

Lizardos Engineering Assoc., P.C.

A. 19′
B. 24′
C. 32′
D. 38′

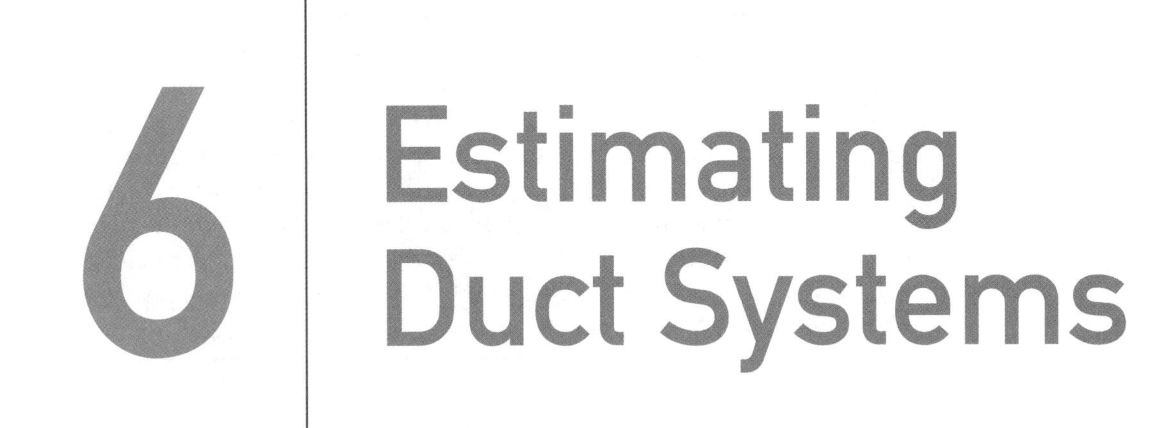

6 Estimating Duct Systems

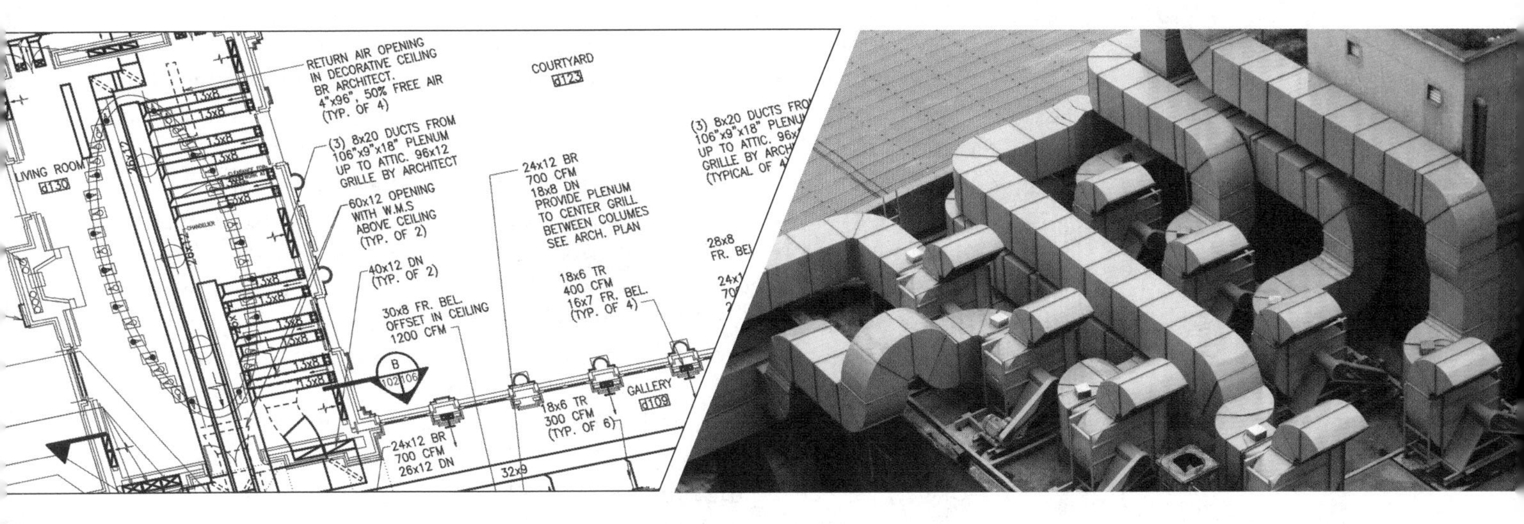

Chapter Outline

6.1 Using an Architect's Scale to Determine Duct System Dimensions and Component Locations

6.2 Total Equivalent Length (TEL)

6.2.1 Total Equivalent Length of a Duct Run with One Air Path

6.2.2 Total Equivalent Length of a Duct Run with Multiple Air Paths

6.3 Determining the Amount of Material Required to Fabricate a Duct Section

6.3.1 Sheet Metal Varieties and Gages

6.3.2 Duct Charts

Learning Objectives

After studying this chapter, you will be able to:

- Use an architect's scale to determine the actual size of duct sections.
- Define the total equivalent length of a duct run.
- Estimate the total equivalent length of a duct run based on print information.
- Calculate the total equivalent length of a duct run with a single air path and multiple air paths.
- Discuss the types of sheet metal, including their advantages and disadvantages.
- Explain sheet metal gages and those most commonly used in the HVACR industry.
- Use a duct chart to determine the amount of material required to fabricate a duct section.

Introduction

Ductwork is an essential component in any heating, cooling, or mechanical ventilation/exhaust system because it is responsible for distributing airflow to, or removing air from, an occupied space. There are many factors to consider when preparing to fabricate ductwork, including size, length, height, and material. Often, math calculations or data interpretation must be completed to determine these factors. This chapter introduces some of the most important calculations required to estimate the cost of fabricating and installing ductwork for a given HVACR project.

Technical Terms

duct chart
easy elbow
gage
hard elbow
sheet metal
square elbow
total equivalent length (TEL)

6.1 Using an Architect's Scale to Determine Duct System Dimensions and Component Locations

Chapter 4, *Measurements and Scales*, introduced the architect's scale and the process for producing scaled drawings. This scale also can be used to read and extract information, such as determining the dimensions and lengths of duct runs. To accomplish this, the scale used to obtain information and the scale used to create the drawing must be the same. Many errors made in extracting information from project plans are due to using an incorrect scale.

Without using the architect's scale, we would need to convert the measurement on the drawing to the actual length of the duct. Consider the straight duct section of 21″ × 12″ shown in **Figure 6-1.** This print was created using a 1/8″ = 1′0″ scale, which means each 0.125″ represents 1′. A traditional ruler shows that the duct section was drawn 1.75″ in length. We then need to divide 1.75″ by 0.125 to determine that the drawing represents a duct section 14′ long.

The architect's scale helps us avoid this long process. First, determine the scale used to create the drawing. Once known, the zero mark on the architect's scale is then lined up with one edge of the duct section, **Figure 6-2.** Next, the opposite end of the duct section is lined up with the scale. Because

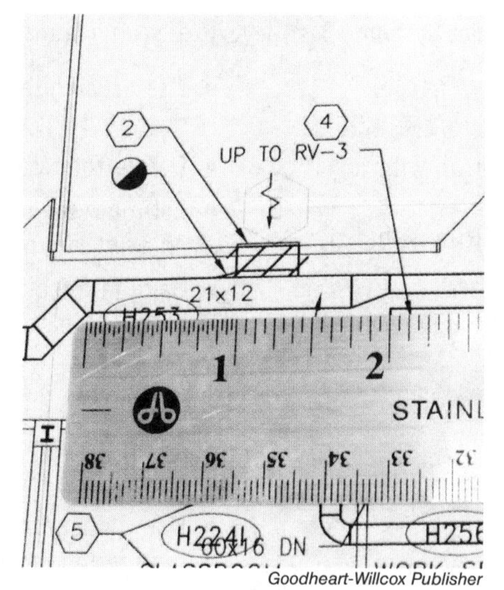

Goodheart-Willcox Publisher

Figure 6-1. The length of the duct section in the drawing is 1.75″ long.

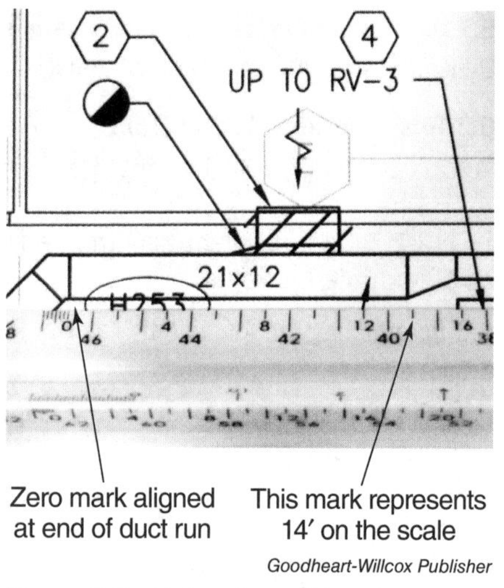

Goodheart-Willcox Publisher

Figure 6-2. Using the architect's scale, the 14′ measurement can be immediately determined.

the 1/8″ = 1′0″ scale starts from the left side, the numbers increase from left to right. The end of the duct section lines up with the line that is midway between the 12′ line and the 16′ line. This shows the duct section is 14′ long. This process can also be used to determine the location or dimensions of an object on a print.

6.2 Total Equivalent Length (TEL)

A properly sized air distribution system ensures that an air-conditioning system's blower can move the required amount of air to and from the conditioned space. As air moves through an air distribution system, the air pressure in the duct decreases. This pressure drop must be considered. Pressure drop in ductwork is affected by the diameter of the duct, the length of the duct, and the number of fittings used.

Pressure drop caused by fittings translates to an equivalent length, which is added to the length of the straight duct run. This determines the total equivalent length (TEL). The ***total equivalent length*** is the equivalent length of each fitting added to the length of the straight duct runs and is used to calculate the total *effect* of the ductwork and fittings on the performance of the entire air distribution system. The equivalent length for duct fittings is determined in part by the construction numbers.

When the system's airstream has only one path to take, all duct sections, fittings, and other associated components are taken into account. When there are multiple possible paths for the system air to take, each possible path must be evaluated and only the highest TEL value is used to size the entire air distribution system. For complete information regarding the equivalent lengths for a wide range of duct fittings, refer to ASHRAE documents or ACCA Manual D, which provides comprehensive information on all aspects of the air distribution system design process.

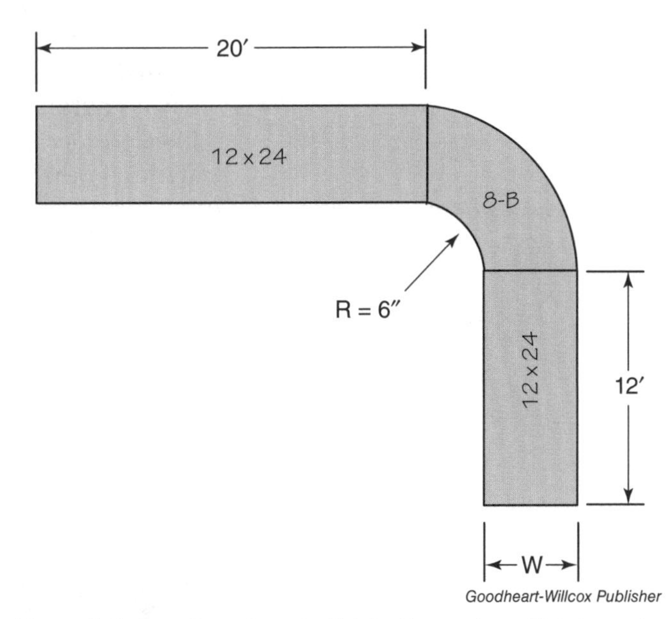

Goodheart-Willcox Publisher

Figure 6-3. A portion of an air distribution system. Not drawn to scale.

6.2.1 Total Equivalent Length of a Duct Run with One Air Path

Consider the portion of the air distribution system shown in **Figure 6-3**. This layout has two sections of ductwork that are 20′ and 12′ long, respectively. The 8-B elbow in this run must also be accounted for. To evaluate the fitting, information regarding its configuration must be found. In the figure, each straight duct section has a cross-sectional measurement of 12 × 24.

The 8-B elbow must then be compared to the three 8-B fitting types to identify its type, **Figure 6-4**. The fitting is an 8-B EASY elbow because elbows fabricated with the *W* measurement as the shorter cross-sectional dimension are classified as ***easy elbows***. Those fabricated with the *W* measurement as the longer cross-sectional dimension are ***hard elbows***. Elbows fabricated with equal cross-sectional dimensions are ***square elbows***.

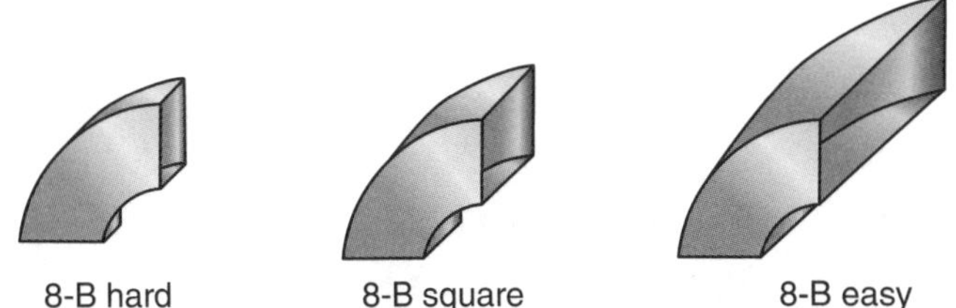

Goodheart-Willcox Publisher

Figure 6-4. The three types of 8-B elbow fittings.

	Vanes	Bend Type	R/W = 0	R/W = 0.25	R/W >0.5
90° Radius Elbow	No (8-B)	Hard	90	35	20
		Square	75	30	15
		Easy	65	25	10
	Yes (8-C)	Hard	30	10	5
		Square	25	10	5
		Easy	40	10	5

Goodheart-Willcox Publisher

Figure 6-5. Equivalent length chart for 8-B and 8-C fittings.

The table in **Figure 6-5** outlines three possible equivalent lengths for the 8-B EASY, depending on the R/W value. The next information needed is the R/W value, which is 6″ (R) / 12″ (W). So, the R/W value is 6″/12″ = 0.5. Using the table, because the R/W value is 0.5, the equivalent length for the 8-B EASY elbow is 10′.

Therefore, the total equivalent length for the portion of the duct run in **Figure 6-3** is 42′, or (20′ + 10′ + 12′).

6.2.2 Total Equivalent Length of a Duct Run with Multiple Air Paths

When an air distribution system contains multiple airstream paths, the TEL of each possible path must be calculated. Only the largest TEL is used to design the duct system. The longest physical duct run is not always automatically the path with the largest TEL.

Consider the portion of the air distribution system shown in **Figure 6-6.** With all other things being equal, one would correctly conclude that the

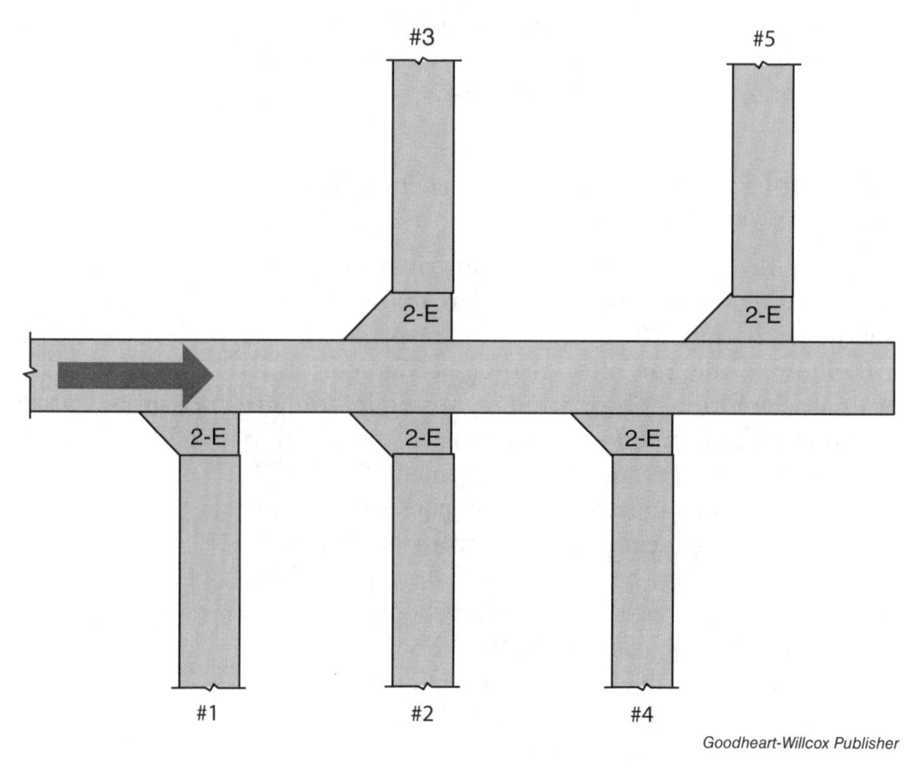

Goodheart-Willcox Publisher

Figure 6-6. Portion of an air distribution system.

path leading to branch #5 would be the one with the largest TEL. This is because the distance to the 2-E takeoff for branch #5 is longer than the others. However, all other things are not equal. Notice the five 2-E takeoffs connected to the main supply duct. The equivalent lengths of each of these takeoff fittings are different, depending on the number of downstream takeoffs that follow it. Any takeoff that is directly opposite of another is considered to be downstream. Because the direction of airflow is from left to right, takeoffs 2, 3, 4, and 5 are all downstream of takeoff 1. Takeoff 5 is downstream of all the rest in the duct run.

To determine the equivalent length of each takeoff, a table like **Figure 6-7** must be used. From the previous figure, we see there are four downstream takeoffs for takeoff 1. From the table, we can determine that the equivalent length for takeoff 1 is 45′. For takeoff 2, there are three downstream takeoffs, which means the equivalent length for that fitting is 40′. Because takeoff 2 and takeoff 3 are directly across from each other, the equivalent length of takeoff 3 is also 40′.

The table also helps determine that the equivalent lengths for takeoffs 4 and 5 are 30′ and 25′, respectively. These lengths are then added together to determine the total equivalent length. Using the same portion of a duct system in **Figure 6-6** in conjunction with the table, it can be determined that the 2-E takeoff for branch #1 has an equivalent length 20′ greater than that for branch #5. So, if these two takeoff fittings are less than 20′ apart from each other, the TEL for branch #1 will be greater than that for branch #5.

Looking at another example, the square 5-H return elbow shown in **Figure 6-8** can have an equivalent length of either 30′ or 45′ depending on the ratio of the W and H measurements. If the H/W ratio is 1, the equivalent length is 45′, and if the H/W ratio is 2, the equivalent length is 30′. If turning vanes are installed in the 5-H return fitting, it then becomes classified as a 5-K fitting, and has an equivalent length of only 10′ regardless of its H/W ratio.

The 5-I radius elbow with the mitered inside corner shown in **Figure 6-9** can have an equivalent length of either 30′ or 45′ depending on the ratio of the W and H measurements, like the 5-H fitting. If the H/W ratio is 1, the equivalent length would be 45′. If the H/W ratio is 2, the equivalent length would be 30′.

Total Equivalent Length Based on Downstream Branches						
Fitting Type	Number of Downstream Branches					
	0	1	2	3	4	5+
2-A	35′	45′	55′	65′	70′	80′
2-D	40′	50′	60′	65′	75′	85′
2-E	25′	30′	35′	40′	45′	50′
2-F	20′	20′	20′	20′	25′	25′
2-G	65′	65′	65′	70′	80′	90′

Goodheart-Willcox Publisher

Figure 6-7. Equivalent length chart for 2-A, 2-D, 2-E, 2-F, and 2-G takeoff fittings.

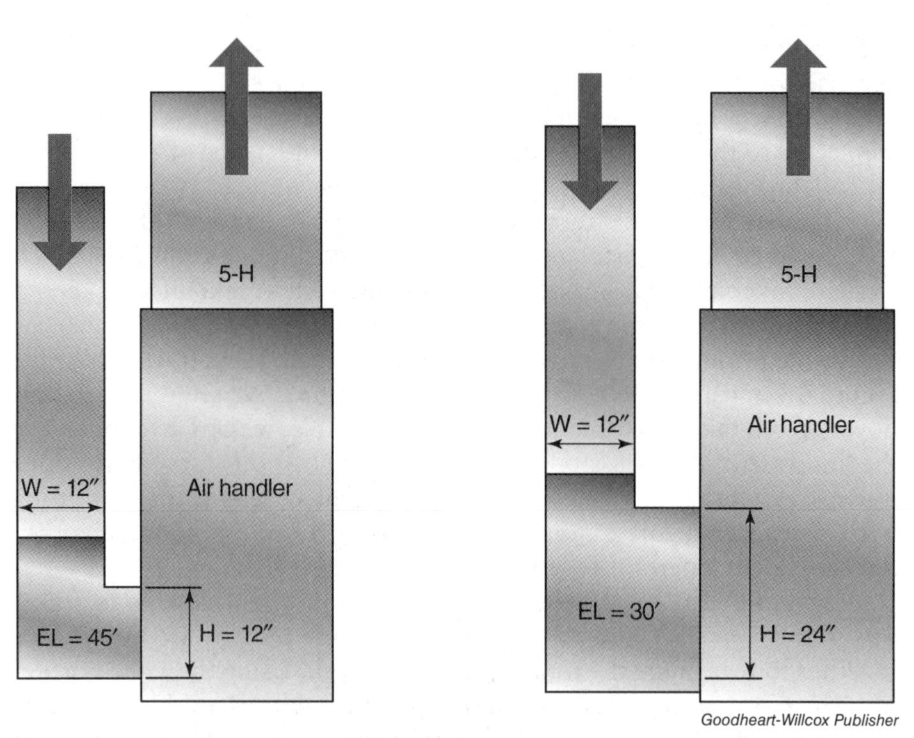

Figure 6-8. Different types of 5-H return fittings at the air handler.

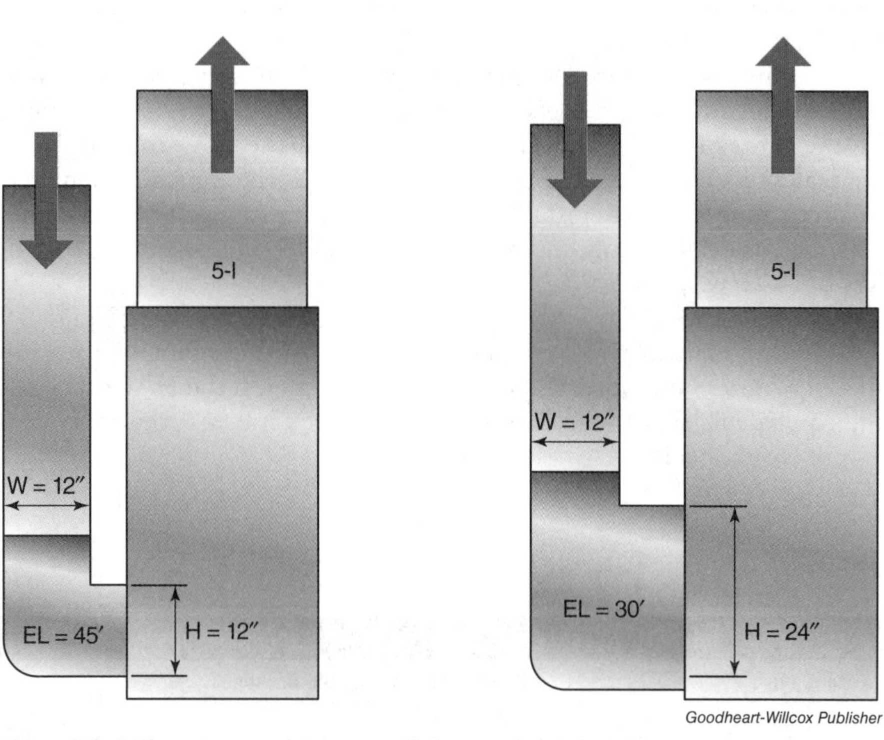

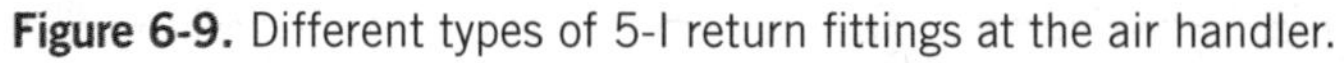

Figure 6-9. Different types of 5-I return fittings at the air handler.

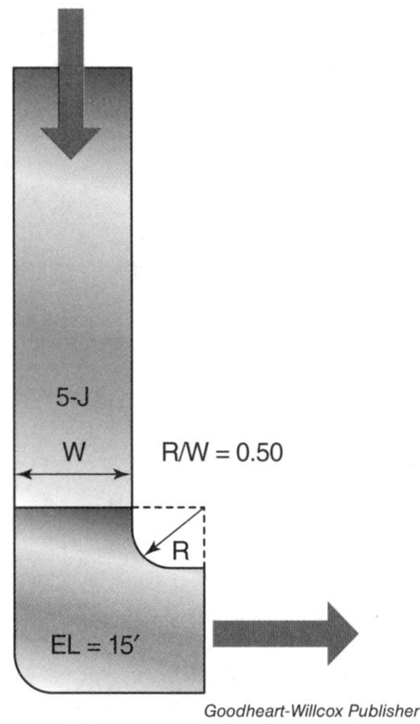

Figure 6-10. The 5-J return fitting at the air handler.

The equivalent length for the radius elbow shown, assigned the 5-J classification in **Figure 6-10**, is determined by a somewhat different method. In this case, the radius, R, of the inside of the fitting is compared to the W measurement. By dividing the radius, R, by the W measurement, the equivalent length can be found. The larger the radius with respect to the W measurement, the lower the equivalent length will be.

Fitting	Type	H/W	Equivalent Length
5-H	Square elbow	1	45′
		2	30′
5-I	Elbow w/ mitered inside corner	1	45′
		2	30′
5-K	Square elbow w/ turning vanes	Any	10′
Fitting	**Type**	**R/W**	**Equivalent Length**
5-J	Radius elbow	0.25	20′
		0.50	15′
		1.00	10′

Goodheart-Willcox Publisher

Figure 6-11. Equivalent lengths for the 5-H, 5-I, and 5-J, and 5-K fittings.

The equivalent lengths for the 5-H, 5-I, and 5-K, and 5-J fittings are shown in **Figure 6-11**.

6.3 Determining the Amount of Material Required to Fabricate a Duct Section

An estimator determines the price for a particular project that should be in line with proposals from other contractors. This individual is also responsible for allowing the company to meets its expenses and make a fair profit. This requires an estimator to extract information from the set of job plans to determine what materials are required for the job and how much of these materials are needed.

In the case of ductwork, the price of the project includes costs for materials, fabrication, and installation. Fabrication and installation costs are often based on the weight of the duct sections, whereas the amount of insulation or lining materials is determined by the surface area of the duct sections. This section will discuss how to calculate the amount of materials needed and how to determine the weight of a particular duct section.

6.3.1 Sheet Metal Varieties and Gages

Sheet metal is a class of materials often used to fabricate duct systems. Common varieties of sheet metal include hot-rolled mild steel, galvanized steel, stainless steel, and aluminum. You must understand the appropriate sheet metal application for a duct system for proper installation.

The most commonly used sheet metal material for ducts is galvanized steel, which is steel that has a zinc coating. This zinc coating helps prevent oxidation from occurring and rust from forming. Hot-rolled steel is the least expensive sheet metal, but it is not protected from rust. Stainless steel and aluminum are much more expensive materials, but are often used for specialty applications. They are both resistant to corrosion, but aluminum is much lighter and easier to work with. Stainless steel is much more durable than aluminum, so it is often used when an abrasion-resistant material is needed.

	Hot-Rolled Steel		Galvanized Steel		Stainless Steel		Aluminum	
Gage	Pound/Square Foot	Thickness (in)	Pound/Square Foot	Thickness (in)	Pound/Square Foot	Thickness (in)	Pound/Square Foot	Thickness (in)
26	0.75	0.0179	0.91	0.0217	0.76	0.0187	0.22	0.0159
24	1.00	0.0239	1.16	0.0276	1.01	0.0250	0.28	0.0201
22	1.25	0.0299	1.41	0.0336	1.26	0.0312	0.36	0.0253
20	1.50	0.0359	1.66	0.0396	1.51	0.0375	0.45	0.0320

Goodheart-Willcox Publisher

Figure 6-12. Chart used to determine the weight per square foot of sheet metal based on gage and metal type.

Sheet metal not only comes in a variety of material types, but in a variety of thicknesses, or ***gages***, **Figure 6-12**. Duct sections with larger cross-sectional measurements require thicker material to provide a more durable system. Thinner materials tend to flex and may generate significantly higher noise levels, especially when the system's blower cycles on and off. The most commonly used gages in the HVACR industry range from 20 to 26. The lower the gage number, the thicker the material.

The weight of a material is important in estimating the cost of a project. Thus, the type of metal used, as well as its gage, must be known. The weight of the section can be calculated once the number of square feet of material required to fabricate a duct section is determined. This weight is found by multiplying the number of square feet of metal by the weight per square foot of the material. For example, a duct section that requires 5 ft^2 of material that weighs 1.5 lb/ft^2 will weigh 7.5 lb.

6.3.2 Duct Charts

A ***duct chart*** provides information such as cross-sectional measurements, gage, sheet metal material weight, and the number of square feet of metal required to fabricate each foot of duct. The information contained in this chart is used to make accurate estimates for duct section fabrication and installation, **Figure 6-13**. For the complete chart, refer to *Duct Chart* in the Appendix. If the duct section dimensions are known, the weight of the completed section, along with how much material will be required to fabricate the section can be determined. This information is extremely valuable to the estimator because it saves a great deal of manual calculations.

The duct chart provided has eight columns. Columns 1 and 2 are used to determine the row from which the material, gage, weight, and square footage values can be obtained. Column 1 represents the sum of two adjacent cross-sectional duct measurements, while column 2 represents the larger of these two measurements. Once the proper line or row has been identified, the estimator can quickly determine the following items:

- The gage metal used to fabricate the duct section (column 3)
- The amount of square feet of material needed to fabricate each 1′ section of duct (column 4)
- Based on the material used to fabricate the duct section, the amount of pounds each linear foot of the completed duct section will weigh (columns 5 through 8)

Duct Chart							
H + W (inches)	Longest Side (inches)	Gage	Square Feet of Material per 1′ of Duct	Hot-Rolled	Galvanized	Stainless	Aluminum
					Weight/Foot (pounds)	Weight/Foot (pounds)	Weight/Foot (pounds)
12	All	26	2.00	1.50	1.81	1.51	0.45
13	All	26	2.17	1.63	1.96	1.64	0.49
14	All	26	2.33	1.75	2.11	1.76	0.52
15	All	26	2.50	1.88	2.27	1.89	0.56
16	All	26	2.67	2.00	2.42	2.02	0.60
17	All	26	2.83	2.13	2.57	2.14	0.63
18	All	26	3.00	2.25	2.72	2.27	0.67
19	0 – 12″	26	3.17	2.38	2.87	2.39	0.71
	13″ – 30″	24	3.17	3.17	3.66	3.20	0.90
20	0 –12″	26	3.33	2.50	3.02	2.52	0.75
	13″ – 30″	24	3.33	3.33	3.85	3.36	0.95
21	0 – 12″	26	3.50	2.63	3.17	2.65	0.78
	13″ – 30″	24	3.50	3.50	4.05	3.53	0.99
22	0 – 12″	26	3.67	2.75	3.32	2.77	0.82
	13″ – 30″	24	3.67	3.67	4.24	3.70	1.04
23	0 – 12″	26	3.83	2.88	3.47	2.90	0.86
	13″ – 30″	24	3.83	3.83	4.43	3.86	1.09
24	0 – 12″	26	4.00	3.00	3.62	3.02	0.90
	13″ – 30″	24	4.00	4.00	4.62	4.03	1.14
25	13″ – 30″	24	4.17	4.17	4.82	4.20	1.18
26	13″ – 30″	24	4.33	4.33	5.01	4.37	1.23
27	13″ – 30″	24	4.50	4.50	5.20	4.54	1.28

Goodheart-Willcox Publisher

Figure 6-13. A portion of a duct chart.

Using a Duct Chart—Example 1

Consider a 4′ section of 10″ × 6″ duct that must be fabricated from aluminum sheet metal. The first step is to add the height and width measurements of the desired duct section together.

$$10'' + 6'' = 16''$$

Locate the 16″ row in column 1 of the duct chart. Since column 2 refers to all measurement combinations, simply follow this row to the right. This duct section, from column 3, is fabricated using 26-gage sheet metal. Continuing to the right and looking under column 4, we see that each foot of this duct section requires 2.67 ft^2 of sheet metal. The desired duct section is 4′ long, so 10.68 ft^2 of sheet metal is needed. This is determined by multiplying square feet by the duct section length.

$$2.67\ ft^2 \times 4 = 10.68\ ft^2$$

Continuing to column 8 for aluminum, each 1′ length of this duct section weighs 0.60 lb. Because the desired duct section is 4′ long, the completed duct section weighs 2.4 lb.

$$0.60 \text{ lb} \times 4 = 2.4 \text{ lb}$$

We can determine that 10.68 ft^2 of duct insulation or duct lining is required if the plan calls for either the insulating or lining of the duct section. Duct lining or insulation material typically covers the entire surface of the duct, so the amount of this material is equal to the amount of metal needed for the section.

Using a Duct Chart—Example 2

A 4′ section of 16″ × 8″ duct needs to be fabricated from stainless steel sheet metal. The first step is to add the height and width measurements of the desired duct section together, giving us 24″ (16″ + 8″).

Locating the 24″ row in the duct chart and following this row to the right, we see there are two rows that correspond to 24″. The row used corresponds to the larger of the two cross-sectional measurements of the duct, namely 16″. The 16″ measurement falls in the 13″ – 30″ range, so that row is used.

Following that row to the right, this duct section will be fabricated using 24-gage sheet metal. Continuing to the right, we see that each foot of this duct section will require 4.00 ft^2 of sheet metal. Because the desired duct section is 4′ long, 16 ft^2 of sheet metal will be needed (4 × 4).

Continuing to the column for stainless steel (column 7), each 1′ length of this duct section will weigh 4.03 lb. Because the desired duct section is 4′ long, the completed duct section will weigh 16.12 lb (4.03 × 4). It can also be determined that 16 ft^2 of duct insulation or duct lining will be required if the plan calls for either the insulating or lining of the duct section.

Summary

- An architect's scale can be used to read and extract information, such as determining the dimensions and lengths of duct runs.
- As air moves through an air distribution system, air pressure drops, affected by the diameter of the duct, the length of the duct, and the number of fittings used.
- Pressure drop caused by fittings is identified as an equivalent length. This length is added to the length of the duct run to calculate the total equivalent length (TEL).
- Construction numbers are used to determine the equivalent length of the fittings.
- Easy elbows are fabricated with the W measurement as the shorter cross-sectional dimension, whereas hard elbows have this measurement as the longer cross-sectional dimension. Elbows fabricated with equal cross-sectional dimensions are square elbows.
- Calculating the total equivalent length of a duct run with multiple air paths requires evaluating the number of downstream fittings and finding their appropriate equivalent length.
- Commonly used sheet metal materials for the HVACR industry are hot-rolled steel, galvanized steel, stainless steel, and aluminum.
- The term *gage* is used to identify the thickness of sheet metal. The lower the gage number, the thicker the sheet metal.
- Duct charts are used to determine the amount of material required to fabricate sections of ductwork.

Chapter Assessment

Name ______________________ Date ____________ Class ______________

Know and Understand

_______ 1. The total equivalent length for the portion of the air distribution system shown in the following figure is _____.

A. 56′

B. 86′

C. 101′

D. 116′

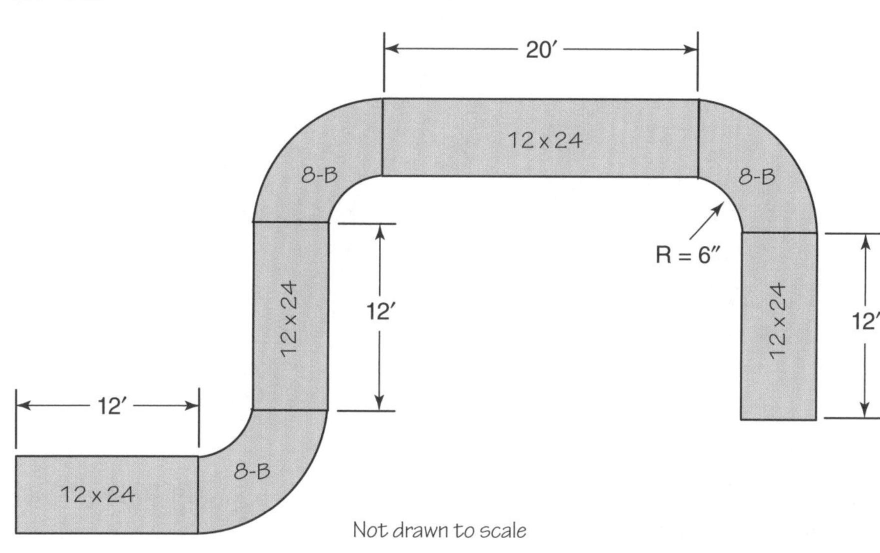

_______ 2. Which of the following materials are most similar with respect to their density?

A. Hot-rolled steel and galvanized steel

B. Stainless steel and aluminum

C. Galvanized steel and stainless steel

D. Hot-rolled steel and stainless steel

_______ 3. Stainless steel is a desirable material for some duct systems because it is _____.

A. lightweight and durable

B. abrasion and corrosion resistant

C. inexpensive and corrosion resistant

D. lightweight and abrasion resistant

_______ 4. Aluminum is a desirable material for some duct systems because it is _____.

A. lightweight and corrosion resistant

B. abrasion and corrosion resistant

C. inexpensive and corrosion resistant

D. lightweight and abrasion resistant

_______ 5. The main drawback of hot-rolled steel ductwork is that it _____.

A. can rust

B. is expensive

C. Both A and B.

D. Neither A nor B.

Apply and Analyze

1. A straight, rectangular duct section requires 10 ft^2 of 22-gage galvanized steel sheet metal. Assuming no waste remains after fabricating the section, how much will the finished duct section weigh?

____

2. A 4′ section of 12″ × 10″ duct needs to be fabricated. What gage metal will be used?

____

_______ 3. A 5′ section of stainless steel ductwork weighs exactly 16 lb. Which of the following is a possible description of the section?

A. Cross-sectional measurements of 19″ × 8″, 24-gage sheet metal
B. Cross-sectional measurements of 14″ × 5″, 24-gage sheet metal
C. Cross-sectional measurements of 19″ × 8″, 22-gage sheet metal
D. Cross-sectional measurements of 14″ × 5″, 22-gage sheet metal

Name ______________________ Date ____________ Class ______________________

4. How many square feet of 1″ thick acoustical lining material is needed to completely line a 6′ section of 14″ × 12″ duct?

____

Critical Thinking

1. Why can the value obtained for the amount of sheet metal, in square feet, to be used for fabrication also be used to determine the amount of material needed to insulate, wrap, or line a duct section?

____

____

____

2. Why does the addition of turning vanes in an elbow or other fitting reduce the equivalent length of the fitting?

____

____

____

3. Why is it *not* desirable to use a metal that is thinner than specified when fabricating and installing ductwork?

____

____

____

Large Prints Activity

Name ______________________ Date ____________ Class ______________

Practice Using Large Prints

Refer to Print-03 in the Large Prints supplement to answer the following questions. Note: The large prints are created at 50% of their original size. When extracting information from the print using an architect's scale, use a scale that is one-half of the one indicated on the drawing.

1. Consider the straight duct section connected to the left side of the heating coil in Section D-102 is fabricated from galvanized steel:

   A. What gage metal should be used?

   ____

   ____

   B. How much will the section weigh?

   ____

   ____

   C. How many square feet of sheet metal will be required for this section? (Hint: Use the *Duct Chart* in the Appendix.)

   ____

   ____

7 Piping Drawings

Chapter Outline

7.1 Piping Systems
- 7.1.1 Piping System Installation

7.2 Piping Abbreviations

7.3 Valves

7.4 Pipe Fittings
- 7.4.1 Elbows
- 7.4.2 Tees
- 7.4.3 Unions

7.5 Gauges and Switches
- 7.5.1 Drains, Strainers, and Cleanouts
- 7.5.2 Pipe Guides

7.6 Piping Symbols and Drawing Concepts

Learning Objectives

After studying this chapter, you will be able to:

- Discuss the various functions of piping arrangements.
- Identify the types of abbreviations used on HVACR plans.
- Describe valves commonly used on HVACR projects.
- Differentiate between directional valves, flow-control valves, and pressure control valves.
- Identify the symbols used to represent various types of valves commonly found in HVACR piping drawings.
- Describe the types of fittings used in piping systems.
- Identify fittings commonly used on HVACR piping drawings.
- Explain the gauges and sensors used in piping systems.
- Explain the function of pipe guides and hangers.
- Identify symbols that express specific piping concepts.

Technical Terms

aquastat

balancing valve

check valve

concentric reducer

eccentric reducer

flanged union

flow switch

mixing valve

pipe guide

pressure switch

pressure-reducing valve

reducer

screwed union

union

valve

Introduction

A system designer or engineer must be able to convey ideas on piping system prints to ensure that the piping systems are installed correctly. On the job site, HVACR-related piping can involve the installation of refrigerant lines, condensate lines, chilled water lines, hot water heating lines, water-cooled condenser lines, fuel lines, drain lines, and make-up water lines. Not only must these lines be properly sized and pitched, they must also be properly supported. Properly designed and installed piping arrangements such as these help ensure the satisfactory operation of the air-conditioning system for years to come. This chapter introduces you to some concepts that are used to extract important information from piping drawings or prints. These skills will prove useful to those who ultimately estimate or install the piping circuits for HVACR equipment.

7.1 Piping Systems

The function of an HVACR system is to provide thermal comfort to the occupants of the space. This requires all of the system's components to function together. Various fluids must move through the system's piping arrangements. These piping arrangements are responsible for a number of tasks:

- Transferring condensate from the air handlers to remotely located drains.
- Carrying high-pressure liquid refrigerant and low-pressure vapor refrigerant between the indoor and outdoor units of a split air-conditioning system.
- Supplying heated water to duct-mounted hot water coils from the boiler.
- Carrying water from hot water coils back to the boiler.
- Supplying chilled water to duct-mounted cooling coils.
- Carrying water from cooling coils back to the chiller.
- Supplying water to water-cooled condenser coils from a cooling tower.
- Carrying water from condenser coils back to the cooling tower.
- Supplying fossil fuels, such as natural gas and oil, to furnaces and boilers.
- Providing make-up water to boilers and cooling towers.
- Transferring water between ground loops/coils and equipment loops/coils in geothermal heat pump applications.

7.1.1 Piping System Installation

Proper piping installation ensures an effective system. There should be little turbulence in the pipes. Turbulence increases pressure drops in the piping runs, which causes reduced fluid flow rates. In addition, turbulent fluid flow often results in excessive noise levels, which can be both distracting and annoying, especially for those in close proximity to the piping arrangements.

BETWEEN THE LINES

Piping Guidelines

Proper installation guidelines must be followed. For example, incorrectly installed or sized refrigerant lines may not allow oil to return to the compressor, which can cause premature pump failure. Improperly sized or installed refrigerant lines can decrease system capacity and operating efficiency.

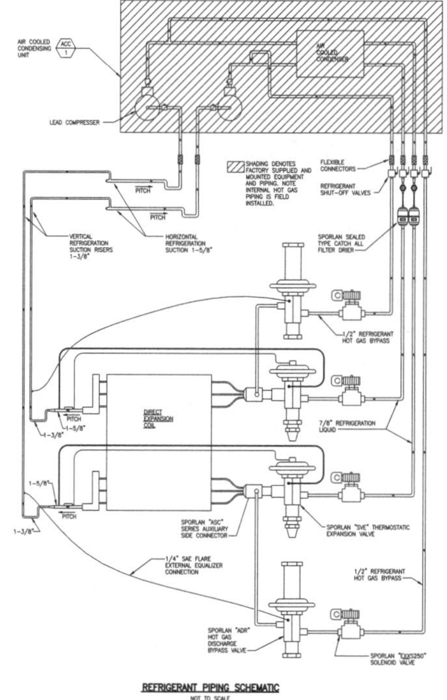

Lizardos Engineering Assoc., P.C.

Figure 7-1. Pictorial diagram of refrigerant piping.

An HVACR piping plan includes guidelines for proper piping installation, **Figure 7-1.** In most cases, the notes on the print state that all work must be performed in accordance with local codes and manufacturers' guidelines, **Figure 7-2.**

Piping systems must be installed with future service and repair in mind. For example, installing isolation valves on heat exchangers, or other system components adds to the initial cost of the job, but saves time and money when those components require servicing. Job plans and prints often account for such service-related features. For example, the piping circuit in **Figure 7-3** shows components labeled BA. These are ball valves that isolate various portions of the piping circuit for system service and maintenance. Without these valves, future system servicing would be difficult.

1. DESCRIPTION OF WORK
THE WORK TO BE PERFORMED UNDER THESE SPECIFICATIONS AND ACCOMPANYING DRAWINGS INCLUDES YEAR-ROUND HEATING AND AIR CONDITIONING FOR THE ENTIRE RESIDENCE, INCLUDING BASEMENT, RECREATIONAL, SERVICE AND STAFF AREAS, AS WELL AS ALL STRUCTURES LOCATED ON, OR REFERENCED IN, THE PLANS.

2. MATERIALS AND WORKMANSHIP
ANY AND ALL MATERIALS AND EQUIPMENT THAT ARE USED IN CONJUNCTION WITH THE EXECUTION OF THE PROJECT AS OUTLINED IN THESE SPECIFICATIONS, PLANS AND DRAWINGS ARE TO BE OF THE HIGHEST QUALITY AVAILABLE AT THE TIME OF CONSTRUCTION. IF, DURING THE COURSE OF CONSTRUCTION, NEW MATERIALS OR EQUIPMENT BECOMES AVAILABLE THAT WILL SIGNIFICANTLY ENHANCE THE FUNCTIONALITY OF THE OVERALL SYSTEM, THE ARCHITECT OR ITS REPRESENTATIVE SHALL RESERVE THE RIGHT TO AMEND THE SPECIFICATIONS TO INCORPORATE SUCH MATERIAL AND/OR EQUIPMENT. THE CONTRACTOR WILL, AS A RESULT, HAVE THE ABILITY TO AMEND THE PORTION(S) OF THE PROPOSAL THAT ARE AFFECTED BY THESE CHANGES, INCLUDING ANY ADDITIONAL TIME AND/OR CHARGES THAT ARE NECESSARY TO INCORPORATE THESE CHANGES. IF, AT THE TIME OF THE PROPOSED CHANGES, A SIGNIFICANT AMOUNT OF THE INSTALLATION HAS BEEN COMPLETED, THE CONTRACTOR CAN, AT HIS DISCRETION, REQUEST, WITH ACCEPTABLE SUPPORTING DOCUMENTATION, THAT THE PROPOSED CHANGES NOT BE INCORPORATED INTO THE PROJECT PLAN. THE ULTIMATE DECISION REGARDING WHETHER OR NOT PROPOSED CHANGES WIL BE INCORPORATED INTO THE FINALIZED AND APPROVED PLAN WILL BE MADE BY THE ARCHITECT OR SYSTEM DESIGNER, ONLY AFTER ALL OPTIONS HAVE BEEN DISCUSSED WITH THE PROPERTY OWNER OR ITS REPRESENTATIVE.

3. RULES AND ORDINANCES
ALL MATERIALS, LABOR AND PRACTICES SHALL BE IN STRICT ACCORDANCE WITH THE RULES AND REGULATION OF THE NFPA AND ANY OTHER LOCAL CODES, RULES, LAWS OR ORDINANCES THAT ARE IN EFFECT AT ANY TIME DURING THE COURSE OF THE PROJECT.

4. DIAGRAMMATIC NATURE OF PLANS
PLANS, AS A WHOLE, ARE DIAGRAMMATIC IN NATURE AND ARE NOT INTENDED TO SHOW COMPLETE DETAILS OF ALL WORK TO BE PERFORMED. THESE PLANS ARE PROVIDED FOR THE SOLE PURPOSE OF ILLUSTRATING THE TYPE OF EQUIPMENT TO BE USED, ALONG WITH IDENTIFYING ANY SPECIAL CONDITIONS THAT MAY NEED TO BE ADDRESSED SO THAT EXPERIENCED INDIVIDUALS CAN SUCCESSFULLY AND COMPLETELY PREPARE AN ACCURATE ESTIMATE FOR THE WORK TO BE PERFORMED. THE CONTRACTOR SHALL BE RESPONSIBLE FOR MAKING MEASUREMENTS, OBSERVING, FIRST-HAND, THE JOB SITE AND ADAPTING THE SCOPE OF WORK TO THE CONDITIONS PRESENT ON THE JOB SITE. UNDER NO CIRCUMSTANCES SHALL THE WORK PERFORMED VIOLATE ANY LOCAL CODES OR THE SPECIFIC INSTRUCTIONS OF THE EQUIPMENT MANUFACTURER. IT IS, THEREFORE, THE RESPONSIBILITY OF THE CONTRACTOR TO BRING TO THE ATTENTION OF THE ARCHITECT OR SYSTEM DESIGNER ANY ISSUES CONTAINED IN THE PROJECT PLANS OR DRAWINGS THAT CONTRADICT ANY LOCAL CODES, INSTALLATION GUIDELINES, OR BEST INDUSTRY PRACTICES.

5. COOPERATION
THE CONTRACTOR AND ITS AGENTS AGREE TO WORK IN CONJUNCTION WITH OTHER CONTRACTORS AND THEIR AGENTS AND SHALL COORDINATE ITS WORK WITH THAT OF THE OTHER CONTRACTORS. ALL WORK WILL BE PERFORMED IN A MANNER THAT WILL NOT UNNECESSARILY DELAY OR PREVENT OTHER CONTRACTORS FROM COMPLETING THEIR WORK IN A TIMELY MANNER.

Goodheart-Willcox Publisher

Figure 7-2. Construction/project-related notes on the plans.

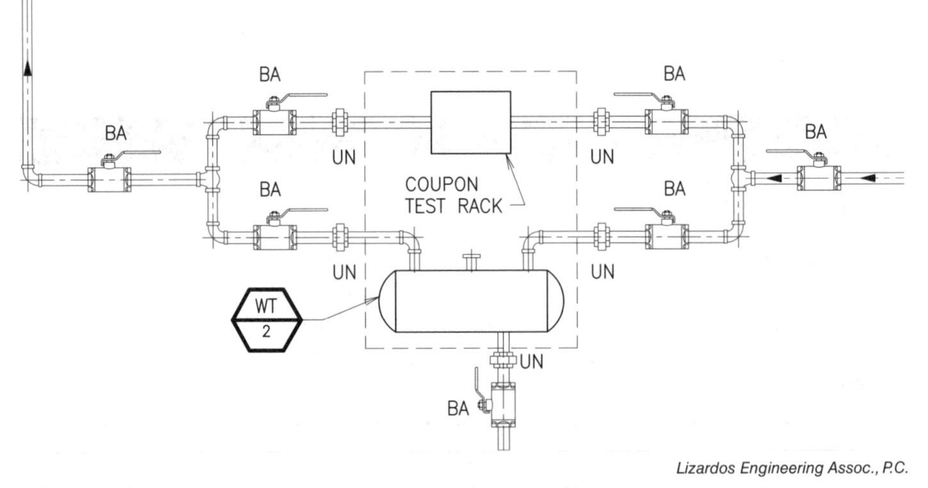

Figure 7-3. A piping circuit print includes ball valves that are used to assist in the future service and repair of components.

A complete circuit requires a start and endpoint, as well as a driving force to push something through. In an air distribution system, the interconnecting ductwork carries air from the occupied space to the air handler and then back to the conditioned space. The driving force is the system's blower. In a piping circuit, a network of interconnecting pipes carry water, refrigerant, fuel, and other fluids through the system. The driving force is often a pump. In some instances, the driving force can be gravity. Like duct systems, control devices, such as sensors and valves, are used to monitor and adjust flow and accomplish basic tasks.

A system design engineer is responsible for combining control devices in a manner that allows the system to operate. This individual must also present this vision in an acceptable manner on the job plans.

7.2 Piping Abbreviations

Piping plans include abbreviations denoting what each pipe is intended to carry. The various types of piping lines required in an HVACR project include refrigerant lines, water and domestic water lines, and fossil fuel lines. See **Figure 7-4**. Refrigerant lines are often identified at the discharge line (hot gas), the liquid line, and the suction line. Water lines installed as part of an HVACR project include chilled and heated water, condenser water, and drain lines. Domestic water lines are identified as hot or cold water. Fuel supply lines are also included when a project deals with fossil fuels.

These abbreviations are often integrated into the lines to avoid confusion with multiple pipes in the same drawing, **Figure 7-5**. Abbreviations can vary from job to job, from print to print, and from designer to designer. Be sure to check the legend of the particular print you are reading.

HVACR Piping Symbols	
Refrigerant Lines	**Symbol**
Hot gas line (discharge line)	— HG —
Refrigerant liquid (liquid line)	— RL —
Refrigerant discharge (discharge line)	— RD —
Refrigerant suction (suction line)	— RS —
Water Lines	**Symbol**
Drain	— D —
Condensate drain	— CD —
Chilled water return	— CHWR —
Chilled water supply	— CHWS —
Condenser water return	— CWR —
Condenser water supply	— CWS —
Heating hot water return	— HHWR —
Heating hot water supply	— HHWS —
High temperature heating water return	— HTWR —
High temperature heating water supply	— HTWS —
Domestic Water Lines	**Symbol**
Hot water	—— – – —— or —— HW ——
Cold water	—— — —— or —— CW ——
Fuel Supply Lines	**Symbol**
Fuel oil return	—— FOR ——
Fuel oil supply	—— FOS ——
Fuel oil vent	- - - FOV - - -
Gas vent	- - - - GV - - - -
High pressure gas	—— HG ——
Low pressure gas	—— G ——
Medium pressure gas	—— MG ——

Goodheart-Willcox Publisher

Figure 7-4. HVACR piping symbols for refrigerant lines, water lines, domestic water lines, and fuel supply lines.

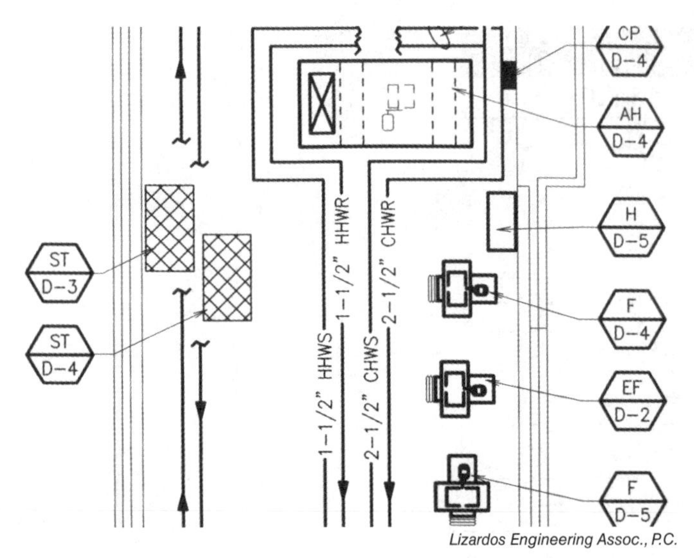

Figure 7-5. Abbreviations are included on lines to represent the types of pipes.

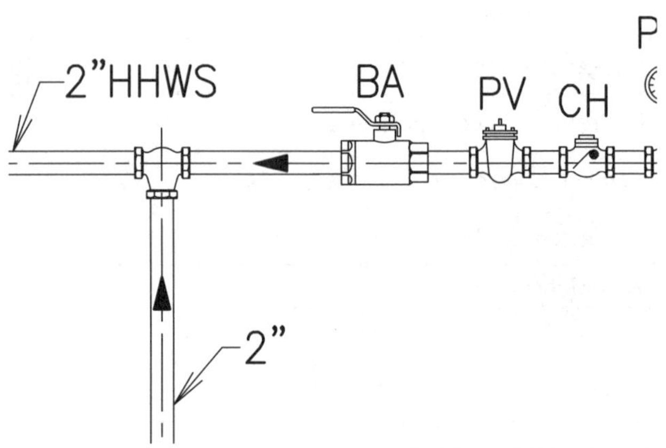

Figure 7-6. Fluid flow direction is indicated on the diagram. The intended direction of flow through the valve labeled BA is from right to left.

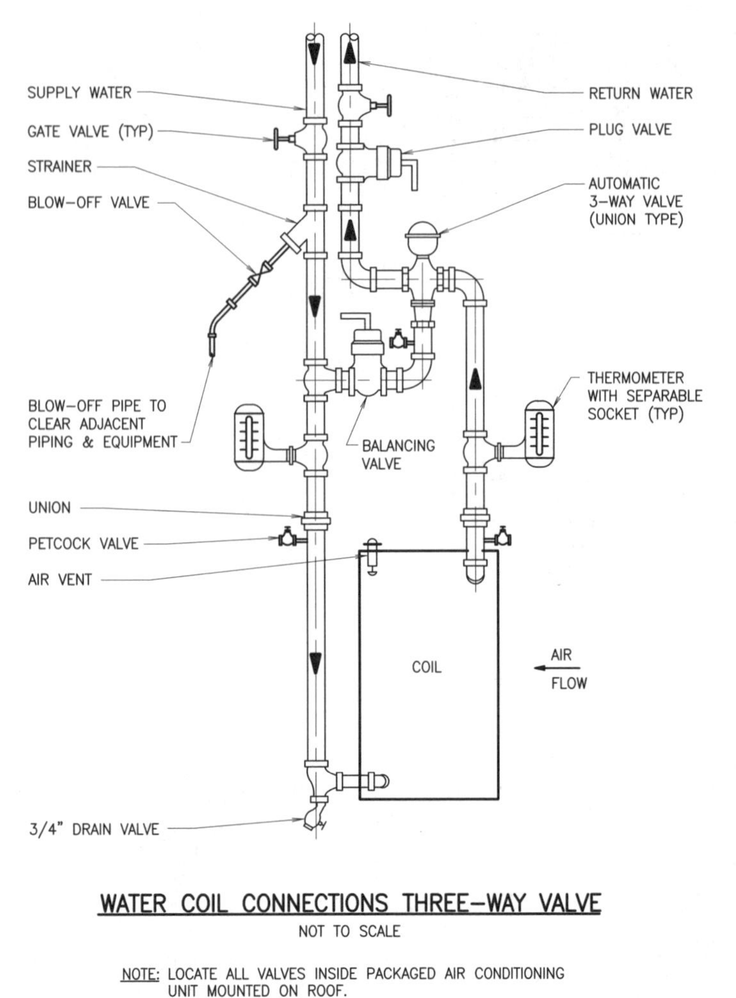

Figure 7-7. A pictorial diagram of a water coil and connections. A balancing valve is used here to control the amount of water passing through the coil.

7.3 Valves

Valves are common piping circuit components. A ***valve*** is a piping system fitting used to control the flow of air or fluid through a pipe. These devices perform a number of different functions and are selected based on the needs of the system being installed.

Many valves are used in HVACR piping systems to control the flow of water through the system. On prints, the direction of flow is indicated, **Figure 7-6.** Common valves used in piping systems include check valves, balancing valves, mixing valves, and pressure-reducing valves.

Check valves allow the free movement of fluid in only one direction. These valves are key for proper system operation in refrigerant systems because they prevent liquid or vapor from flowing in the wrong direction.

Balancing valves set the desired flow through a particular branch in a multi-circuit piping arrangement. Balancing valves are typically of the ball, gate, or globe variety. Consider a hot water heating coil circuit, as shown in **Figure 7-7.** Hot water flows downward through the supply pipe, as indicated by the arrows in the diagram. The water passes a tee fitting connected to a balancing valve before entering the coil. The balancing valve allows only a portion of the water to flow through the coil, while permitting some water to bypass the coil completely. Without the balancing valve, all the water in the loop would flow through the coil.

Mixing valves typically have three ports—two inlet ports and one outlet port. The inlet ports carry hot and cold water streams into the valve, while the stream at the outlet of the valve is made up of a combination of the cold and hot water. The temperature of the mixed water is determined by the position of the valve, which varies depending on the desired water temperature. These valves are usually electronically or electromechanically controlled. Thus, the position of a particular valve is determined by the mode of system operation or is positioned in response to a change in a measurable parameter.

Several valves are used to limit the amount of pressure sent through a system. ***Pressure-reducing valves*** control pressure in a system. They change a higher inlet pressure to a lower pressure at the valve's outlet. With pressure-reducing valves, the valve's outlet pressure can never be higher than its inlet pressure. In cases where the inlet pressure is lower than the maximum predetermined outlet pressure, the valve's inlet and outlet pressures will be the same. Therefore, if the valve's set point at the outlet is 12 psig and the valve's inlet pressure is 15 psig, the pressure at the outlet will be 12 psig. However, if the valve's set point is 12 psig and the valve's inlet pressure is only 10 psig, the pressure at the outlet of the valve will be 10 psig.

Symbols for these valves, as well as other valves common in HVACR piping, are shown in **Figure 7-8**.

Common Valves for HVACR		
Valve	**Symbol**	**Description**
Ball valve	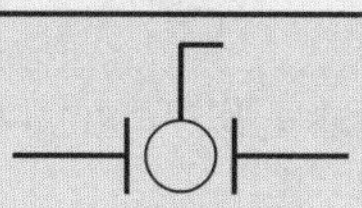	Ball with bored passageway; includes handle that requires only ¼ turn to adjust from fully open to fully closed
Gate valve		Provides straight-line flow through an open gate
Globe valve	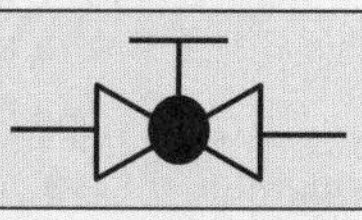	Adjusts fluid flow rate by turning handle to different positions
Plug valve	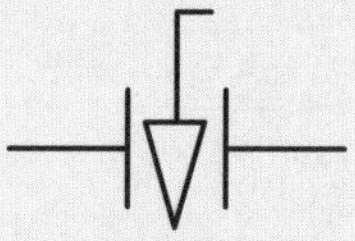	Cone shaped; rotates within the body of the valve to allow for adjusting fluid flow
Check valve	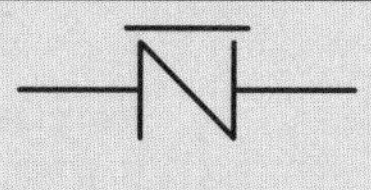  	Ensure fluid flows in only one direction; without handles or operating devices
Mixing valve	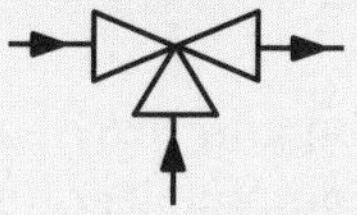	Three-port, two inlet and one outlet; valve modulates to adjust amounts of hot and cold water entering to produce desired temperature

Goodheart-Willcox Publisher

Figure 7-8. Common symbols for HVACR piping systems.

(*continued*)

Common Valves for HVACR		
Valve	**Symbol**	**Description**
Diverting valve		Three-port, one inlet and two outlet; adjusts amount of water supplied to each outlet branch
Electronically controlled valve	S	Position of valve is determined by mode of system operation in response to an increase or decrease of a quantity
Balancing valve		Sets desired flow through a branch in a multi-circuit piping system
Butterfly valve		Rotating disc to modulate fluid flow
Angled valve		Two-port valve, 90° configuration; facilitates horizontal turns in piping system
Gas valve		Manually stops the flow of gas; uses flame safety mechanism to shut valve when unsafe conditions are present
Gas cock		Manually-operated gas valves that provide positive shutoff for gas supply
Pressure-relief valve		Opens if system pressure rises to an unsafe level
Pressure-reducing valve		Two-port valve; provides maximum outlet pressure

Goodheart-Willcox Publisher

Figure 7-8. (*continued*).

7.4 Pipe Fittings

It is common for the direction and size of a piping arrangement to change. You may also need to make equipment connections, create takeoff branches, and route the piping through walls and other barriers. This must be done while ensuring the integrity of the arrangement. Pipe fittings, which include elbows, tees, reducers, and unions, are used to facilitate these changes.

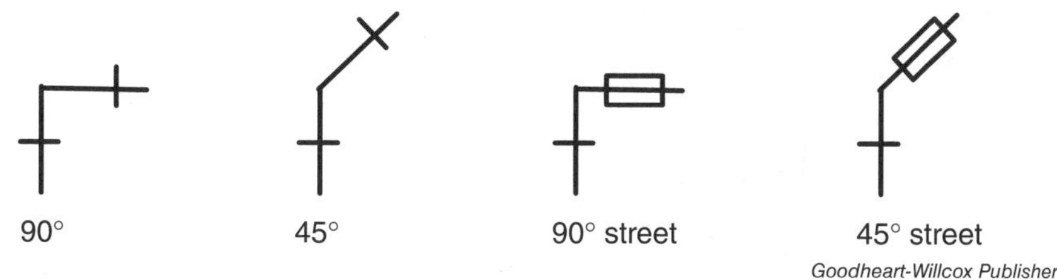

Goodheart-Willcox Publisher

Figure 7-9. Piping symbols for an elbow and street elbow.

7.4.1 Elbows

Elbows are one-piece, two-port pipe fittings used to change the direction of a piping run. They are available in 45° and 90° configurations, **Figure 7-9**. Traditional elbows have two female ports, but street elbows are configured with one male port and one female port. Street elbows are commonly used when a pipe connected to a system component, such as a valve, needs to make a quick turn.

Elbows used for horizontal turns in a piping circuit are easily seen on plan view drawings, but vertical turns are not easily identified. On plan drawings, symbols are used to indicate vertical turns for elbows turned up and turned down, **Figure 7-10.**

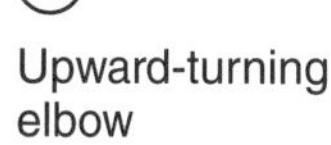

Goodheart-Willcox Publisher

Figure 7-10. Piping symbols for an upward-turning elbow and a downward-turning elbow.

7.4.2 Tees

Tees, **Figure 7-11**, are one-piece, three-port pipe fittings that split a piping run into two separate paths. The two ports that are in-line with each other are referred to as the "run," while the port perpendicular to the run is referred to as the "bull" of the tee. All three ports may be the same size, but will often differ depending on the specific application.

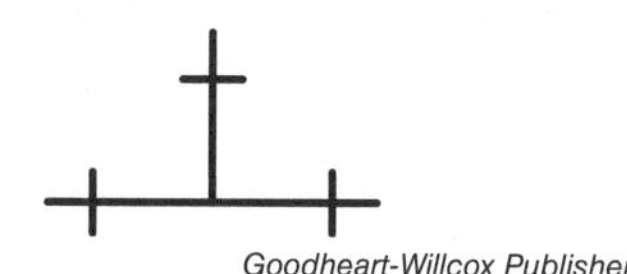

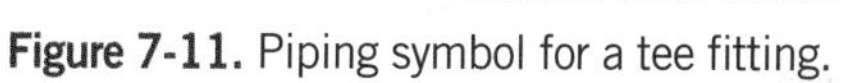

Goodheart-Willcox Publisher

Figure 7-11. Piping symbol for a tee fitting.

A tee fitting is identified by three measurements. First, the two run measurements are listed and then the bull. For example, a tee that has a 3/4 run and a 1/2″ bull will be classified as a 3/4″ × 3/4″ × 1/2″ tee. If the connection sizes are the same, only one measurement is required.

On a plan drawing, pipe measurements are shown on the lines that represent each individual pipe. Like elbows, tees can be positioned horizontally, vertically, or at another angle, and are represented with specific symbols, **Figure 7-12.**

Reducers are one-piece, two-port pipe fittings that change the size of a pipe. Reducers are often configured as in-line devices, with the inlet and outlet ports of the fitting following the same path. These reducers can be either ***concentric reducers*** or ***eccentric reducers***. Concentric reducers have inlet and outlet ports aligned at their center points.

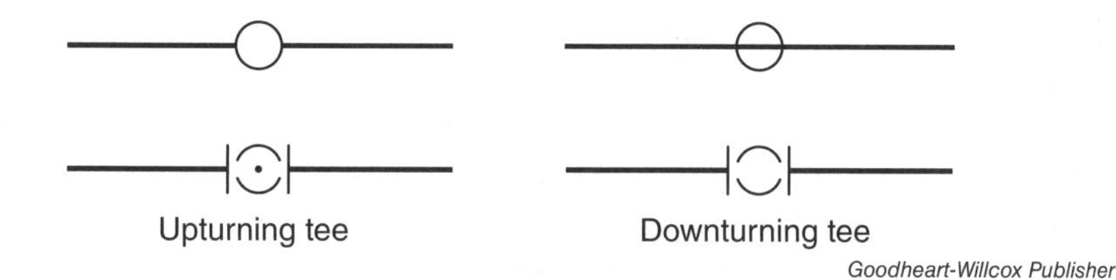

Goodheart-Willcox Publisher

Figure 7-12. Piping symbols for upturning tees and downturning tees.

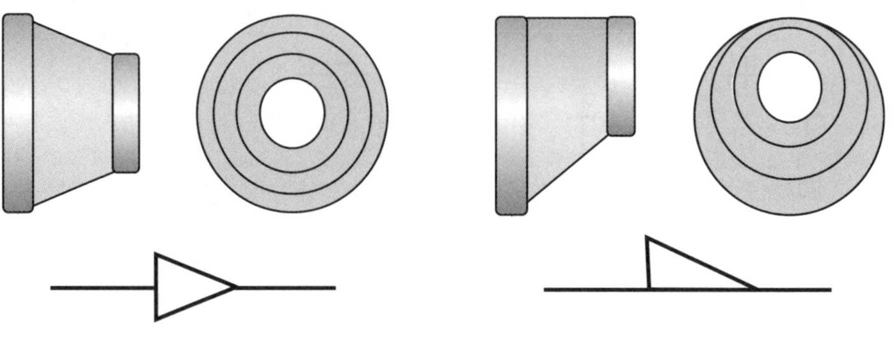

Concentric reducer

Eccentric reducer

Goodheart-Willcox Publisher

Figure 7-13. Concentric reducers and eccentric reducers are depicted differently on a print based on the view shown.

Eccentric reducers have offset inlet and outlet ports. These two reducers and their piping symbol are shown in **Figure 7-13**. An elbow and reducer can be combined into a single fitting to reduce the number of fittings used in a run. This fitting is called a reducing elbow.

7.4.3 Unions

Unions connect sections of piping material that may need to be disconnected at a later time. Different types of unions are based on the number of pieces that make up the fitting. Screwed unions and flanged unions are common types, **Figure 7-14**.

A ***screwed union*** has three pieces, and is referred as a three-piece union. The male and female ends of the union are secured, either by threaded or soldered connections, to the pipes being joined. The nut is positioned on the pipe where the male portion of the union is connected, and is screwed onto the female portion. A ***flanged union*** is a two-piece connector that is bolted together rather than threaded together.

Although unions do not affect fluid flow through a piping arrangement or the operation of the system, they are extremely important for system servicing and maintenance and should be installed wherever indicated on a print. The piping symbols for unions are shown in **Figure 7-15**.

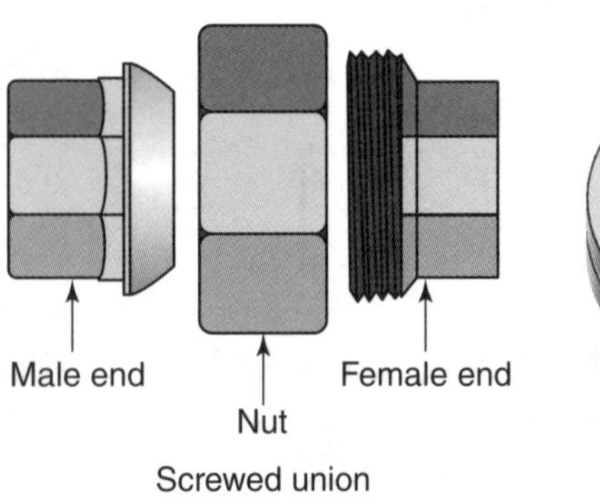

Flanged union

Goodheart-Willcox Publisher

Figure 7-14. Two common types of unions are screwed unions and flanged unions.

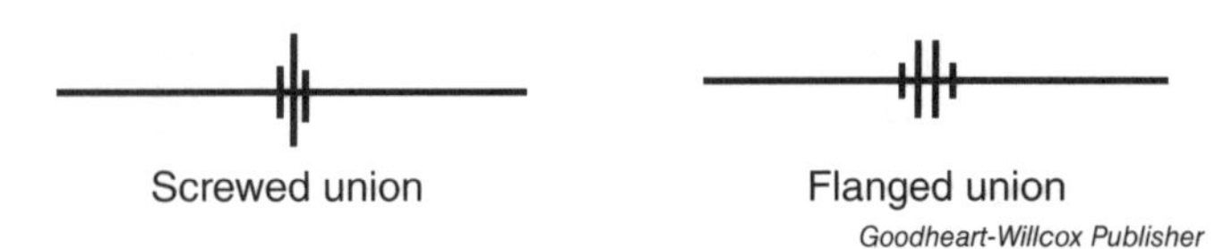

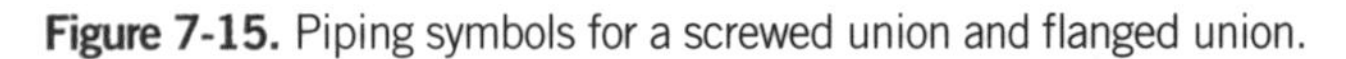

Figure 7-15. Piping symbols for a screwed union and flanged union.

7.5 Gauges and Switches

Sensors are used to monitor and ensure proper system operation. These devices can measure a number of different system conditions including pressure, temperature, and fluid flow.

Thermometers and pressure gauges, **Figure 7-16**, are examples of sensors used by a system operator to observe the present operating conditions. Other devices ensure that the system functions within acceptable operating ranges and opens and closes electrical contacts based on the operating ranges.

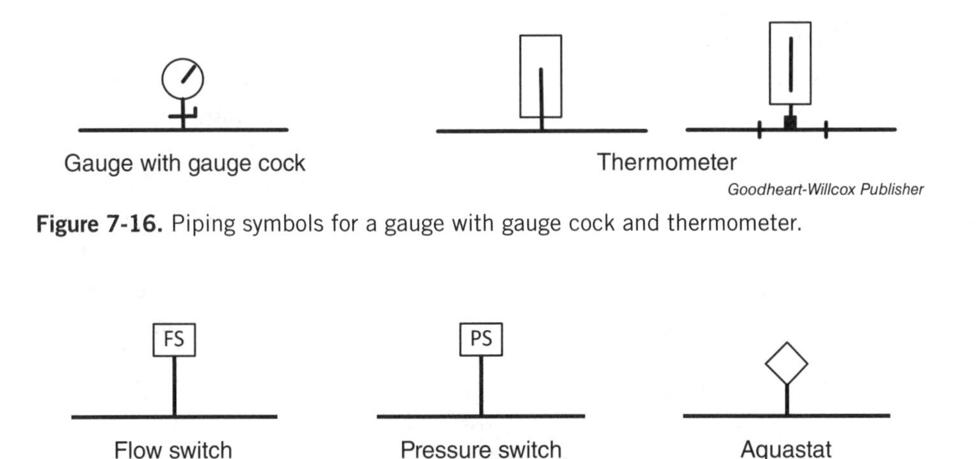

Goodheart-Willcox Publisher

Figure 7-16. Piping symbols for a gauge with gauge cock and thermometer.

Goodheart-Willcox Publisher

Figure 7-17. Piping symbol for a flow switch, pressure switch, and aquastat.

Flow switches, pressure switches, and aquastats are examples, **Figure 7-17**. A ***flow switch*** detects when fluid flow is detected in a pipe, while a ***pressure switch*** detects when the sensed pressure reaches a predetermined level. ***Aquastats*** open or close one or more sets of electrical contacts when the sensed water temperature reaches a predetermined level. These devices can be utilized as either operational or safety components. They must be selected and installed as outlined in the project plans.

7.5.1 Drains, Strainers, and Cleanouts

Floor drains and strainers are components that also improve system operation. Condensate drips from the evaporator coil of an air-conditioning system and must be removed from the system to prevent damage. A floor drain is used to remove it from the structure. Common symbols for floor drains are provided in **Figure 7-18**.

Any impurities, organic matter, or other foreign substances in a water circuit must be removed using strainers and cleanouts. These two components help prevent clogging and keep the water chemistry stable. Some common symbols for strainers and cleanouts are shown in **Figure 7-19**.

Floor drain

Floor drain with P-trap

Goodheart-Willcox Publisher

Figure 7-18. Piping symbol for a floor drain and symbols for floor drains equipped with P-traps.

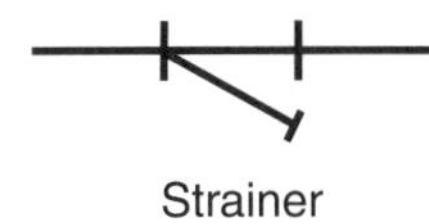

Strainer

Cleanout plug

Strainer with cleanout plug

Goodheart-Willcox Publisher

Figure 7-19. Piping symbol for a strainer, cleanout plug, and strainer with a cleanout plug.

7.5.2 Pipe Guides

A ***pipe guide***, **Figure 7-20**, allows a pipe to move in one direction. Pipe guides facilitate thermal expansion or contraction of a pipe, while reducing the possibility of damage.

The permitted movement of a pipe is in a direction parallel to the pipe itself. A guide is made of two parts. One part is stationary, while the other moves back and forth within the stationary portion of the guide. The stationary portion is secured to the support structure. The symbol for this component, **Figure 7-21**, accurately depicts its function.

Pipe guides are not designed to support the weight of the pipe and the fluid flowing through it, thus pipe supports are used. The details of the pipe supporting hardware are provided on detail drawings on the print, **Figure 7-22**. Depending on the structural elements on the project, multiple types of piping supports may be used. Details of how pipe supports should be attached to the structure are also included in the plan drawings, **Figure 7-23**.

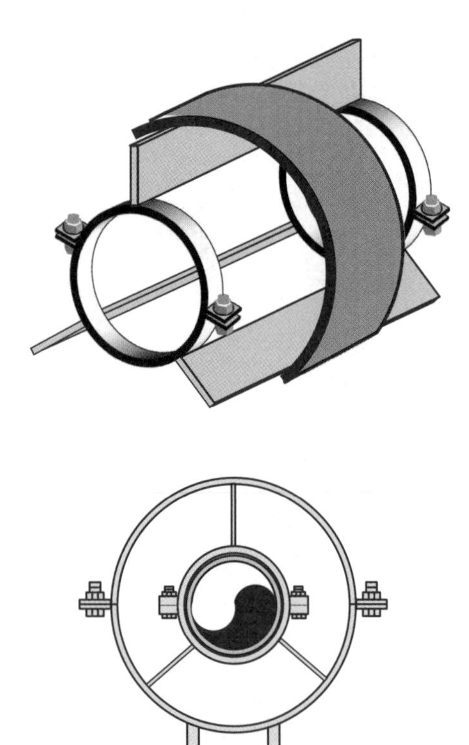

Goodheart-Willcox Publisher

Figure 7-20. A pipe guide and the end view of a pipe guide.

Goodheart-Willcox Publisher

Figure 7-21. Symbols used to represent a pipe guide on a print.

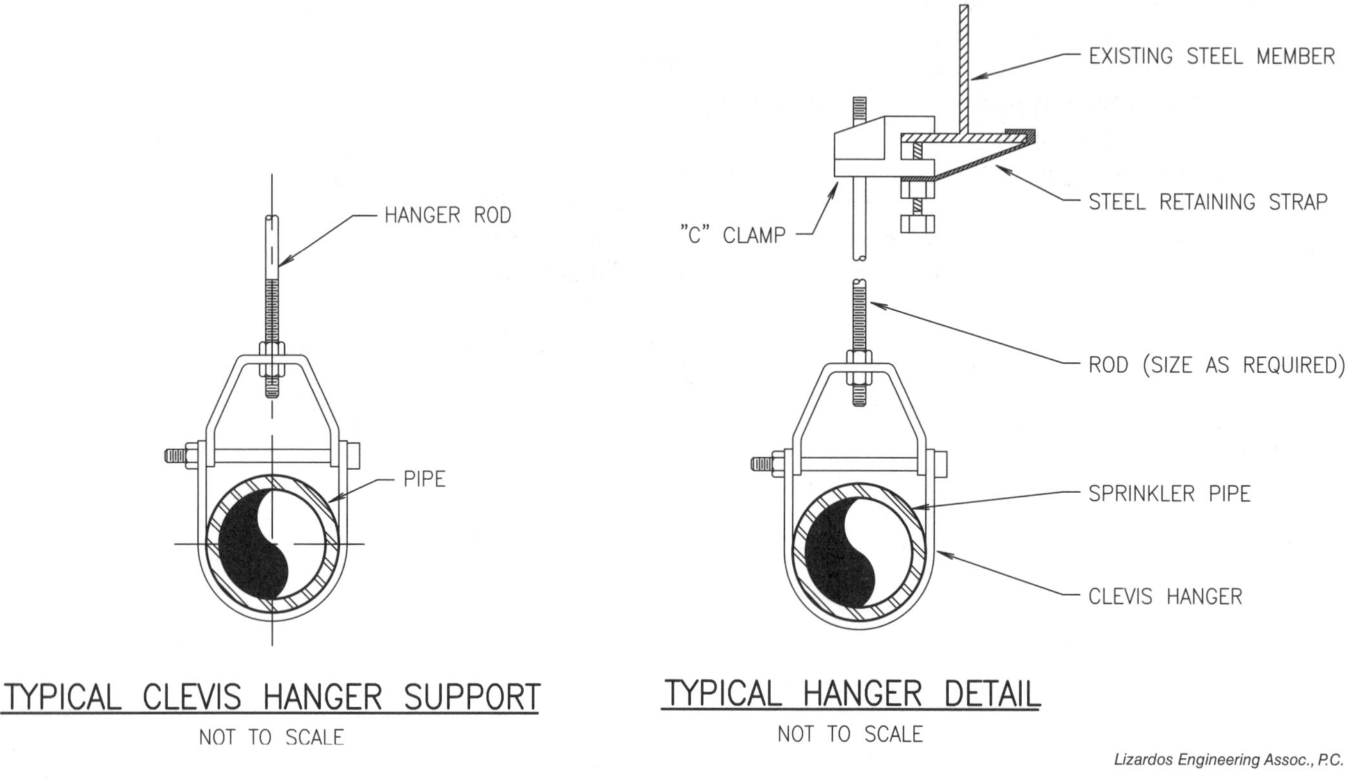

Lizardos Engineering Assoc., P.C.

Figure 7-22. Pipe hanging details.

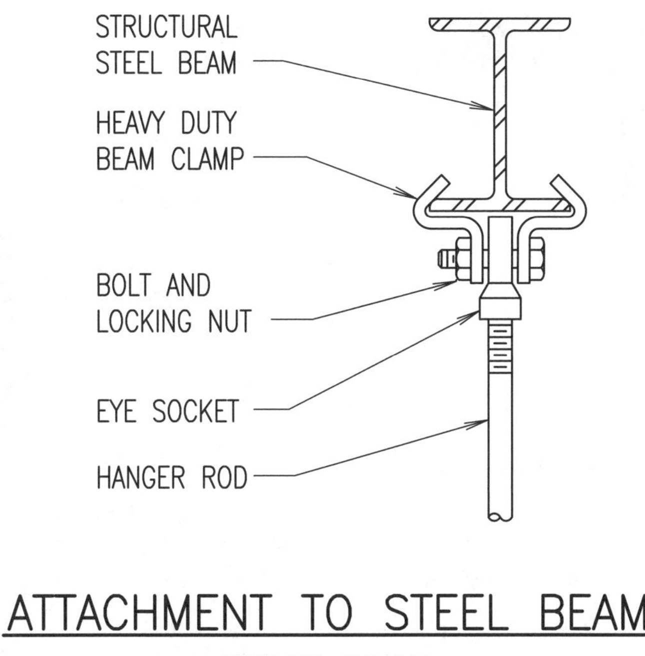

Lizardos Engineering Assoc., P.C.

Figure 7-23. Structural mounting details.

7.6 Piping Symbols and Drawing Concepts

Additional piping symbols include pumps, coils, water heaters, water meters, hose bibs, flexible connectors, expansion joints, and air vents, **Figure 7-24**. Other items are included on a print but do not represent a system component. These symbols are intended to help the print reader to better understand and execute the project. They may include new piping to be installed, existing piping that is to remain in place, or existing piping that is to be removed.

On a piping plan or drawing, new piping is identified as a solid, bold line. The piping line may also include an arrow that indicates the direction of fluid flow, **Figure 7-25**. Existing piping that is to remain in place (for renovation projects) is indicated by a series of long and short dashed lines.

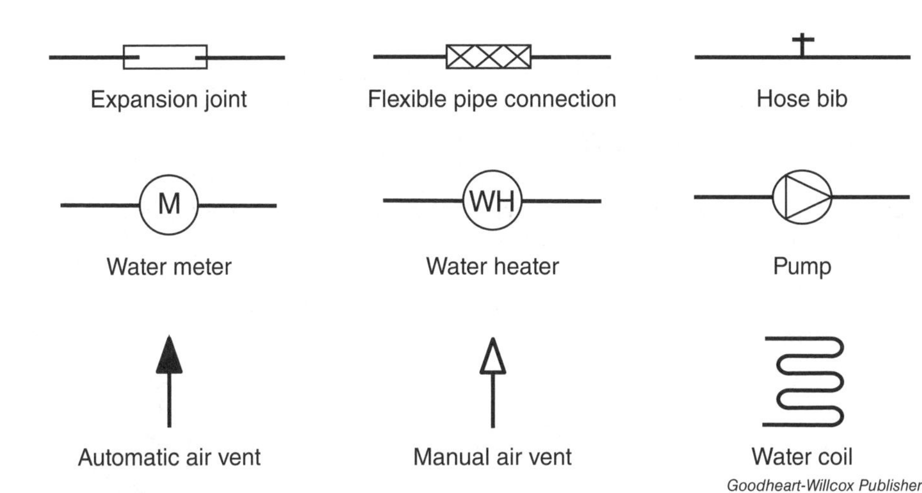

Goodheart-Willcox Publisher

Figure 7-24. Additional piping symbols are shown.

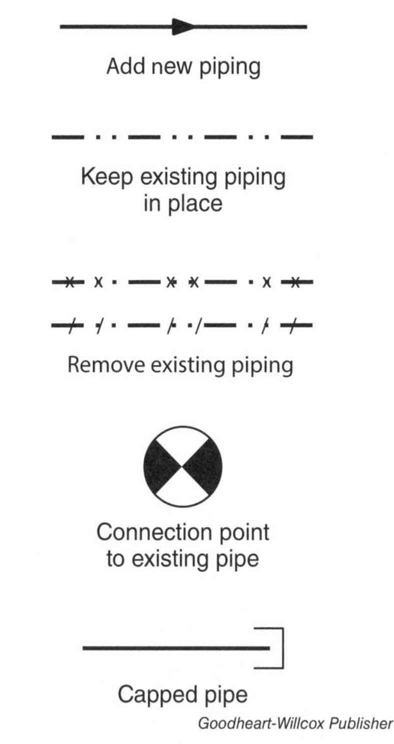

Goodheart-Willcox Publisher

Figure 7-25. Piping symbols used to provide details about current or new piping installation.

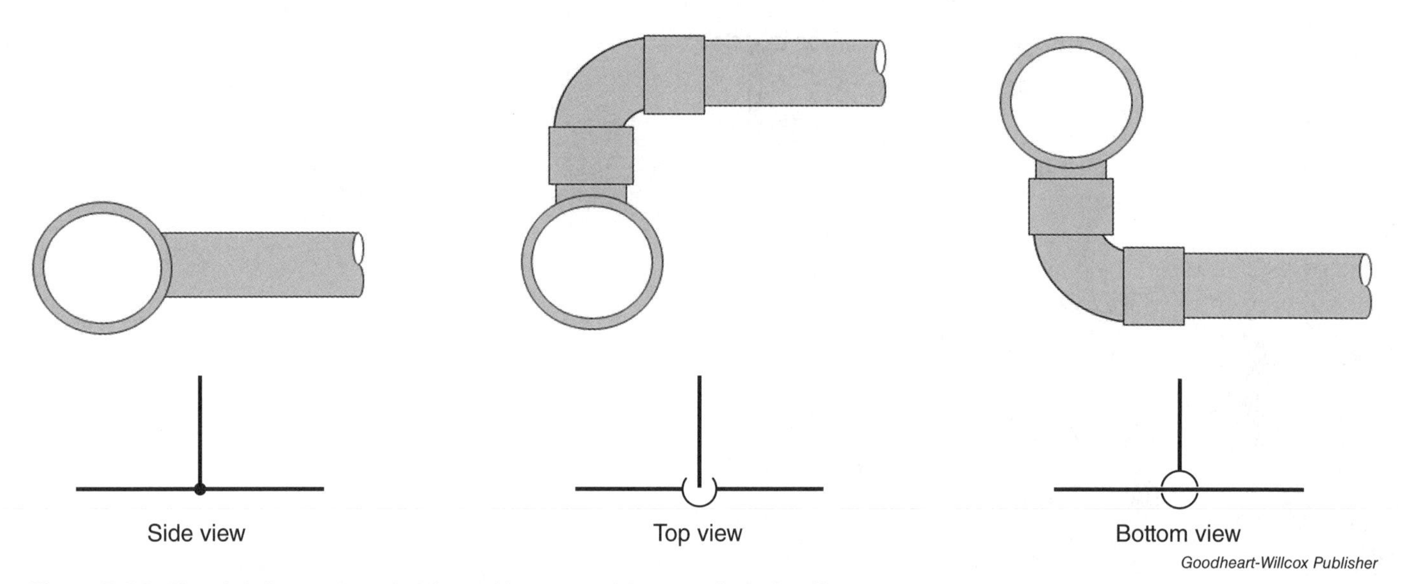

Figure 7-26. Pictorial view and symbol for a side, top, and bottom pipe takeoff.

Existing piping to be removed (also for renovation projects) is denoted by a series of crosses along the length of the line(s). When a connection to existing piping is to be made, the connection symbol is used. There is also a symbol to represent when a pipe is capped.

Like duct systems, branch circuits are common in piping arrangements. Sometimes, these branches are taken from the side of a pipe, the top, or the bottom, **Figure 7-26.** The symbol that most closely resembles a tee is used to indicate a branch side takeoff. Branch top and bottom takeoffs are also shown. Tee fittings, which are similar to takeoffs, can also be on the top, side, or bottom of a main piping run.

Job specifications may require that pipes be pitched either upward or downward. Pitches in pipe runs are difficult to identify on a piping plan or drawing, especially because these pitches are often very slight, such as 1/4″ per running foot of pipe. The letters *U* and *D* indicate an upward or downward pitch, respectively, in conjunction with a directional arrow, **Figure 7-27.**

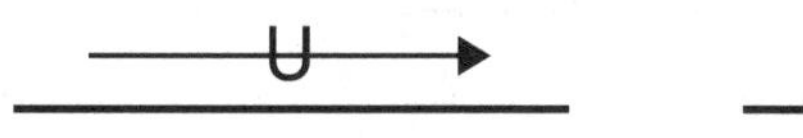

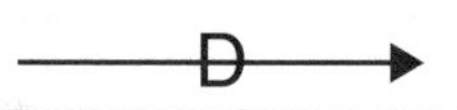

Figure 7-27. Piping symbol for an upward-pitched pipe from left to right, and a downward-pitched pipe from left to right.

Summary

- Piping systems perform many tasks to ensure the proper function of an HVACR system and provide thermal comfort to occupants of a space.
- An HVACR piping plan includes guidelines for proper installation. Piping systems must be installed with future service and repair in mind.
- Drawings use piping abbreviations to identify the fluid being carried in the various pipes.
- Always read any notes presented on the prints to ensure compliance with all codes and regulations.
- HVACR piping circuits can carry refrigerant, water, or fuel, so the circuits must be properly designed and installed to safely contain these substances.
- Many different types of valves are used in piping systems. Valves in HVACR piping systems typically regulate the flow of air or limit the pressure through a system.
- Valves can be manually operated or automatically controlled devices.
- Always ensure valves are installed according to the installation guidelines to ensure proper operation of the device.
- Pipe fittings change the direction and size of a pipe. Common pipe fittings include elbows, tees, reducers, and unions.
- Gauges and switches monitor and ensure proper system operation. Examples include thermometers, pressure gauges, flow switches, pressure switches, and aquastats.
- Floor drains, strainers, and cleanouts are components used to improve system operation and prevent damage in the system.
- Pipe guides allow a pipe to move in one direction and facilitate thermal expansion or contraction.
- Other items are included in prints that do not represent a system component but help execute the project. These symbols can indicate new piping to install, existing piping to remove, specific connections, and up and downturning fittings.

Chapter Assessment

Name ______________________ Date __________ Class ______________

Know and Understand

Refer to the figures below to answer questions 1–4.

_______ 1. The following symbol represents a _____ valve.

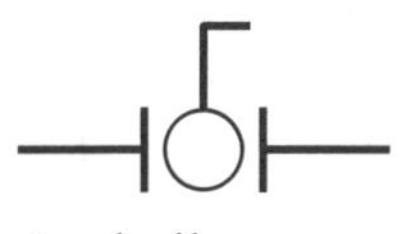

A. ball
B. gate
C. butterfly
D. plug

_______ 2. The following symbol represents a _____ valve.

A. ball
B. check
C. globe
D. plug

_______ 3. The following symbol represents a _____ valve.

A. butterfly
B. gate
C. globe
D. check

_______ 4. The following symbol represents a _____ valve.

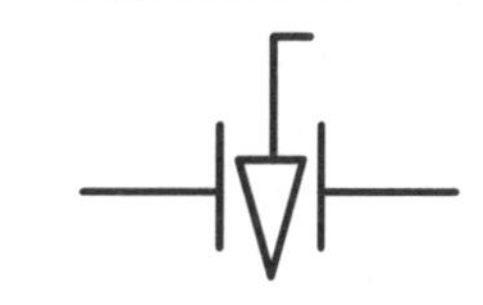

A. check
B. gate
C. globe
D. plug

_______ 5. _____ valves set the desired flow through a particular branch in a multi-circuit piping arrangement.

A. Balancing
B. Mixing
C. Pressure-reducing
D. Butterfly

_______ 6. The valve that can be moved from a fully open to a fully closed position by moving the valve's handle only 90° is the _____ valve.

A. plug
B. ball
C. gate
D. check

_______ 7. The valve that must be fully installed with its directional arrow pointing in the direction of intended fluid flow is the _____ valve.

A. check
B. ball
C. plug
D. globe

_______ 8. The abbreviations HHWS and HHWR would most likely be used in conjunction with which type of system?

A. Hydronic heating system
B. Chilled water cooling system
C. Domestic hot water system
D. Hydronic cooling system

_______ 9. The symbol for a diverting valve is comprised of _____.

A. one inlet port and one outlet port
B. one inlet port and two outlet ports
C. two inlet ports and one outlet port
D. two inlet ports and two outlet ports

_______ 10. The symbol for a mixing valve is comprised of _____.

A. one inlet port and one outlet port
B. one inlet port and two outlet ports
C. two inlet ports and one outlet port
D. two inlet ports and two outlet ports

_______ 11. The device that allows for pipe movement in the direction of fluid flow and not in any other direction is the pipe _____.

A. hanger
B. strap
C. guide
D. support

_______ 12. The pipe guide should be selected and installed so it is _____.

A. able to support 100% of the weight of the pipe and the fluid it carries
B. able to support 200% of the weight of the pipe and the fluid it carries
C. able to support 300% of the weight of the pipe and the fluid it carries
D. not needed to support the weight of the pipe and the fluid it carries

Refer to the figure below to answer questions 13–19.

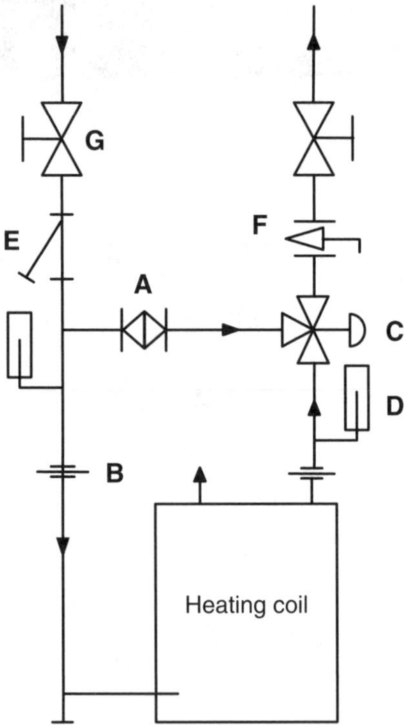

_______ 13. The piping component identified by letter *A* in the figure is a _____ valve.

A. balancing
B. gas
C. butterfly
D. gate

_______ 14. The piping component identified by letter *B* in the figure is a _____.

A. cleanout port
B. flanged union
C. strainer
D. screwed union

_______ 15. The piping component identified by letter *C* in the figure is a(n) _____ valve.

A. diverting
B. automatic 3-way
C. automatic 2-way
D. balancing

Name ______________________________ Date ____________ Class ____________________

_______ 16. The piping component identified by letter *D* in the figure is a _____.

A. pressure switch

B. temperature switch

C. thermometer

D. drain valve

_______ 17. The piping component identified by letter *E* in the figure is a _____.

A. strainer

B. sound trap

C. drain valve

D. pipe guide

_______ 18. The piping component identified by letter *F* in the figure is a _____ valve.

A. gate

B. ball

C. globe

D. plug

_______ 19. The piping component identified by letter *G* in the figure is a _____.

A. gate

B. ball

C. globe

D. plug

_______ 20. A threaded street elbow has _____.

A. only female pipe threads on both ends of the fitting

B. only male pipe threads on both ends of the fitting

C. both female and male pipe threads on the fitting

D. threads on only one end of the pipe fitting

_______ 21. Which of the following best describes a screwed union?

A. A three-piece fitting used to join two sections of pipe at right angles to each other

B. A three-piece fitting used to join two pipe sections that are in-line with each other

C. A two-piece fitting used to join two sections of pipe at right angles to each other

D. A two-piece fitting used to join two pipe sections that are in-line with each other

Apply and Analyze

Refer to the figure below to answer question 1.

1. Explain the purpose of the four unions shown in the following figure.

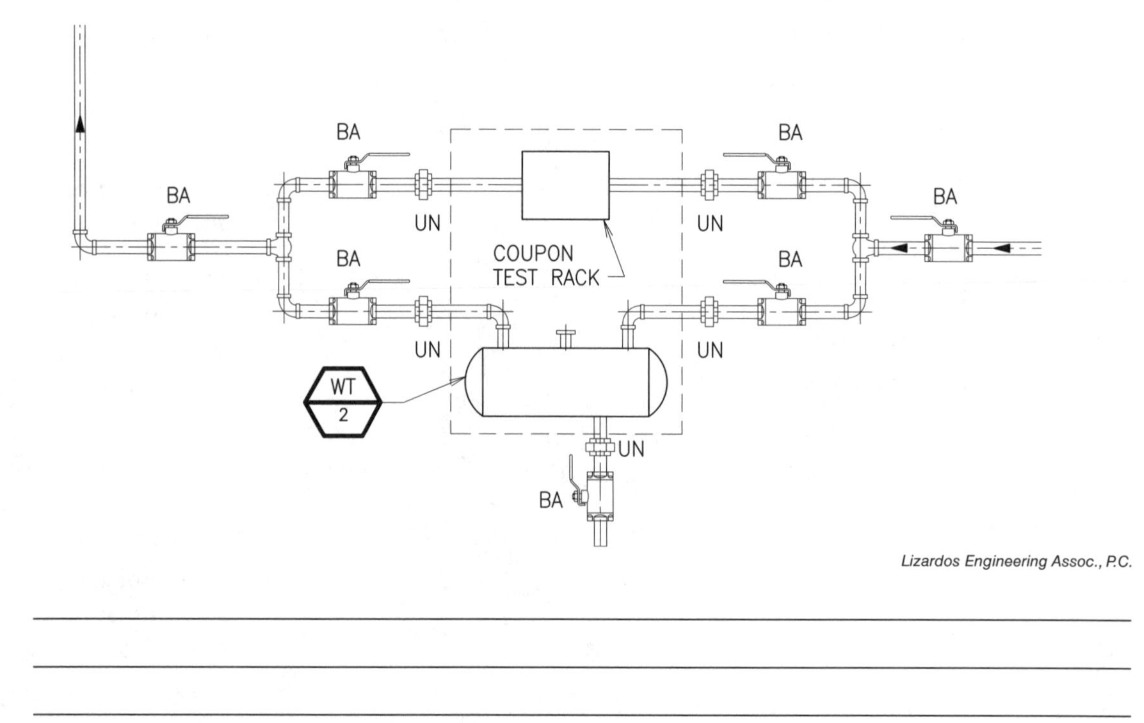

Lizardos Engineering Assoc., P.C.

____

____

____

____

Name ______________________________ Date ____________ Class ____________________

Refer to the figure below to answer questions 2–3.

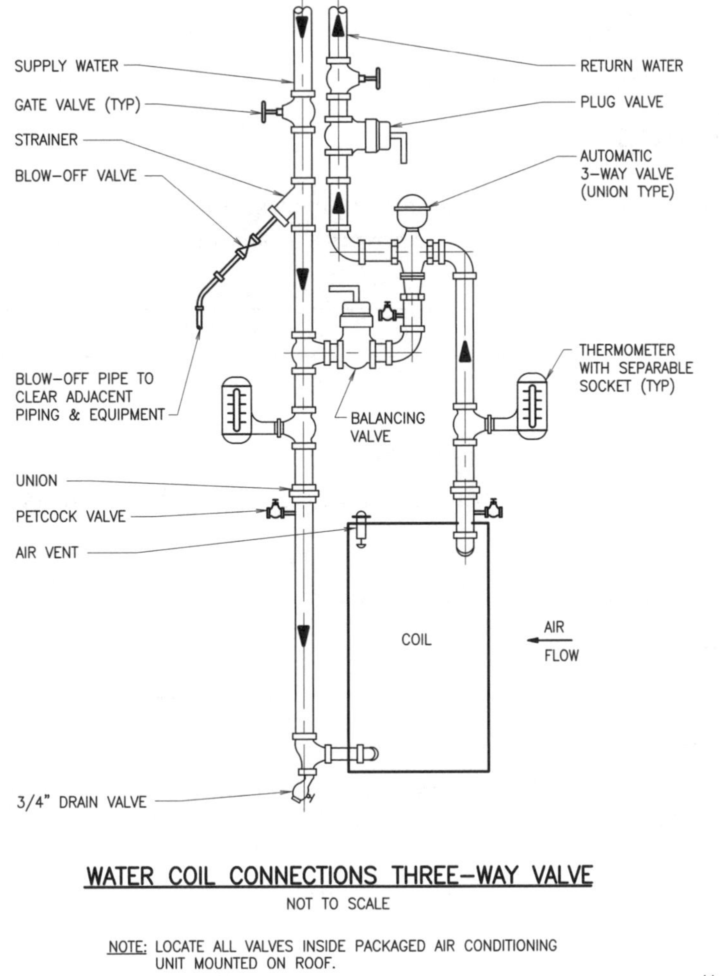

WATER COIL CONNECTIONS THREE-WAY VALVE

NOT TO SCALE

NOTE: LOCATE ALL VALVES INSIDE PACKAGED AIR CONDITIONING UNIT MOUNTED ON ROOF.

Lizardos Engineering Assoc., P.C.

2. Explain why it would *not* be desirable for all of the hot water to flow through the heating coil, as shown below.

____

____

____

____

3. Explain how the balancing valve could be replaced with a three-port mixing valve if the water entering the heating coil was to be maintained at a set temperature.

____

____

____

____

____

4. Can three-port mixing valves and three-port diverting valves be used interchangeably?

Critical Thinking

1. Explain why ball valves and other isolation devices can help prevent system downtime in the event of required system service or a partial system malfunction.

2. Reevaluate your response from critical thinking exercise #1 if a particular system were installed without isolation valves.

3. Why are permanently mounted thermostats and pressure gauges desirable on larger equipment installations?

4. Why are valve cocks positioned between permanently mounted thermostats or pressure gauges and the main system?

5. Who ultimately benefits from the "install with service in mind" mentality?

8 Equipment Installation Considerations

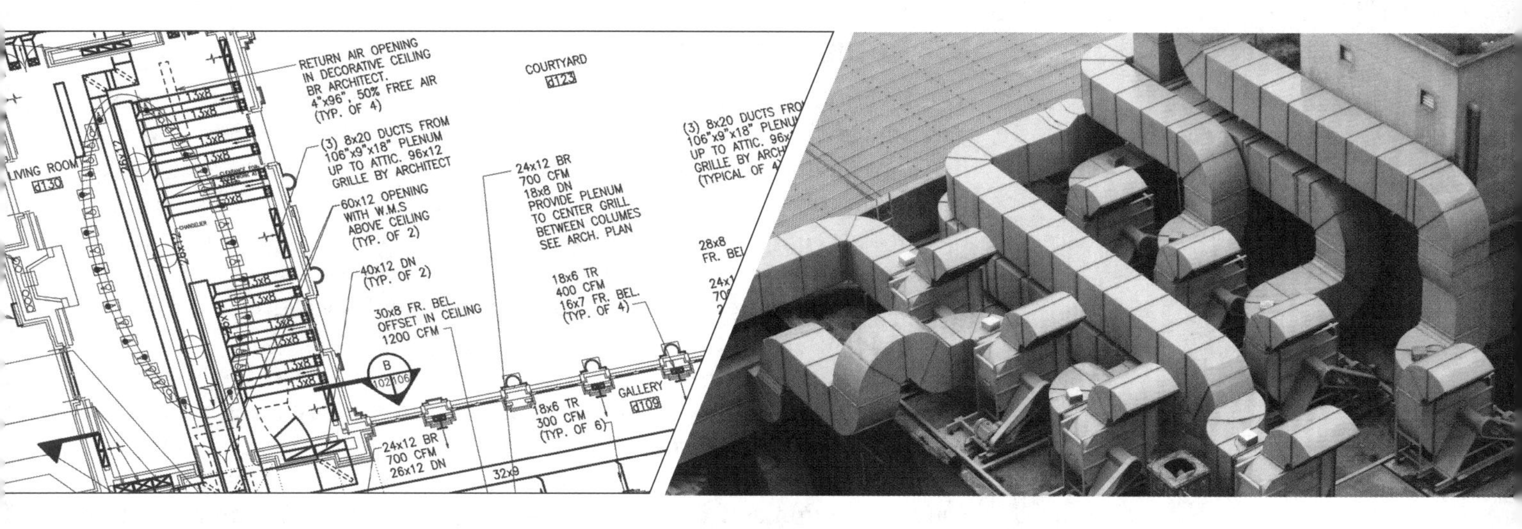

Chapter Outline

Learning Objectives

After studying this chapter, you will be able to:

- Define a split air-conditioning system and its components.
- Identify the considerations for determining the location of a condensing unit.
- Discuss the factors that determine the location of an air handler.
- Describe system components that affect the efficiency of ductwork.
- Explain installation considerations for supply and return duct runs.
- Understand the types of piping used in HVACR systems.
- Identify installation considerations for refrigerant, condensate, and water lines.

Technical Terms

air handler

auxiliary drain pan

bar slip

building setback

C-clamp

clevis-type hanger

condensate

condensate drain line

condensate pump

condensate trap

condensing unit

cross-breaking

drive

pitch

refrigerant trap

safety float switch

slip

split air-conditioning system

variance

Introduction

A complete set of project plans consists of many different elements. These elements include plumbing, electrical, carpentry, masonry, decorative landscaping elements, structural elements, HVACR, interior decorative elements, design finish elements, project utilities, and site topography. Although the goal is for plans to be as accurate and complete as possible, it is likely that neither the architect, who prepared the plans, nor the project owner are experts in all of the factors that must be addressed in the plan set. Thus, you must be able to interpret a set of job plans to install HVACR equipment and understand these plans, while being aware of the industry's best practices and the various manufacturers' installation instructions. This chapter addresses some important aspects of HVACR system installation and provides references to project drawings as they relate to these installation practices, processes, and procedures.

8.1 Equipment Location

The location of various pieces of air-conditioning equipment is usually determined during the concept and layout stages of the project. It is preferred to have HVACR equipment located in close proximity to the areas being conditioned. However, this is not always possible or desirable. Each option for equipment placement must therefore be evaluated against other options to decide what is best for the project. When evaluating specific options, factors such as aesthetics, system operation, and serviceability must be considered.

These factors are integral when determining the best location for a split air-conditioning system. ***Split air-conditioning systems*** are a common air-conditioning configuration and divide their systems into two separate locations. Split air-conditioning systems are different from package-type equipment, which has its system in a single location and all its components in one cabinet. Package units are commonly installed on the roof of a structure or alongside the structure at ground level. A split system has a condensing unit that is typically located outdoors, while its air handler is located indoors. There are many considerations to determine their exact locations.

8.1.1 Condensing Unit Location

A ***condensing unit*** includes both the condenser coil and the compressor of an air-conditioning system. Factors for considering the location of a condensing unit include the following:

- Airflow restrictions
- Local building codes
- Sound/noise transmission
- Location of electrical power
- Proximity to the indoor unit
- Ground slope
- Serviceability

Airflow Restrictions

In split air-conditioning systems, airflow through the condenser coil must be able to mix freely with the outside air. Effective outside air mixing helps ensure that the coil's discharge air does not recirculate back through the coil. If it does recirculate through, as shown in **Figure 8-1**, excessively high system discharge pressures occur, and the system will not function as intended. Therefore, the unit should not be located under any overhangs or porches. The manufacturer's installation literature should always be consulted when evaluating the clearances for a specific piece of equipment. Under no circumstances should the outdoor unit be positioned against a structure.

If the proposed location of the condensing unit, or even a package-type unit, contradicts the manufacturer's specifications or facilitates air recirculation through the outdoor coil, these discrepancies must be brought to the attention of the general contractor, architect, or system designer. An alternative location can then be selected. Identifying these discrepancies early helps minimize the cost of making a change further along in the project.

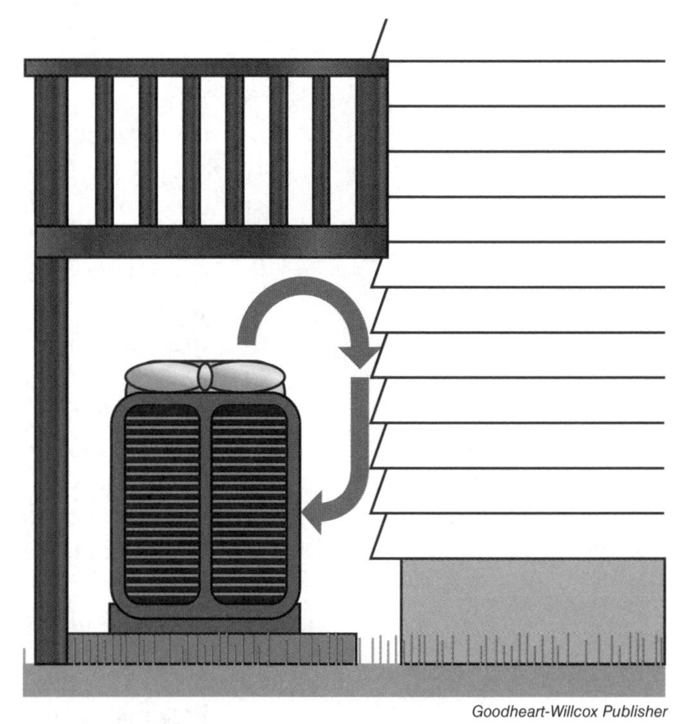

Goodheart-Willcox Publisher

Figure 8-1. Discharge air from the condensing unit should never recirculate back through the condenser coil.

Local Building Codes

Local building codes must be referenced and reviewed to ensure that the equipment is not too close to the property line or that other guidelines are not being violated. If the contractor installing the HVACR system identifies any concerns about possible property boundary encroachments or other violations upon inspection of the project plans, they should be immediately conveyed to the project managers.

BETWEEN THE LINES

Building Setbacks

A ***building setback*** is used to identify the required distance between a property line and any structures on the property, or the distance between two property structures. Building setbacks vary from town to town and from state to state. Therefore, the local planning division is contacted for information on a particular project. The planning division is the local agency that oversees the zoning issues for a particular neighborhood or group of neighborhoods. Because building setbacks can vary greatly by geographic area, zoning designation, and lot size, it is recommended you contact the planning division early in the project.

Sound/Noise Transmission

The condensing unit of an air-conditioning system contains both a compressor and a fan motor. These system components generate noise that can be distracting to those individuals near the equipment. Although newer, high-efficiency equipment has the ability to increase or decrease compressor and condenser fan motor speeds, which lowers the noise levels, there are still times when the level of noise may be unavoidable. To avoid excessive noise during the night, the outdoor unit can be positioned away from any bedrooms or other living areas.

To maximize system efficiency, the outdoor unit should be located close to the indoor unit, thus close to the structure. Yet this can also affect noise levels. To address both the noise and efficiency issues, the condensing unit should be located along the side or back of the house, as far from windows as possible. In addition, the unit should be situated in a manner that minimizes any inconvenience.

Location of Electrical Power

The outdoor portion of a split air-conditioning system requires a separate electric power supply, which must be protected by fuses, circuit breakers, or both. Although the distance between the power distribution panel and the outdoor unit does not affect system operation, capacity, and efficiency, the installation cost is lower if the run is shorter. Although it may be ideal to have the outdoor unit located close to the electrical service panel, it is not the most critical compared to the other factors discussed.

Proximity to the Indoor Unit

The location of the condensing unit with respect to the air handler affects the success of installation. The distance between these two units determines the length of interconnecting refrigerant lines that are installed. Longer refrigerant lines can lead to a decrease in the system's operating efficiency because longer lines can reduce the mass flow rate, or the amount of refrigerant circulated through the system.

There are additional problems with locating the outdoor unit away from the structure. The cost of running the electrical power to the unit increases with distance because trenches must be created to install electrical and refrigerant lines underground. The buried refrigerant and electrical lines make equipment servicing much more difficult, especially if a refrigerant leak develops.

When extracting information from a set of project plans, verify that the length of the refrigerant lines is within the manufacturer's guidelines. If the distance between the system's indoor and outdoor units poses a potential problem, alternative locations must be selected. In situations where alternate equipment locations cannot be identified, contact the equipment manufacturer and request that the installation be evaluated. Upon evaluation, it may be determined that the equipment can be installed as indicated on the plans. A variance can then be requested from the manufacturer. A ***variance*** is a document prepared by the manufacturer that states the installation is acceptable and does not affect system operation, effectiveness, or efficiency. This document thus removes a great deal of liability from the architect and contractor.

Ground Slope

The outdoor unit must be positioned so it remains level, **Figure 8-2.** A level outdoor unit helps prevent any unnecessary strain on the outdoor fan motor and reduces the amount of noise created by equipment vibrations. A level outdoor unit also can ensure adequate compressor

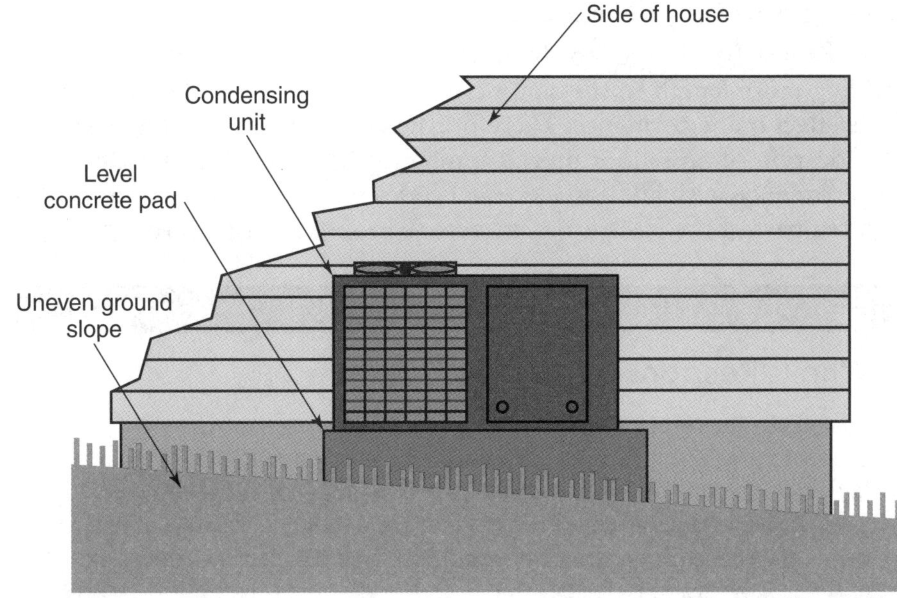

Figure 8-2. The condensing unit must always be installed level, even when the ground is not.

lubrication and makes the installation look more professional. The outdoor unit must never be set directly on the ground because excessive sinking and settling can occur.

Prefabricated outdoor unit composite pads are available, but are only used when instructed in project plans. Some projects require permanent concrete pads to be poured for the installation. The support pad should be larger than the outdoor unit. The outdoor unit is never connected to the structure. Putting the outdoor unit on a patio or deck may cause vibrations and noise transmission into the structure.

Serviceability

The outdoor portion of a split air-conditioning system requires periodic routine maintenance and repair. The unit must be located and positioned so service personnel can access the unit, remove all service panels, and be able to perform all service-related procedures in a safe manner.

8.1.2 Air Handler Location

An ***air handler*** is the indoor portion of a split air-conditioning system that houses the evaporator coil, blower, and metering device. As with a condensing unit, a number of factors must be considered when choosing its location:

- Distance to the electric power supply
- Serviceability
- Ease of condensate removal
- Noise level
- Return air
- Location of space to be conditioned

Distance to the Electric Power Supply

Power must be brought to the indoor unit from the distribution panel. Adding more length to the run increases the cost of installation but has little effect on performance. Therefore, it is not a major factor in deciding the location of an indoor unit. Running of the power line must be done in accordance with all local electrical codes by a licensed electrician. This is often part of the electrical contractor's proposal for the project.

BETWEEN THE LINES

Who Is Responsible?

Sometimes the responsibilities for individual trades overlap. The HVACR contractor, for example, is typically responsible for running the electrical lines from the unit's service disconnect switch to the unit itself. You must refer to all of the notes on the project plans to determine which contractor is responsible for each portion of a project. For example, consider the portion of the plan shown in **Figure 8-3**. Although this drawing is part of the HVACR contractor's plan, the installation of the domestic water heater (DWH) is the responsibility of the plumbing contractor.

Serviceability and Accessibility

The indoor unit must be serviced periodically. A service technician must have access to the unit, remove all access panels, and work on all of the components in the unit. Serviceability is not often addressed in a set of HVACR prints. It is the responsibility of an HVACR contractor to identify any potential problems with the future servicing of the equipment before the installation is completed. Thus, it is in the HVACR contractor's best interest to identify these concerns before they become real problems.

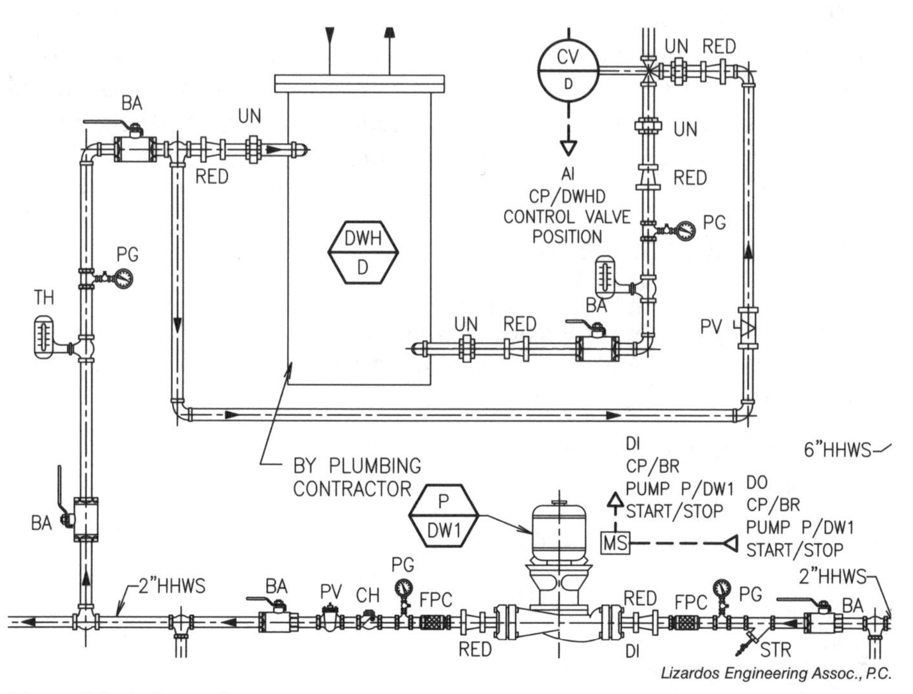

Figure 8-3. A domestic water heater is the responsibility of the plumbing contractor.

An indoor unit is usually located in an attic or crawl space. The unit should be positioned close to the access of the crawl space to make servicing the equipment easier. When located in an attic, a path leading to and under the unit must be provided.

Many local codes require different types of accessibility to the unit:

- A solid, permanent access path must be built to the indoor unit when it is installed in an attic or crawl space.
- A permanent platform at least 30″ wide must be provided at the unit location.
- A service disconnect must be located on or next to the indoor unit.
- A permanently mounted light fixture must be in place, with the light switch located within arm's reach of the attic entrance.

If project plans are missing these items, whether or not they are required by codes, the contractor must make these concerns known.

Ease of Condensate Removal

Because air-conditioning systems have both dehumidification and cooling functions, a project design always includes a plan on how to carry condensate to an appropriate location, such as outside the structure or to a drain. Improper removal of condensate can result in water damage. To reduce the possibility of water-related damage, the length of the condensate drain line should be as short and direct as possible. Excessively long condensate lines have more piping material and fittings, and as such, the chance of a leak is greatly increased.

For an indoor unit located in an attic or overhead area, the condensate drain line is routed to the outside of the structure. If an outside wall is not accessible, the condensate should be directed to a nearby waste line or drain, as long as the piping work complies with all local plumbing codes. For example, if the condensate line runs into a waste line, a trap must be installed to prevent waste gases from entering the space. Even if the project plans do not indicate a downward pitch on the condensate line, there should be a pitch of at least 1/4″ per foot of run. When the indoor unit is located in the basement and no floor drain is accessible, a condensate pump is needed to pump the condensate up to an overhead line, which is routed to a waste line or utility sink.

Always check the installation literature that is supplied with a unit before running the condensate line. Many indoor units require that the condensate line be trapped at the unit's condensate connection.

BETWEEN THE LINES

Project Drawings and Notes

Check project drawings and notes carefully to ensure no project element is overlooked. An example of a project remarks section is shown in **Figure 8-4**. Looking at note #3, condensate pumps are required at each air handler, regardless of whether or not a floor drain or other viable option is available.

UNIT SECTIONS:
CHILLED WATER COIL
HOT WATER COIL
MIXING BOX/FILTER
SUPPLY FAN (PLUG TYPE)
ACCESS SECTION

REMARKS

1. UNITS SHALL BE DISASSEMBLED AT SITE FOR ENTRY INTO BUILDING AND REASSEMBLED ON CONCRETE PADS.
2. PROVIDE HOT WATER AND CHILLED WATER PIPING CONTROL VALVES, SHUT OFF VALVES, ETC. TO EACH AHU CHILLED WATER COIL AND DUCT HOT WATER HEATING COIL.
3. PROVIDE CONDENSATE PUMP FOR EACH AIR HANDLING UNIT TO BE PUMPED INTO CONDENSATE MAIN.
4. BOOK SPECIFICATIONS TO FOLLOW WHEN ALL WINGS ARE ISSUED FOR BID.

Lizardos Engineering Assoc., P.C.

Figure 8-4. Item #3 calls for the use of condensate pumps on all air handlers.

Noise Level

When located in an attic, the indoor unit should never be located directly above a bedroom. In a commercial office setting, the unit should be located away from private offices and conference rooms when possible. Stock areas and basements are appropriate locations for the indoor unit. If the unit is located in a basement, position it so noise distractions and inconveniences will be minimized. When the unit is located in a utility closet, the walls of the closet should be covered with an acoustical material to reduce the amount of noise transmission to the adjacent areas. Even though noise transmission was most likely addressed when the project was drawn up, it is good practice to check the unit's location for possible problems.

Return Air

An HVACR equipment installation plan must include a method of returning air from the occupied space to the indoor unit. The source of the return air should be a common location, such as a hallway. When the unit is installed in a utility closet, place a return grille in the door of the closet. When there is a configuration where no ductwork is used to connect the return grille in the closet door to the air handler or create a path for return air, the term "natural return" is used.

In some instances, such as when the air handler is located in a utility closet, there can be more noise transmission to the occupied space. When the system is located in the attic, the return grille is often set in the ceiling of a common hallway. The return grille is connected to the indoor unit via a return duct, which physically connects the return grill to the unit. The indoor unit should never be located directly above the return grille because the noise levels will be higher. If the unit must be located close to the return grille, you can create a large loop in the return duct, which dampens the unit noise, creating less effect on the occupied space.

Goodheart-Willcox Publisher

Figure 8-5. Ductwork located within the conditioned space.

Distance to the Conditioned Space

The indoor unit should be closely located to the conditioned space. Long duct runs have a negative effect on the operation of the system. The interior of ductwork provides resistance to airflow, reducing the velocity of the air moving through it. Also, heat exchange takes place between the air surrounding the duct and the air passing through the duct. One way to help minimize duct leakage and heat transfer between the air in and around the duct is to run the duct within the conditioned space, **Figure 8-5**. The losses associated with air leakage and heat transfer are eliminated for the portions of the duct system located within the occupied space.

8.2 Ductwork Location and Considerations

When systems are installed as part of a new construction project, ductwork installation is straightforward. Problems are often small and easy to resolve because the duct system is designed along with the structure. When the walls and ceilings are open, the installation crew has free access to these spaces. Duct sections can be installed before any gypsum board or other wall or ceiling covering material is set in place. On new construction projects, wall registers are popular because duct sections can be easily positioned, mounted, and sealed in between the wall studs before any drywall is installed. An example of a wall register illustrated in print is shown in **Figure 8-6**.

Notes and drawings must be carefully studied to ensure that all project elements are addressed. For example, consider an air handler drawing that requires the installation of a duct collar with a flexible connection on all air handlers and an extended return air duct plenum, **Figure 8-7**. Be aware that information can be presented in different forms, such as visually in drawings, complete sentences in note form, or a callout on a project plan. Consider Note #2 in **Figure 8-8**, for example. For sound attenuation, it is required that 2′ of ductwork be provided at the discharge of all fans. Other prints may note additional requirements. Always note all requirements and additions because overlooking any items can affect the price quoted for the job.

A duct system should be constructed of low-resistance sections to maintain proper airflow rates. For example, flexible or spiral duct offers a great deal of resistance to flow and should be avoided when possible. National codes limit the amount of flexible duct material that can be used on any single duct run, so be sure to compare the job specifications to the code limits. Limiting the number of transition sections, offsets, and elbows decreases additional resistance to airflow. Most project plans indicate that all completed duct systems be evaluated, using a duct pressurization test, **Figure 8-9**, to determine that the duct system installation meets local codes and guidelines.

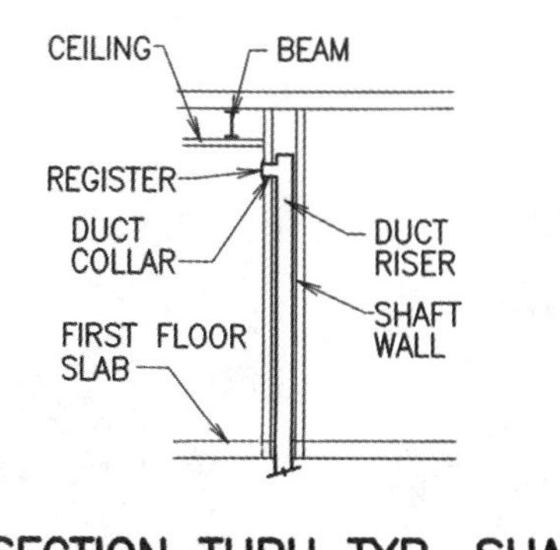

Lizardos Engineering Assoc., P.C.

Figure 8-6. Cross-sectional view of a wall register.

BETWEEN THE LINES

Reducing Air Leaks in Duct Systems

Leaking duct systems can account for up to 30% of the money spent to heat and cool a structure. Shorter, more efficiently designed and installed duct systems have the potential to save significant amounts of energy. One way to ensure a duct system is as tight and leak-free as possible is to use mastic to seal the duct's seams and joints. Mastic is a non-hardening material used to reduce air leaks in a duct system. It should be applied to all seams and joints.

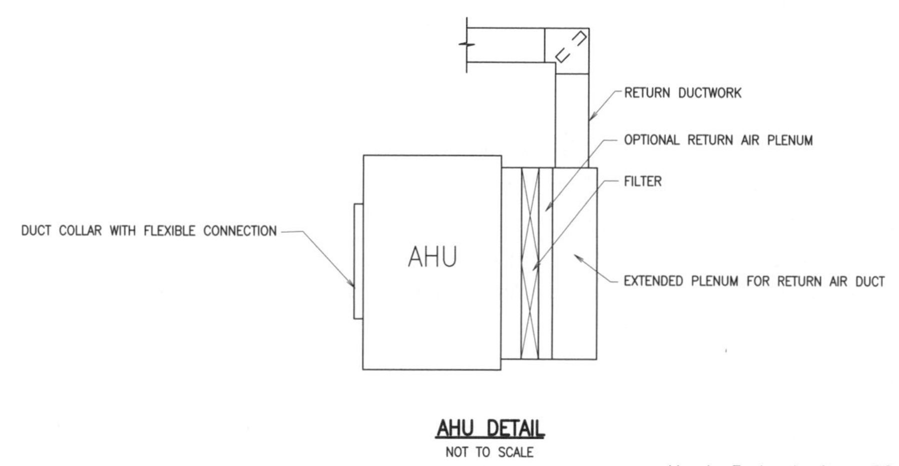

Lizardos Engineering Assoc., P.C.

Figure 8-7. Detail of air handler unit duct connections.

NOTES:

1. ALL RETURN FANS IN SERVICE WING CELLAR SHALL BE CENTRIFUGAL, BELT DRIVEN UTILITY FANS WITH NON-OVERLOADING BACKWARDLY INCLINED ALUMINUM WHEELS AS MANUFACTURED BY PENN VENTILATION CO., INC. FANS SHALL BE SINGLE INLET, SINGLE WIDTH, AMCA ARRANGEMENT 10 WITH CLOCKWISE (OR CCW) ROTATION. AIR DISCHARGE POSITION SHALL BE THD UNLESS SPECIFIED OTHERWISE.

MODEL:	D12	D13	D15
CFM:	1,000	1,250	1,900
RPM:	3020	2855	2600

2. PROVIDE TWO FEET OF DUCTWORK ON ALL FAN DISCHARGE OUTLETS FOR SOUND ATTENUATION.
3. PROVIDE DRIP PAN UNDER ANY AND ALL PIPES THAT CROSS OVER ELECTRICAL BUS DUCT IN SERVICE WING CELLAR AREA. PAN SHALL HAVE TWO INCH LIP ON EITHER SIDE AND SET AT 5° ANGLE TO ALLOW ANY CONDENSATION OR SWEATING TO SLIDE HARMLESSLY ACROSS ONTO THE CONCRETE FLOOR AND AWAY FROM THE ELECTRICAL BUS DUCT.

Lizardos Engineering Assoc., P.C.

Figure 8-8. Notes on sound attenuation.

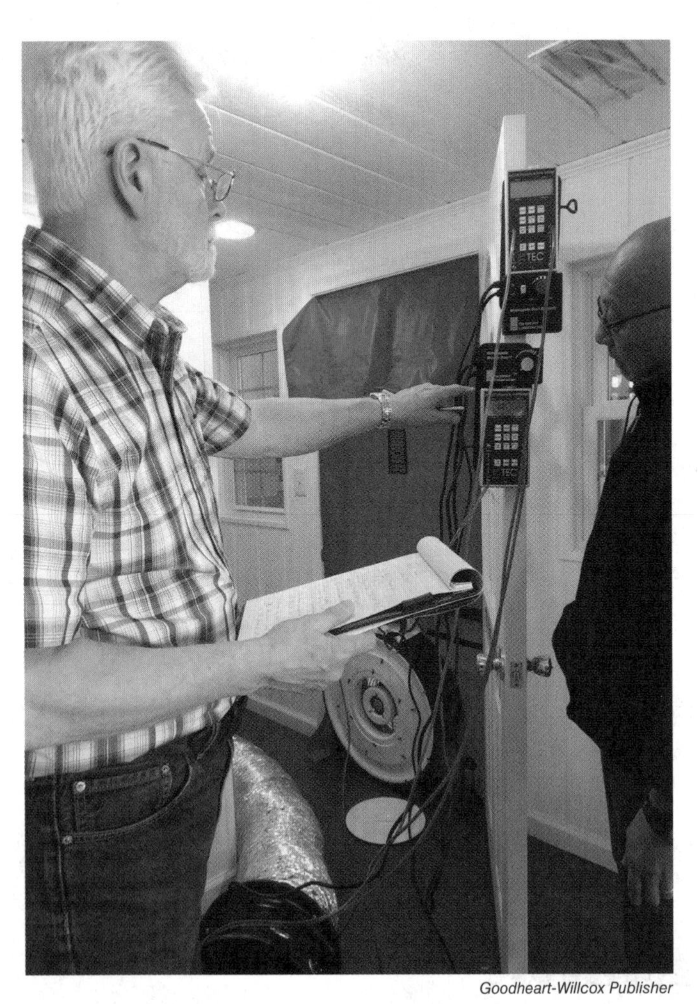

Goodheart-Willcox Publisher

Figure 8-9. Performing a duct pressurization test.

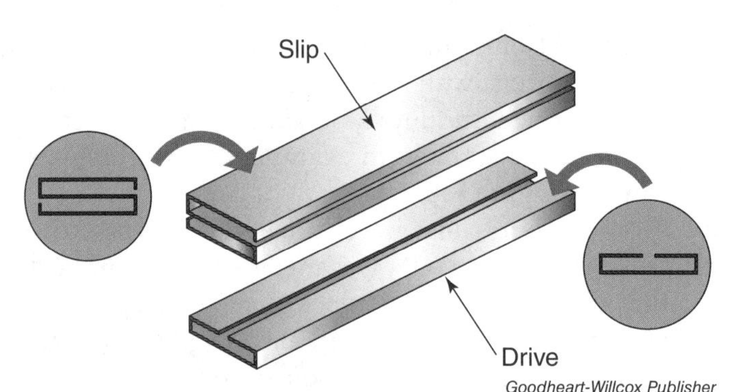

Goodheart-Willcox Publisher

Figure 8-10. Slips and drives.

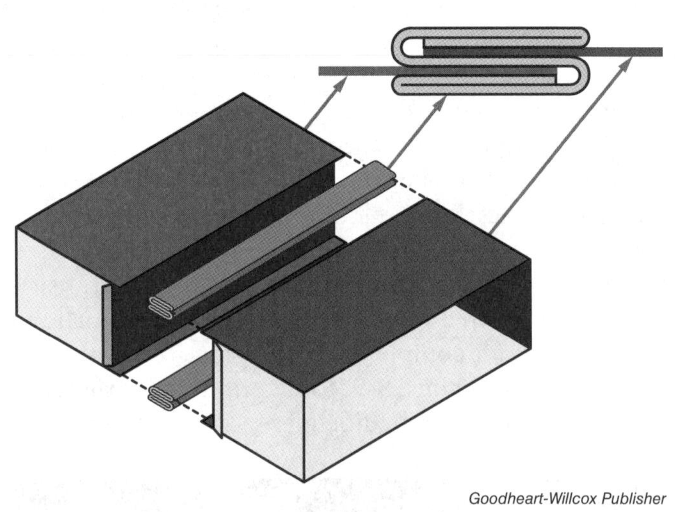

Goodheart-Willcox Publisher

Figure 8-11. Slips installed on the longer sides of the duct sections being joined.

8.2.1 Ductwork Fastening

Depending on the construction of the duct, as well as the material used, various techniques can be used to fasten duct sections together. Common methods for duct fastening include slips and drives, bar slips, and flanged connectors.

Slips and Drives

When installing a duct system made of galvanized sheet metal duct sections, the most common method for joining the sections is with slips and drives, **Figure 8-10**. ***Slips*** are strips of metal formed into an S-shape, **Figure 8-11**, whereas ***drives*** are formed into a flattened C-shape, **Figure 8-12**. The longer edges of the duct slip into the openings in the slip. When the duct sections are pushed together, a drive is used to secure the sections.

To use slips and drives, the shorter edges of the duct must be bent over to form ears. The drive cleat is bent at the top and bottom, around the edges of the duct, to keep it in place to complete the connection.

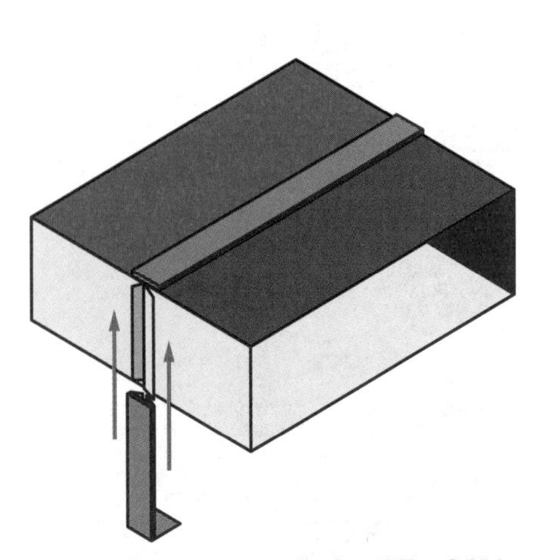

Goodheart-Willcox Publisher

Figure 8-12. Drives installed on the shorter sides of the duct sections being joined.

Bar Slips

Wide duct sections tend to flex and bend when the indoor fan cycles on and off. To reduce the noise generated by this, a modified slip called a ***bar slip*** is used. See **Figure 8-13**. Similar to a slip, the bar slip has an extra perpendicular tab that prevents the slip, and the duct it is connecting, from bending and flexing during blower startup and shutdown.

As the ductwork dimensions increase, the ducts may start to flex upon blower startup and shutdown. This flexing can cause loud banging that is transmitted through the entire air distribution system. Another commonly used method is called cross-breaking. ***Cross-breaking*** is the process of creating small bends between the diagonal corners on each side of a duct section. By cross-breaking the sides of the duct, the sides are less likely to flex when system blowers cycle on and off.

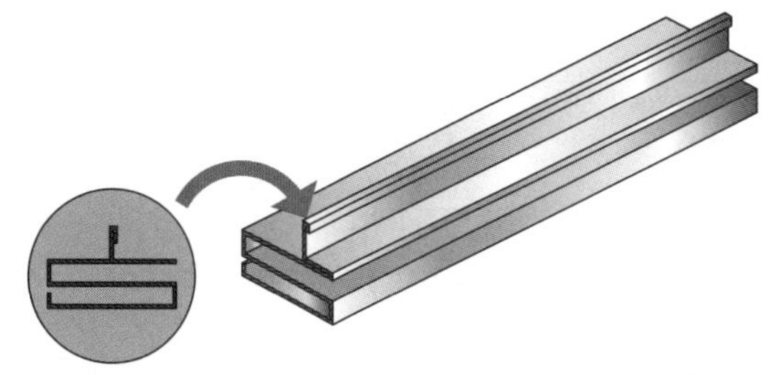

Goodheart-Willcox Publisher

Figure 8-13. Bar slips are used instead of traditional slips when larger duct sections are being connected.

Flanged Connectors

Flanged connectors are more expensive than the slip and drive method. Although primarily used in commercial and larger residential applications, flanged duct connectors are becoming popular in smaller residential projects as well. Flanged connectors are made up of specially fabricated angle connectors, which fit over the edges of the duct section, and matching corner connectors that fit into the angle connectors, **Figure 8-14.** The corner and angle connectors fit together to form a frame that is installed on the edge of the duct section. The angle connectors have an adhesive inside the channel, which sticks to the sheet metal once inserted into the flange. After placing a sealing gasket material between the two mating duct sections, they are bolted together, **Figure 8-15**, completing the connection.

The type of connector intended for a project is often indicated on the project plans. If no specific type is specified, ask for clarification before bidding on the job. The costs associated with using different duct-joining methods are very different.

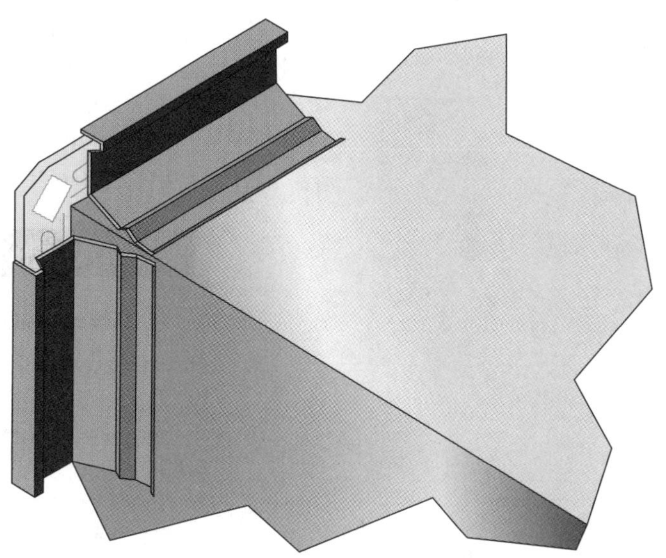

Goodheart-Willcox Publisher

Figure 8-14. The two sections of flanged duct sections are bolted together.

8.2.2 Supporting Sheet Metal Ductwork

Ductwork must be properly supported when installed. Strapping material, or strips of heavier-gage sheet metal, is commonly used to support a duct. These straps are often secured to the rafters or overhead structure and then screwed into the sides and bottom of the duct sections. On smaller-sized ducts, there must be one set of duct straps per duct section. Larger ducts require at least two sets per section.

A number of different construction materials may make up the overhead structure, so different methods may be needed to secure the ductwork. Common overhead materials include wood, steel, and concrete. Detail drawings illustrate how supporting members are attached to overhead structural materials.

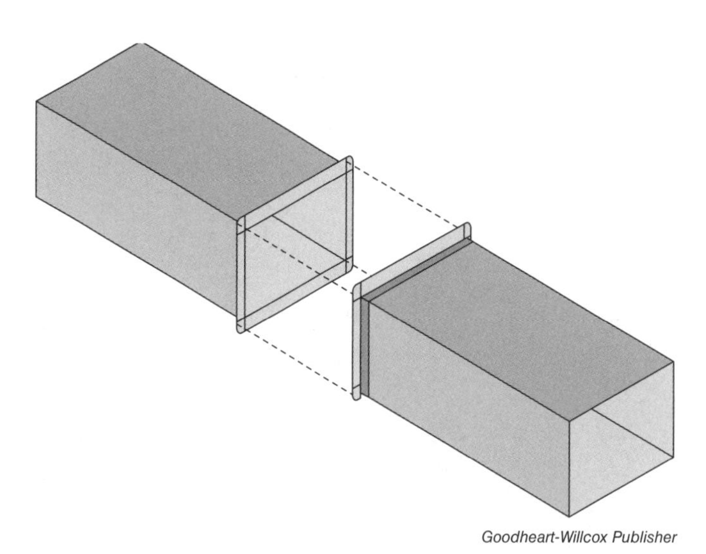

Goodheart-Willcox Publisher

Figure 8-15. Detail of attachment to steel beams.

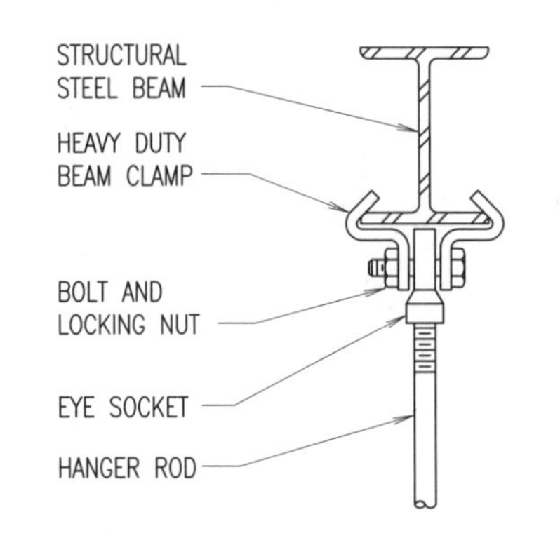

Figure 8-16. Detail of attachment to steel beams.

They provide valuable information regarding proper mounting, supporting, and installing of the various pieces of equipment.

For example, **Figure 8-16** demonstrates a method of how a hanger rod, also referred to as a threaded or all-thread rod, is secured to a structural steel beam using a heavy-duty beam clamp. The threaded rod is secured to the steel beam using an eye socket into which the threaded rod is secured. The eye socket is then secured to the beam clamp using a bolt and locking nut.

In another example, **Figure 8-17**, ductwork is supported with metal straps using the hanging method. A support strap is used, which must be a minimum of 1″ wide and 1/8″ thick if the cross-sectional area of the duct is greater than 2 ft^2. If the cross-sectional area of the duct is less than 2 ft^2, the support strap must be a minimum of 1″ wide and 1/16″ thick. The detail drawing also requires that, if the duct is more than 49″ wide, the support strap must be bent under and fastened to the duct as shown.

When installing ductwork to the air handler or blower outlet, support the ductwork independently of the air handler or blower. Never rely on the blower section to support the weight of the ductwork.

BETWEEN THE LINES

Canvas Connectors

Consider the case of an air handler installed with canvas connectors between the air handler and the supply and return ductwork. Canvas connectors, made of flexible material, can tear, and are not intended to bear weight. When installing canvas connectors, the metal portions of the connector should not touch each other. If they do, noise and vibrations will emanate from the air handler and transmit throughout the air-distribution system. There should be absolutely no metal-to-metal contact between the air handler and the duct system.

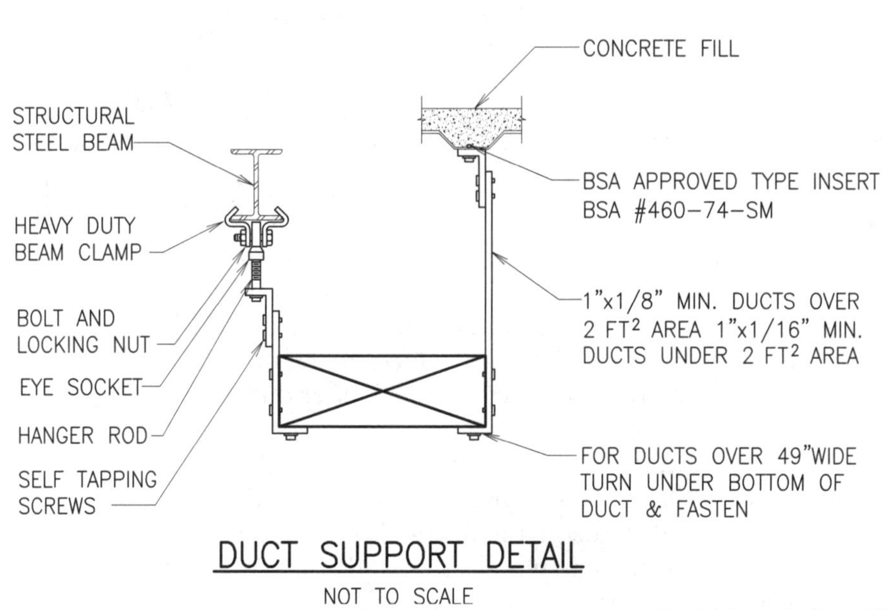

Figure 8-17. Detail of duct support.

8.2.3 Duct Leakage

Duct leakage can be measured by performing a duct pressurization test. This test involves sealing and pressurizing the duct system. To perform a duct pressurization test, all service panels must be taped on the air handler, along with sealing all air registers and grilles. A separate, external blower then pressurizes the air distribution system. With this done, the duct system is evaluated based on the amount of airflow, in cfm, needed to maintain the required pressure in the system. This type of test is performed once the entire duct system has been installed and all registers and grilles are in place.

8.2.4 Duct Insulation

Two major concerns must be addressed when working with sheet metal duct systems: heat transfer and noise transmission. These two issues can be resolved through different methods of insulation.

In sheet metal duct systems, heat transfer takes place between the air in the duct and the air surrounding the duct. When in cooling mode, heat from the air surrounding the duct is transferred to the cooled air inside the duct. If air in the duct system is heated, heat transfer occurs from the heated air to the surrounding air. In both cases, a loss of system effectiveness and efficiency results. Metal ductwork sweats during cooling mode if the surface temperature of the duct is below the dew point of the air surrounding the duct. This can cause water damage to the structure.

To alleviate problems such as these, sheet metal air distribution systems can be insulated. When insulating ductwork, a few options are available:

- Wrapping the duct with foil-covered fiberglass insulation
- Lining the duct with acoustical lining
- Wrapping and lining the ductwork

Once the duct system is installed, the duct can be wrapped with foil-covered fiberglass insulation. This insulation is secured around the duct with reinforced duct tape. The entire duct run must be insulated to help prevent any condensation from forming on the duct surface. Also, the duct system must be wrapped tightly to prevent large air pockets from forming between the duct and the insulation, which can result in condensate accumulation.

To prevent noise transmission, duct sections can be lined with a material called acoustical liner. Duct lining is installed on the sheet metal before the duct section is fabricated. The lining is held in place with pins and contact adhesive. It must be secure on the inside surface of the duct and able to withstand the effects of the high-velocity air moving across it. If the lining comes loose during operation, the ducts can become blocked, causing system malfunction. The decision to line the duct system must be made when the system is initially designed. It is typically shown on project drawings as a dashed line within the confines of the duct section, **Figure 8-18**.

Symbol	Abbreviation	Description
		CEILING SUPPLY AIR DEVICE
		CEILING RETURN OR EXHAUST AIR DEVICE
	AL	ACOUSTICAL LINING
		SIDE CONNECTED SUPPLY, RETURN OR EXHAUST AIR DEVICE

Lizardos Engineering Assoc., P.C.

Figure 8-18. Plan symbol for acoustical lining.

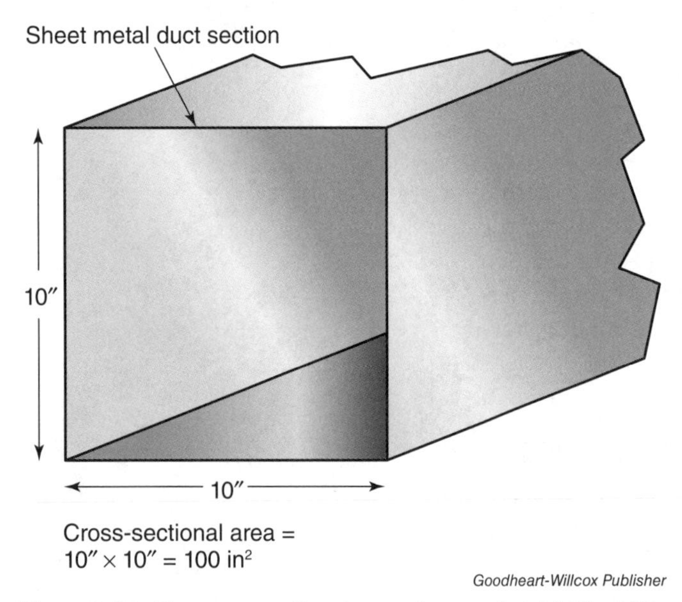

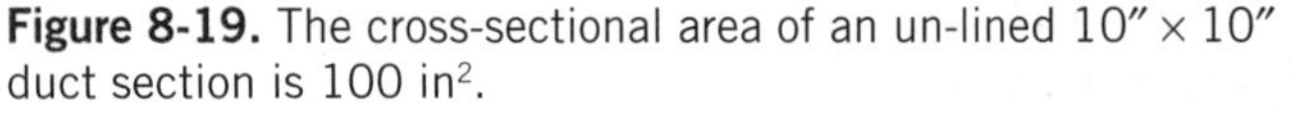

Goodheart-Willcox Publisher

Figure 8-19. The cross-sectional area of an un-lined 10″ × 10″ duct section is 100 in^2.

If a duct section is lined, the duct dimensions must be larger to accommodate the lining. Acoustical lining, depending on the application, is approximately 1″ thick and reduces the effective cross-sectional area of the duct run. For example, a 10″ × 10″ unlined duct has a cross-sectional area of 100 in^2, **Figure 8-19**. A 10″ × 10″ duct lined with 1″ thick acoustical lining will have a cross-sectional area of only 64 in^2.

8.2.5 Heat Exchangers in the Ductwork

An HVACR system installation project usually requires installing a heat transfer surface, coil, or other air-side component in the ductwork. In this case, it is common for the cross-sectional measurements of the air-side device to differ from the cross-sectional measurements of the duct.

It is best to gradually increase and decrease the duct size to fully and completely accommodate the air-side device. See **Figure 8-20**. This figure indicates that, for this specific application and installation, the sides of the duct should be at an angle no greater than 30° to the main duct on the device-inlet side, and no greater than 45° to the main duct on the device-outlet side. This helps minimize the air turbulence in the duct, while increasing the effectiveness and efficiency of the air-side device being installed. In addition, whenever an air-side component is installed in a duct, access doors should be installed to facilitate future component servicing.

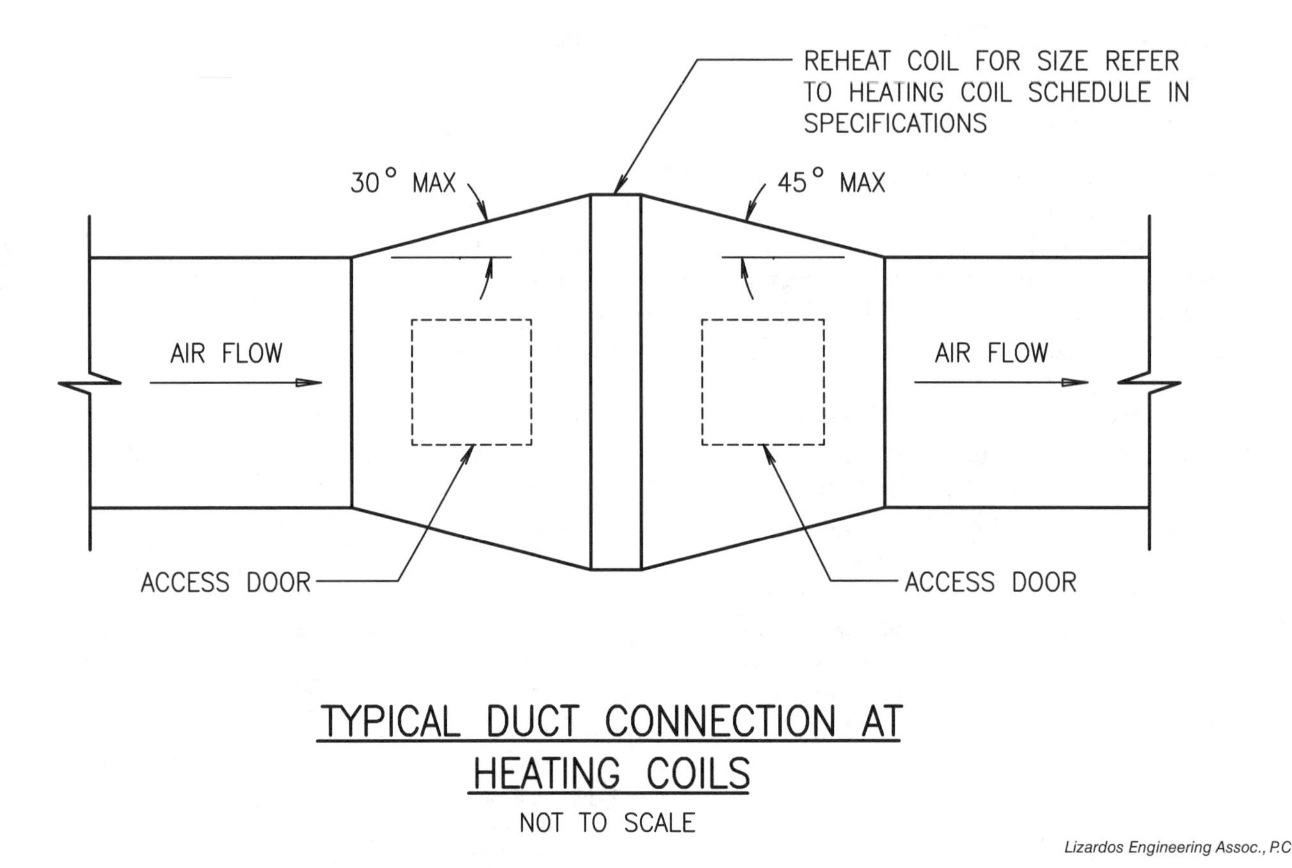

Lizardos Engineering Assoc., P.C.

Figure 8-20. Detail of duct connection to air-side components.

8.3 Plumbing Line Location and Considerations

Plumbing is an integral component in air-conditioning system installation. All air-conditioning systems rely on an interconnecting piping arrangement to allow the system to operate as intended. For instance, refrigerant must circulate through the various system components, and moisture removed from the air must be deposited somewhere. Chilled and heated water coils are often installed in ductwork located far from the point where the water is treated.

HVACR system piping includes:

- Condensate piping
- Refrigerant piping
- Condenser water piping
- Chilled water piping
- Hot water piping

8.3.1 Condensate Piping

An air-conditioning system performs two main functions: cooling and dehumidifying. The cooling and dehumidifying processes are accomplished in the air handler. When air is cooled, its temperature is lowered, and moisture is removed from the conditioned air when it is dehumidified. The moisture, also called ***condensate***, taken from the air must be safely removed from the structure. Thus, a ***condensate drain line*** must be installed to carry the condensate away. Condensate, if not properly managed, can cause major damage to a structure, especially if the air handler is located above the conditioned space, such as in an attic.

Consider a scenario where a condensate line becomes clogged and the condensate overflows the condensate pan in an attic-mounted air handler. The occupant of the space may be unaware of the problem until the water starts dripping from the ceiling. By this time, the ceiling, likely constructed from gypsum board, has absorbed a significant amount of water and is damaged. To help ensure that damage from improperly disposed of condensate is minimized, several factors must be considered when installing condensate piping.

Drain Line Size

Air handlers in an air-conditioning system are equipped with internal drain pans and female pipe connections, which connect to drain lines. The size of a drain line must be the same size or bigger than the fitting provided with the component. A drain pipe must never be less than 3/4″ in diameter. When condensate is removed from the air, water accumulates. The condensate drain line must be able to handle this water flow. In areas that produce large amounts of condensate, it may be necessary to increase the size of the condensate line.

Drain Line Material

A drain line is constructed of the materials specified on the project plan and in accordance with local plumbing codes. Polyvinyl chloride (PVC) piping is acceptable for most residential applications and some commercial applications. When installing a PVC condensate drain line, all fittings and pipe ends must be cleaned first with a PVC primer and then joined using PVC cement. Some local codes require the use of colored primers so it is easy to verify that primer was used. Using cement not designed for use with PVC pipe can result in a weak joint that may cause leaks.

Larger installations and instances where a condensate line is located within a wall or ceiling may require copper piping. When installing a copper drain line, all solder joints must be checked for leaks before the system is put into operation. Other lightweight rigid plastic materials should typically be avoided, but they can be used if the length of the line is very short and the materials are code-compliant.

Drain Line Pitch

A drain line ***pitch*** refers to the downward slope in the direction of water flow. Regardless of material, a drain line must be pitched away from the air handler toward the termination point of the line. The minimum pitch on the line is 1/4″ per 1′ of horizontal run. For example, a 10′ horizontal drain line should be pitched a minimum of 2 1/2″.

Drain lines must be supported to ensure the proper pitch is maintained. To properly drain the condensate, the pitch should be made as large as possible. Improperly pitched drain lines can result in water damage to both the system and the structure.

Condensate Drain Line Traps

Air flowing up the drain line can affect the condensate from draining properly. It also increases the possibility of water-related damage. To prevent this, ***condensate traps*** are often used. Condensate traps are installed in condensate lines to help prevent air from flowing up through the drain line.

These traps are not required on all air handler installations. When the refrigerant coil is located downstream of the blower, so the air is being pushed through the coil, a trap is not needed. Because the drain line inlet is at a higher pressure than the outlet, the condensate will drain properly without a trap.

Condensate traps are used where the blower is located downstream of the refrigerant coil. In instances like this, the air is pulled through the coil and up through the drain line. This causes the pressure at the inlet of the drain line to be lower than the pressure at the outlet, which pulls air in the direction opposite to the direction of condensate flow. If the drain line is under a negative pressure and no trap is installed in the line, the condensate will accumulate in the drain pan, causing overflow and water damage. To help determine whether a trap is needed refer to **Figure 8-21**.

Piping Designs That Require Condensate Traps
Vertical upflow unit with blower above coil
Vertical downflow unit with blower below coil
Left-to-right horizontal unit with blower to right of coil
Right-to-left horizontal unit with blower to left of coil
Piping Designs That Do Not Require Condensate Traps
Vertical upflow unit with blower below coil
Vertical downflow unit with blower above coil
Left-to-right horizontal unit with blower to left of coil
Right-to-left horizontal unit with blower to right of coil

Goodheart-Willcox Publisher

Figure 8-21. A table identifying when a condensate trap is used based on piping design.

Consider the air handler shown in **Figure 8-22.** Because this is an upflow air handler and the blower is above the evaporator coil, a condensate trap is required. Be sure to follow the manufacturer's guidelines with respect to trapping condensate lines. If it is uncertain whether a trap is required, a trap should be installed. Project notes, such as those shown in **Figure 8-23**, require that all air handlers be mounted 8″ above the finished floor line to give adequate clearance for drain traps.

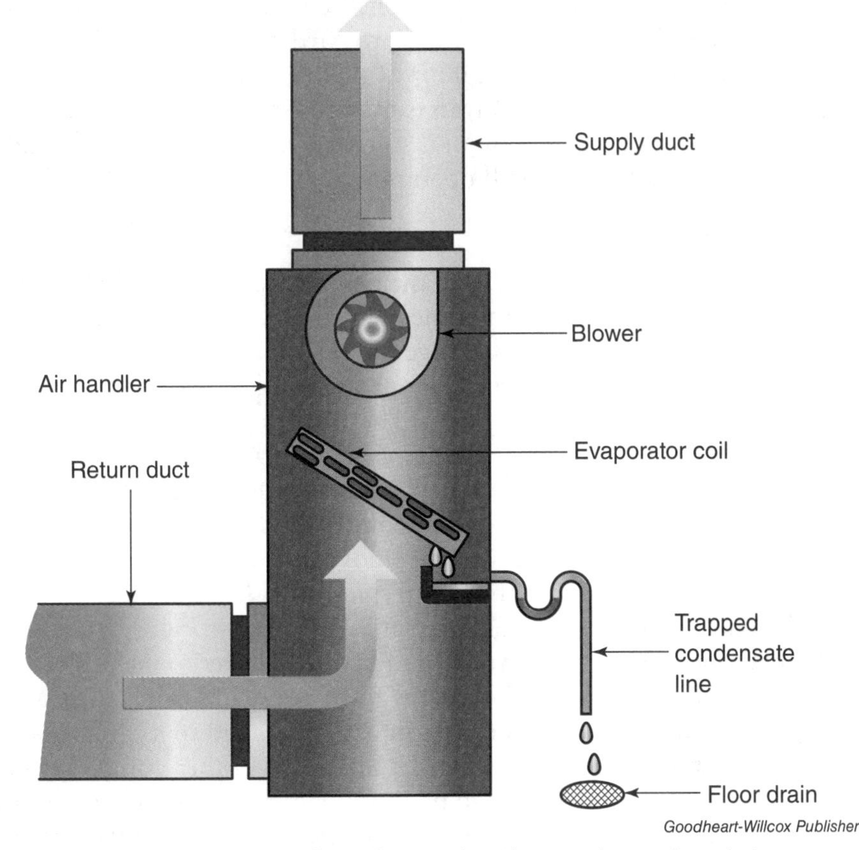

Goodheart-Willcox Publisher

Figure 8-22. This air handler configuration requires the use of a condensate trap.

GENERAL NOTES:

1. ALL AIR HANDLING UNITS SHALL BE MOUNTED 8" ABOVE FINISHED FLOOR TO ALLOW FOR CONDENSATE PUMP AND DRAIN TRAP.
2. PUMP ALL CONDENSATE TO SANITARY DRAIN PIPE.
3. FIRST FLOOR DRAWINGS WILL BE ISSUED UPON FINAL LIGHTING PLANS FROM CBB.

Lizardos Engineering Assoc., P.C.

Figure 8-23. Note on air handler positioning and condensate line water routing.

Condensate removed from the system must be deposed of, usually outside the structure or down a waste line. If the unit is located in an attic or a utility closet, the drain line commonly leaves the structure through the same building penetration as the refrigerant lines and low-voltage wiring. The drain line should be as short as possible to reduce the possibility of a blockage that could cause the line to back up.

If running the line to the outside of the structure is not possible, the line should be run to a nearby waste line or utility sink. In this case, the condensate line must be installed with a trap to prevent waste vapors from entering the space. If the indoor unit is located in a basement, running the condensate line to the outside may not be as easy. In a basement, the unit is below ground level, so a gravity-type drain to the outside is not possible. In such cases, a floor drain can be installed or the condensate can be removed using a condensate pump.

If the unit is installed in an unfinished basement and a floor drain is nearby, the condensate can be directed toward the drain. The line should be run from the unit, along the floor of the basement, which is shown in **Figure 8-24**. Whether or not a trap is required depends on the configuration of the unit. Before the system is started up, the floor drain should be tested to make certain it is still operational.

Condensate Pumps

If a floor drain is not available, or if the basement is finished, a ***condensate pump*** may be necessary to pump the condensate from the location. Condensate pumps come in a wide range of styles based on voltage, holding capacity, and body. Low-profile pumps are used when not much clearance space is available for the pump. These are popular on units installed in ceilings where no termination point is nearby for the condensate. In operation, the pump is normally in the off position and is controlled by a float switch. Once the water in the pump's reservoir reaches a predetermined level, the pump switches on, removing the water from the pump. Once the water level drops, the pump switches off.

The pump is located in close proximity to the unit, **Figure 8-25**. Water is then pumped either outside or to a utility sink. Refer to the project plan for details on where the condensate is to be routed. For instance, a plan note might state that all condensate must be pumped to a sanitary drain pipe. The piping run from the unit to the pump is therefore very short. The need for a trap is determined by the configuration of the indoor unit.

Some condensate pumps are equipped with overflow switches. This will prevent water damage if the pump fails. This overflow control is a normally closed switch that opens if the water level rises above the maximum level permitted by the pump. It is normally wired in series

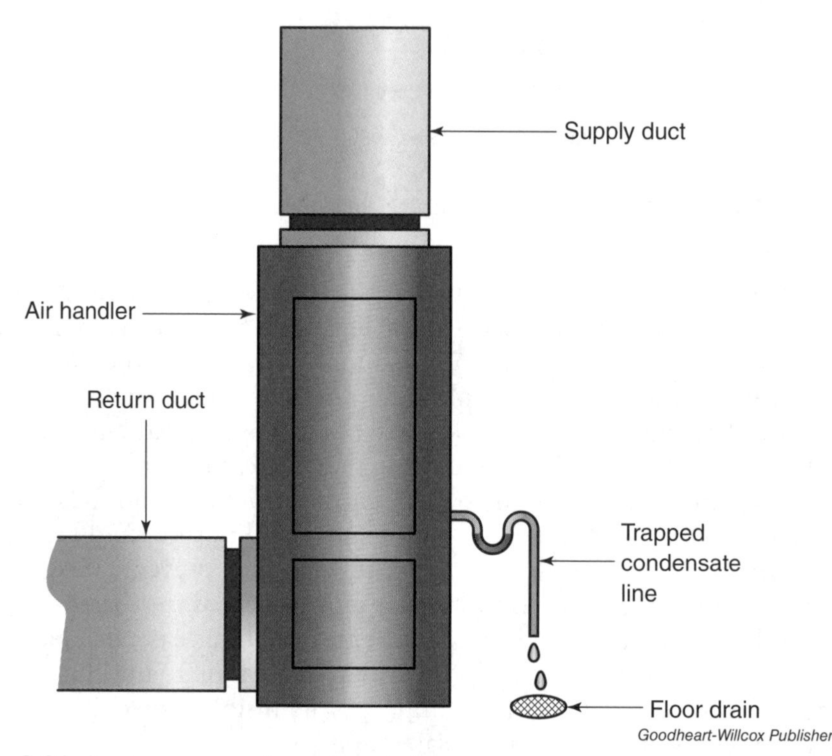

Figure 8-24. Condensate routed to a floor drain.

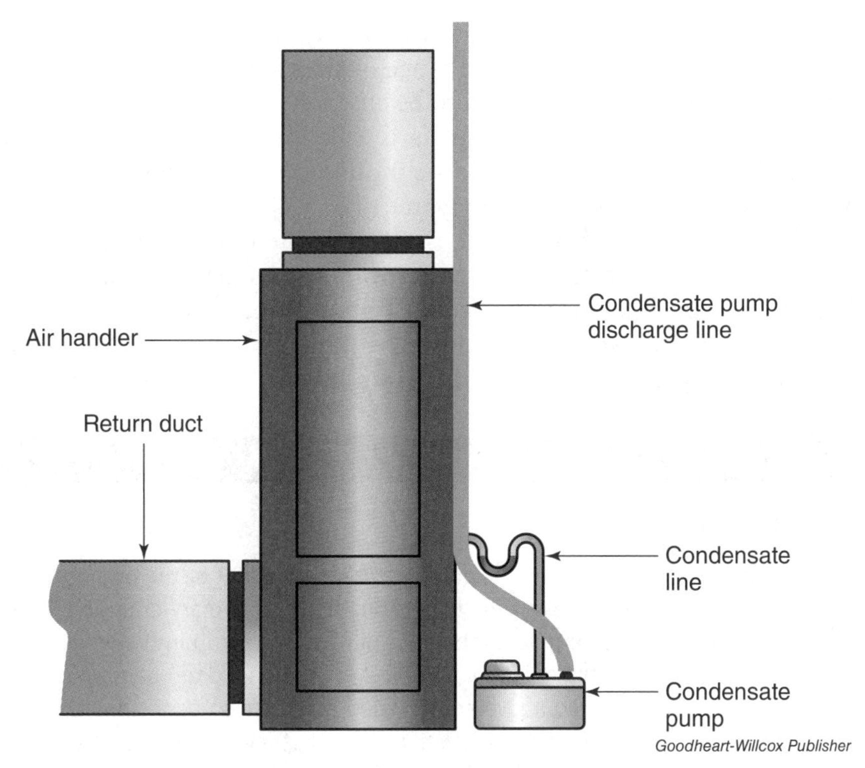

Figure 8-25. Condensate directed to a condensate pump.

with the "R" wire coming off the control transformer. When the overflow control opens its contacts, the system shuts down, preventing any further moisture removal from the conditioned space. The customer then places a service call indicating that the system is not operating. Shutting down the system prevents the water from overflowing the pump and causing damage.

Wiring the overflow safety control through the hot wire that connects the control transformer to the thermostat's "R" terminal disables the entire system. This includes the indoor fan. By disabling the indoor blower, it is likely the equipment operator will become aware of a system problem sooner. Some technicians, however, choose to interrupt only the compressor's control wire, which allows the indoor blower to run. Breaking the "R" conductor is the most effective way to ensure the problem is identified faster.

Auxiliary Drain Pans

Auxiliary drain pans are placed under the air handler to eliminate the possibility of water damage due to an overflow. The purpose of an auxiliary pan is to catch any overflow condensate that may result from a blocked or clogged condensate line. The piping from the auxiliary drain pan directs the condensate to an often conspicuous location, such as along a wall at the exterior of the structure. If water is spotted flowing from this pipe, it is most likely a problem with the primary drain line.

An auxiliary drain pan differs from the pan manufactured in the air handler. The pan that is manufactured along with the air handler is much smaller than the auxiliary drain pan and is located directly under the evaporator coil. The auxiliary drain pan must be sized so it can catch water leaking from the unit, which means the dimensions of the pan must be larger than the footprint of the indoor unit. Prefabricated plastic drain pans can be purchased in a wide range of sizes suitable for many systems. Larger pans, however, may need to be custom-made. Pans can be made of sheet metal, as long as the corners are sealed to ensure the pan can hold water. The auxiliary pan should be mounted securely under the unit in the event it actually has to hold water.

The primary condensate line is piped to the outside of the structure, or to a waste line, in such a way that water is disposed of discretely. The auxiliary drain line does not need a trap, as long as it is not being tied into a waste line. No pressure difference exists between the two ends of the auxiliary drain line, so water flows freely from the pan.

BETWEEN THE LINES

Auxiliary Drain Line

The auxiliary drain line must be completely independent of the primary drain line. Never join the auxiliary drain line and the primary drain line into a common line. This eliminates the possibility of a backup in the auxiliary drain line. Project plans should always call for auxiliary drain pans whenever the air handler is located above an occupied space. In some instances, auxiliary pans are called for when water pipes that are positioned above electrical devices or wiring have the potential to sweat.

Safety Float Switches

In the event of a drain line problem, auxiliary pans can fill up, causing water to overflow and potentially damage the area below. To prevent this, a ***safety float switch***, **Figure 8-26**, can be installed in the auxiliary pan. A safety float switch is a normally closed device that is wired in series with a low-voltage control circuit. It opens and shuts down the system if the level of water in the auxiliary drain pan reaches a predetermined level. Once the system shuts down, the customer calls the service company to report that the system will not operate.

Many contractors opt to eliminate the drain line connections from the auxiliary drain pan altogether and instead install a safety float switch. This eliminates the leak potential in the auxiliary drain pan fittings and the piping associated with the auxiliary drain line. Because there is no piping connection to the pan in this case, there is the potential for water damage in the event the auxiliary pan fills with water. The pan must be securely mounted to prevent any pan flexing and movement to minimize this possibility.

Goodheart-Willcox Publisher

Figure 8-26. Safety float switch.

8.3.2 Refrigerant Lines

Proper installation of refrigerant lines is critical to ensure the long-term satisfactory operation of an air-conditioning system. Poorly installed refrigerant lines can result in premature compressor failure, as well as a system that does not operate effectively and efficiently.

There are a number of factors to consider when installing refrigerant lines:

- Run the refrigerant piping as short, straight, and direct as possible to avoid excessive piping.
- Keep the number of fittings to a minimum.
- Use tubing benders when possible.
- Use long radius elbows whenever elbow fittings are needed.
- Ensure all brazed joints are as perfect as possible, using a dry-nitrogen purge.
- Pitch the suction line back toward the compressor to aid oil return.
- Install refrigerant traps when necessary and/or indicated.
- Always insulate the suction line.
- Install flexible connectors and shutoff valves as required for future service.
- Follow all project guidelines regarding piping location.

Length of the Piping Run

Consider the long-standing remmendation that the indoor and outdoor units be located as close to one another as possible. If this was done, the length of the refrigerant lines would also be as short as possible, which would increase the system's operating efficiency. In addition, the piping run should be as direct as possible, with no excessive solder joints or pipe sections.

Choice of Pipe Fittings

When piping is installed, a number of pipe fittings are likely installed as well. These fittings include 90° elbows, 45° elbows, and couplings. With 90° elbows, be sure to select those with a long radius. Long radius elbows provide less resistance to refrigerant flow than those with a shorter radius. Shorter radius elbows are typically used on plumbing piping arrangements and not on refrigerant piping systems. Using long radius elbows helps maintain a constant refrigerant velocity.

Each fitting added to the refrigerant piping circuit increases the resistance encountered by the refrigerant, so the number of fittings should be kept to a minimum. Reducing the number of fittings in the piping circuit reduces the number of brazed joints required to connect the indoor and outdoor units. This decreases the chance of refrigerant leaks. When possible, use 45° fittings instead of 90° elbows when piping in refrigerant lines. Fittings with a 45° angle offer far less resistance to flow than 90° fittings.

Brazing and Brazed Joints

High-quality materials, including those employed for brazing, must be used to ensure a high-quality installation and continued satisfactory system operation. For example, all refrigerant lines vibrate to some degree. Silver-bearing filler materials better withstand these vibrations and ensure leak-free connections more so than products with no silver.

All brazed joints must be carefully inspected before a new system is put in operation. Any joint that appears faulty or weak must be rebrazed. When brazing, pass 1–2 psig of dry nitrogen through the refrigerant lines being brazed. The nitrogen used to pressurize refrigerant lines during the brazing process is classified as medical-grade, dry nitrogen. It is not the nitrogen typically used in welding applications.

Some equipment manufacturers require that field-installed filter driers not be used in conjunction with their equipment. Thus, a nitrogen purge must be introduced to the piping circuit during brazing. This helps prevent oxidation from forming on the interior surfaces of the piping and also reduces the possibility of metering device and other system component clogging. R-410A and the POE and PVE oils used in conjunction with R-410A have a strong scrubbing effect on the interior surfaces of the copper pipes. Any oxidation on the pipes will be scoured away and end up in the metering devices and other system components.

Soft-Drawn Copper Tubing

One way to minimize the number of solder joints and fittings required is to use soft-drawn copper tubing. Soft-drawn tubing can be bent around corners and shaped to form offsets, reducing the number of fittings. Soft-drawn copper tubing comes in rolls, typically 50′ in length. When using soft-drawn tubing make certain that the tubing does not kink.

The best way to ensure that the tubing does not kink is to place the loose end on the ground, holding it in place with your hand or foot. Then, carefully unroll the tubing, keeping the roll on the ground. Pulling the tubing off the coil can twist the tubing, creating waves in the line. This may form small refrigerant traps if the run is horizontal. An excessive number of refrigerant traps can affect system operation, refrigerant flow rates, and hinder the process of compressor oil return. Do not use soft-drawn copper tubing if hard-drawn or rigid piping is specified for the project.

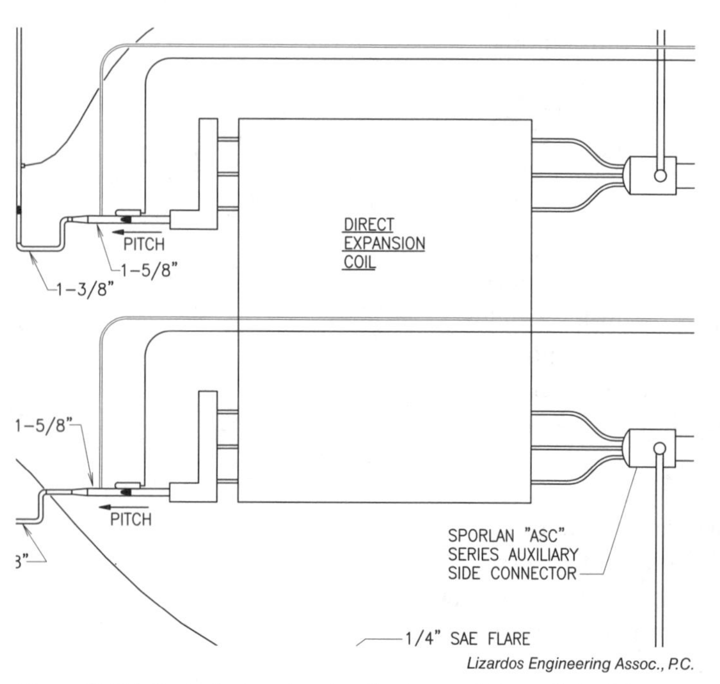

Figure 8-27. Project plan showing the traps at the outlet of the evaporator coils.

Refrigerant Traps

Refrigerant traps are a component of the refrigerant piping installation, but are not often included in project plans. They are responsible for helping return oil to the system's compressor. Equipment installation literature is specific regarding the use of refrigerant traps. Traps must not be left out if specified in the plans.

A refrigerant trap allows system oil not in the compressor to accumulate in one location. When the line becomes blocked or the trap becomes liquid-locked, the compressor's suction pulls the oil back to the crankcase. Refrigerant traps are typically installed in systems where the indoor unit is located below the outdoor unit. An example is when the air handler is located in the basement, and the condensing unit is on the roof. In this case, a refrigerant trap should be located at the bottom of the vertical suction line run, **Figure 8-27**. In addition, an inverted trap should be placed at the top of the vertical run on the roof to ensure the oil is able to return to the compressor, **Figure 8-28**. The suction line connecting the inverted trap to the compressor must be pitched to the compressor to allow the oil to flow freely. An example of pitching suction lines is shown in **Figure 8-29**.

The exact measurements of the trap depend on the capacity of the system and the line sizes. Refer to the manufacturer's installation literature for exact trap sizing information. On larger installations where the air handler is located below the outdoor unit, install one refrigerant trap for every 15′ of vertical suction line run, **Figure 8-30**, or as required by the manufacturer. Because the installation of suction line refrigerant traps requires more time and materials, this must be taken into account when the project cost is initially determined. If not included in the original proposal, a change order is required to ensure the installation complies with the equipment manufacturer's installation guidelines.

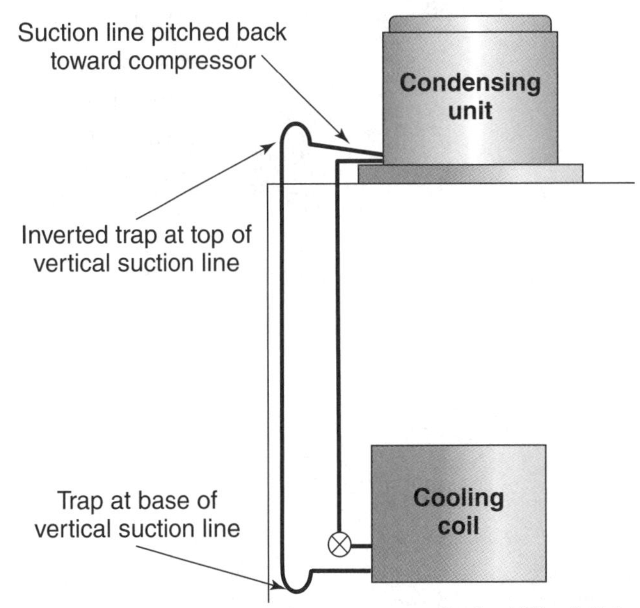

Figure 8-28. Refrigerant trap and inverted trap located at the bottom and top of the vertical suction line run. Notice the pitch on the suction line returning to the compressor.

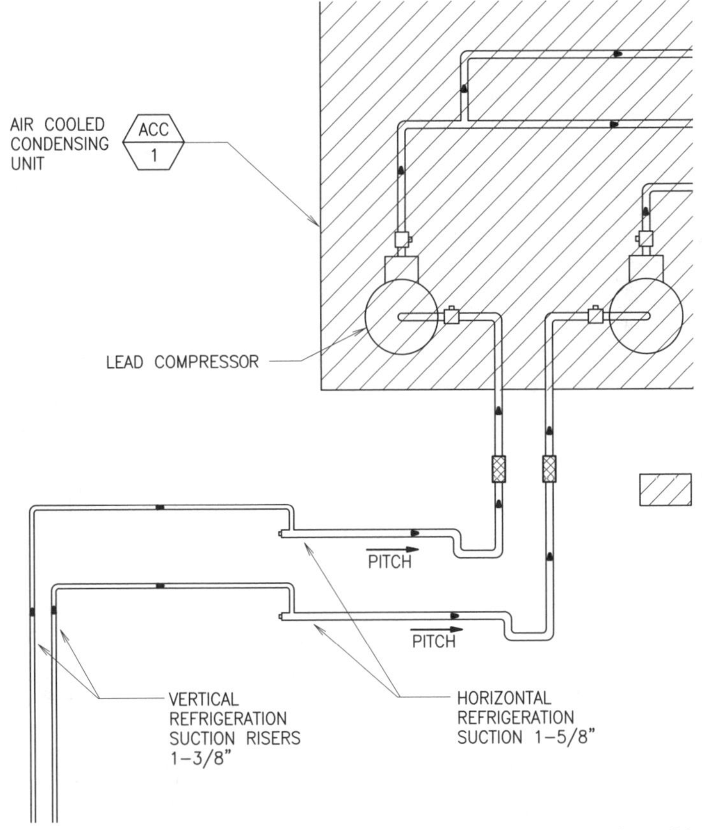

Figure 8-29. Drawing indicating that the suction line is to be pitched back to the compressor.

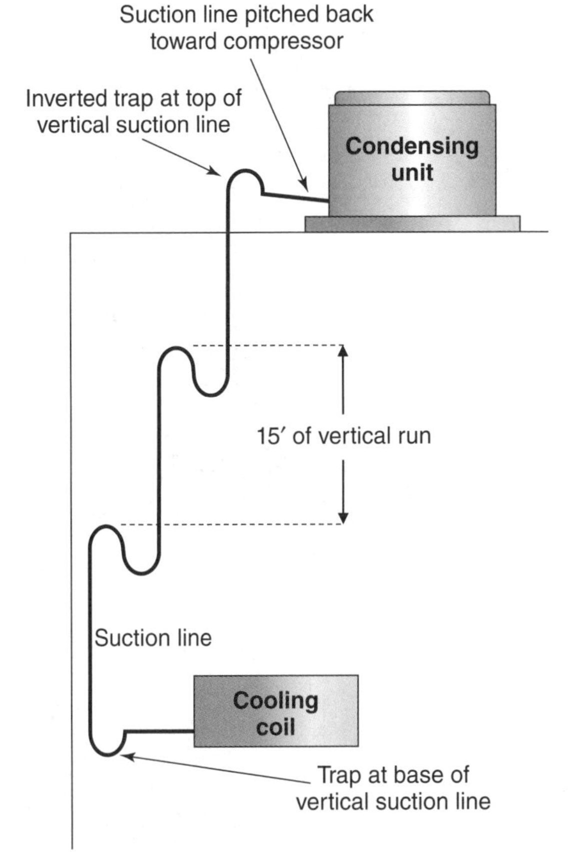

Figure 8-30. One refrigerant trap is installed for each 15′ of vertical suction line.

Install Flexible Connectors and Shutoff Valves

Air-conditioning systems should always be installed with future service in mind. All of the major system components should be accessible, and the various pieces of equipment should be able to be isolated from the rest of the system. This not only makes service and maintenance easier, but can also help minimize the loss of refrigerant in the event of a leak. The portion of the project shown in **Figure 8-31**, for example, calls for the installation of flexible connectors and refrigerant shutoff valves.

Follow All Project Guidelines Regarding Piping Locations

Follow all project guidelines regarding the location and positioning of all piping arrangements. An important goal is to have all piping accessible for future servicing, but also be out of way and not impede access to other system components.

A print can, in a few ways, show clear instructions to the one executing it. Consider the ductwork and piping arrangement shown in **Figure 8-32**. System return and supply ductwork is positioned neatly against the wall, while the piping, which is mounted against the same wall, is located parallel to, and under, the ductwork. Another example of this appears in **Figure 8-33**. Instructions are provided to locate all HVACR piping along the wall.

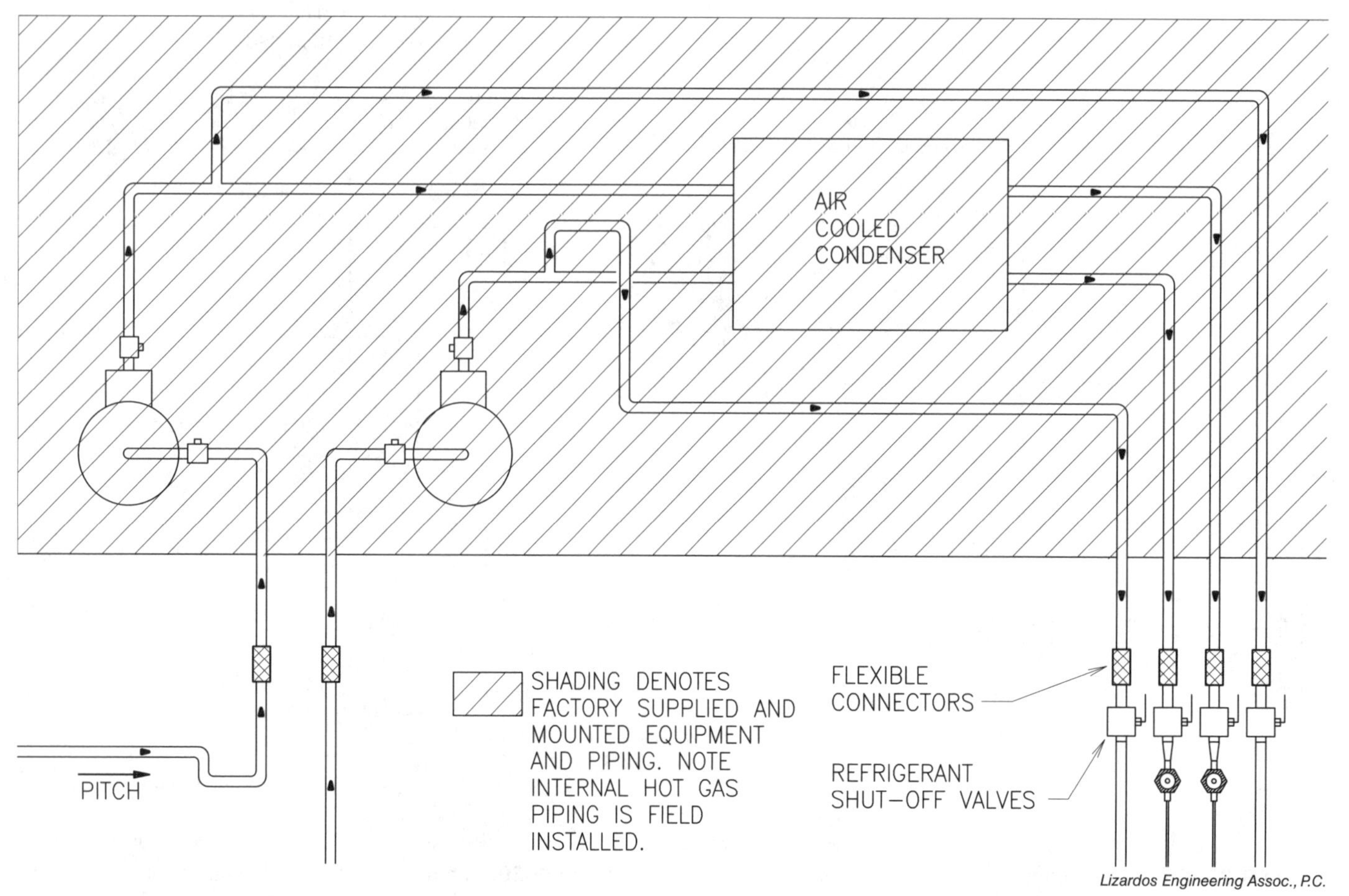

Figure 8-31. Flexible refrigerant line connectors help reduce noise transmission through the structure, while refrigerant shut-off valves facilitate future system service.

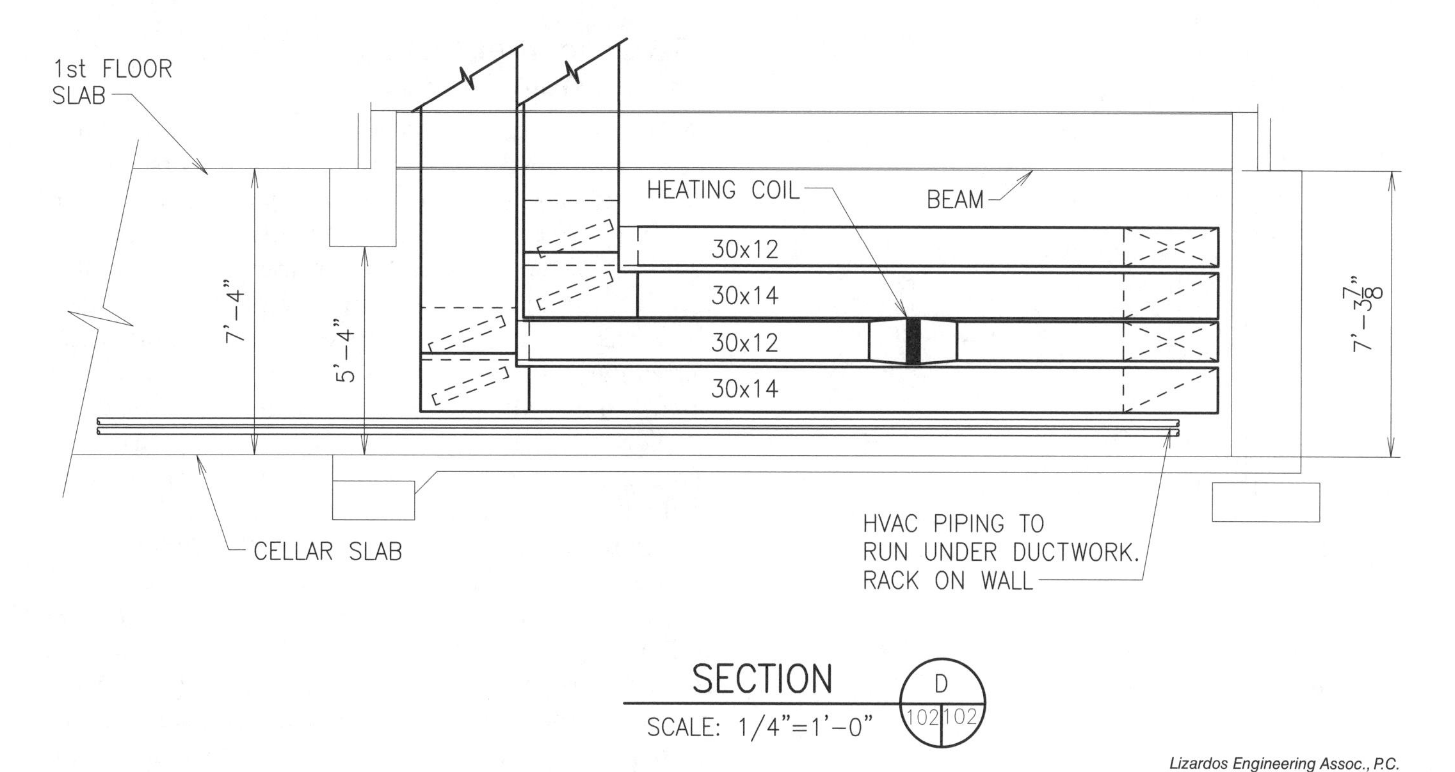

Lizardos Engineering Assoc., P.C.

Figure 8-32. Notes on HVACR piping locations.

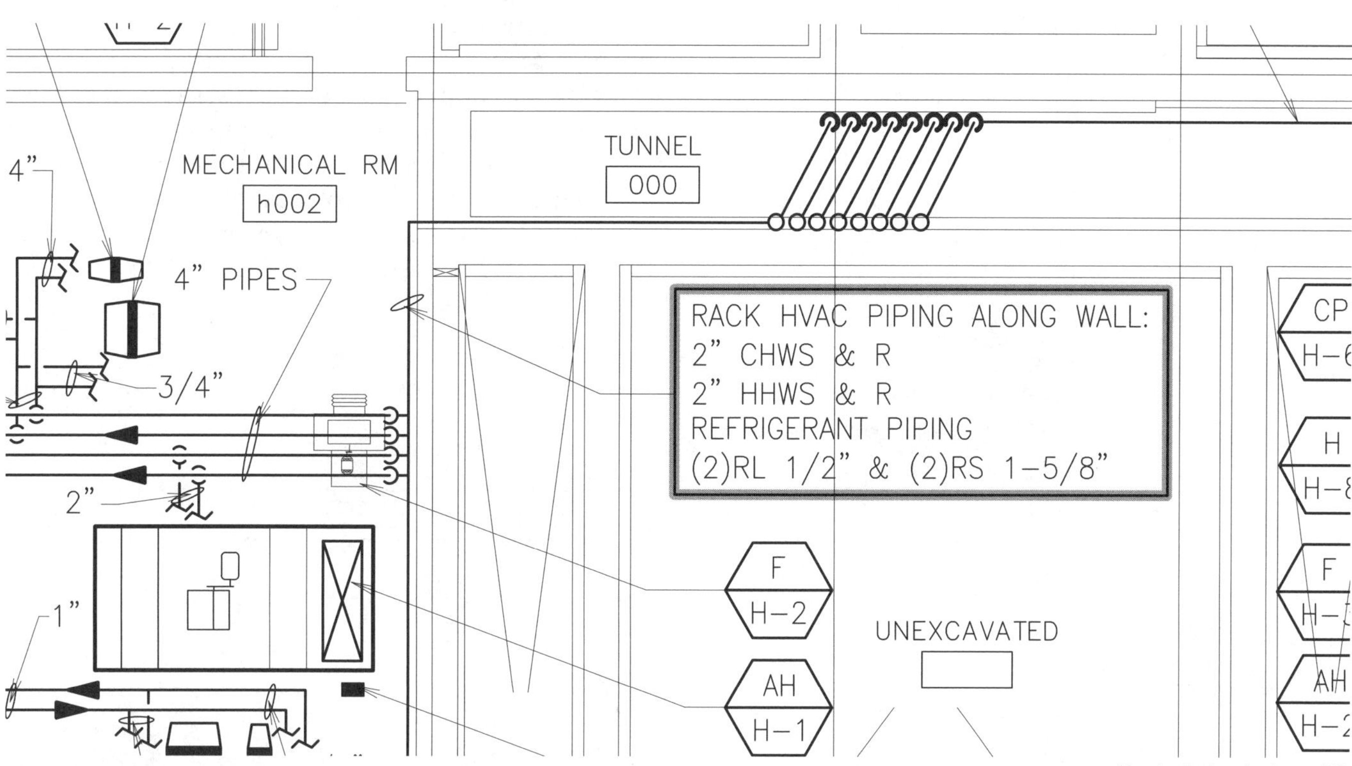

Lizardos Engineering Assoc., P.C.

Figure 8-33. Note on location of hot and chilled water supply and return lines (highlighted). Notice that although eight lines are referenced, only one line is drawn.

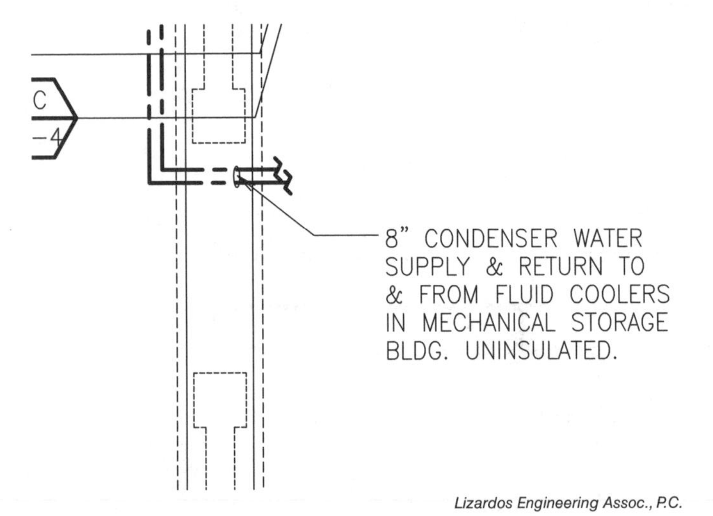

Lizardos Engineering Assoc., P.C.

Figure 8-34. Note on condenser water piping insulation.

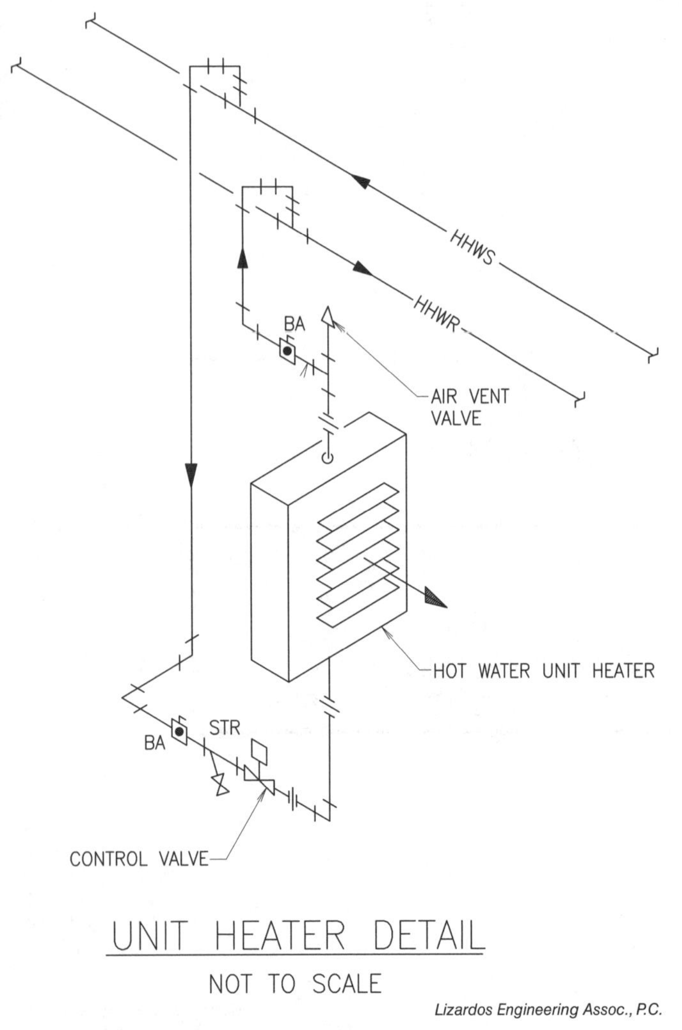

Lizardos Engineering Assoc., P.C.

Figure 8-35. Detail of unit heater piping installation. Refer to drawing MN101 for the complete image.

8.3.3 Condenser, Chilled, and Hot Water Lines

Besides condensate and refrigerant lines, other lines must be installed to carry water to and from remote cooling towers. Water lines that connect chilled water and hot water coils located in the ductwork are also required. When working with prints during installation, you should be able to understand what purpose the water flowing in the pipes is serving. This helps determine whether or not the pipes should be insulated and how the piping installation is performed.

For example, when chilled water is supplied to a remote cooling coil, the water should reach the coil at the lowest possible temperature because this water is used to cool an occupied space. If the chilled water lines are not insulated, the water picks up heat as it flows to the coil. The water's cooling effect on the conditioned space is then reduced. This is also true for hot water supplied to a hot water heating coil from a boiler. If these hot water lines are not insulated, the heated water gives up heat to the surrounding air and reaches its destination at a lower temperature than desired. Condenser water lines, however, should remain uninsulated, as shown in **Figure 8-34.** This helps cool the water and increase the efficiency of the air-conditioning systems.

When working with hot water lines, provide the necessary service valves, strainers, and balancing valves required for proper system operation. **Figures 8-35** illustrates detailed piping diagrams on the fittings and valves required for the project. Piping arrangements that carry hot water have a tendency to expand and contract at greater rates, something which often requires expansion loops, **Figure 8-36.**

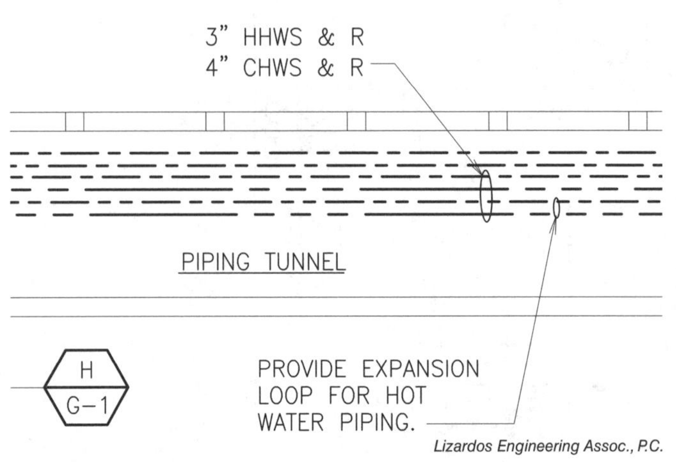

Lizardos Engineering Assoc., P.C.

Figure 8-36. Although not drawn, the note indicates a need to account for pipe expansion.

8.3.4 Pipe Penetrations

Piping that runs through walls, partitions, or floors must be in accordance with local fire codes, especially when they are fire-rated, or fire-resistance rated. Fired-rated walls and construction members are intended to help prevent the spread of fire while providing structural support. Fire-rated walls should not be confused with fire-barrier walls, which only prevent the spread of fire.

Fire-rated walls can be rated for either one side or both sides. If the wall is an exterior wall, the assembly can be rated for one side, as a fire is likely to occur only on the inside of the structure. Interior fire-rated wall assemblies often require rating on both sides. The most common fire rating for a wall is one hour, which indicates the wall can resist exposure to fire for that duration. Depending on the particular project, walls and other construction members can have ratings of two, three, or four hours.

BETWEEN THE LINES

Fire Wall Penetration

Penetrating a fire wall is done only when absolutely necessary. When it is necessary, the penetration point must be sealed according to local codes to ensure the wall can still adequately protect the occupants of the structure from fire.

An example of how a project print addresses a fire wall penetration is shown in **Figure 8-37**. The penetration requires that the pipe and insulation be passed through a sleeve in the wall, **Figure 8-38**. The space between the sleeve and pipe is filled with fire-stopping material and special insulation material. A cover, or escutcheon, is called for on both sides of the penetration.

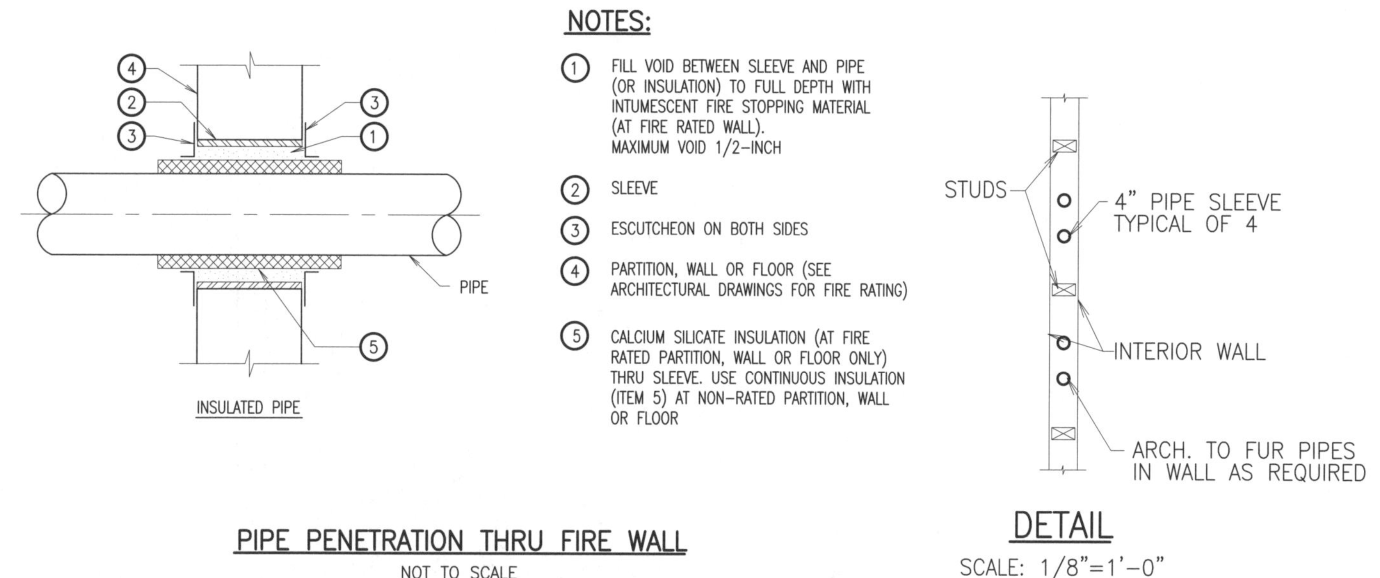

Figure 8-37. Detail of pipe penetration through a fire wall.

Figure 8-38. Detail of wall cutaway showing pipe sleeves.

STRUCTURAL PENETRATIONS		
TAG NO.	STRUCTURAL OPENING SIZE	DESCRIPTION \ REMARKS
FP/1	2"	HEATING HOT WATER SUPPLY
	2"	HEATING HOT WATER RETURN
	2"	REFRIGERANT PIPING SUCTION 1-1/8"
	2"	REFRIGERANT PIPING LIQUID 1/2"
FP/2	10"	CONDENSER WATER SUPPLY
	10"	CONDENSER WATER RETURN
FP/3	2"	HEATING HOT WATER SUPPLY
	2"	HEATING HOT WATER RETURN
	2"	REFRIGERANT PIPING SUCTION 7/8"
	2"	REFRIGERANT PIPING LIQUID 3/8"

Lizardos Engineering Assoc., P.C.

Figure 8-39. Locations of structural penetrations.

Local codes vary greatly, so always adhere to the codes in effect for the project area. Because of the extra work involved in preparing pipes for structural penetrations, a separate list is often created and supplied as part of project plans to ensure all penetrations are accounted for, **Figure 8-39.**

8.3.5 Piping Support

Project plans usually give a contractor some flexibility on how to support piping materials because more than one support method might be required as a piping run makes its way through the structure. The construction materials into which the pipe supports are fastened might be different in different locations within the structure. For example, the ceiling structure might be wood in one part of a building and concrete in another. In addition, the weight of a piping run will vary based on the size of the pipe and the fluid the pipe is intended to carry.

When securing a section of threaded, or hanger, rod to a structural steel beam, a heavy-duty beam clamp arrangement can be used, **Figure 8-40.** The specifics of this mounting method are precise and should be followed unless otherwise instructed by the engineer or architect. If alternative instructions are given, it is strongly recommended these instructions be provided in writing, especially if they contradict the print details.

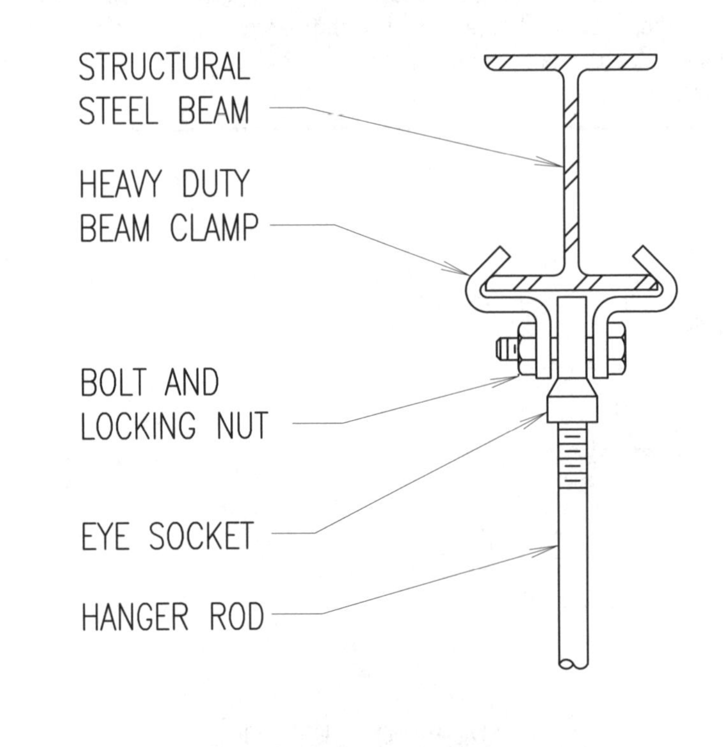

Lizardos Engineering Assoc., P.C.

Figure 8-40. Detail of attachment to steel beams.

A ***clevis-type hanger*** can be used as support for piping material, **Figure 8-41**. This is a popular three-piece, pipe-hanging device. The clevis is the U-shaped strap onto which the pipe rests. The top portion of the hanger connects to the supporting rod, while the third part, the clevis pin, joins the top and bottom parts of the hanger together.

A ***C-clamp*** can also be used for securing a section of threaded (hanger) rod to a structural steel beam. This clamp is fastened to an existing steel beam or member, **Figure 8-42**. The threaded rod is then threaded into the C-clamp. Because vibration from the system can cause the clamp to loosen, steel retainer straps are often specified in drawings.

Be sure to follow all project plan instructions, whether expressed in words or depicted visually in diagrams. These are the standards and guidelines you will be held to. The contractor is ultimately responsible for any damage or loss that arises as a result of improperly followed or executed instructions.

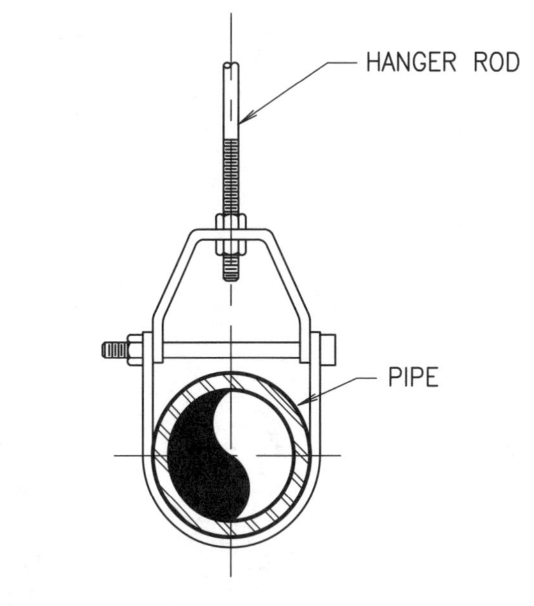

Figure 8-41. Clevis hanger support detail.

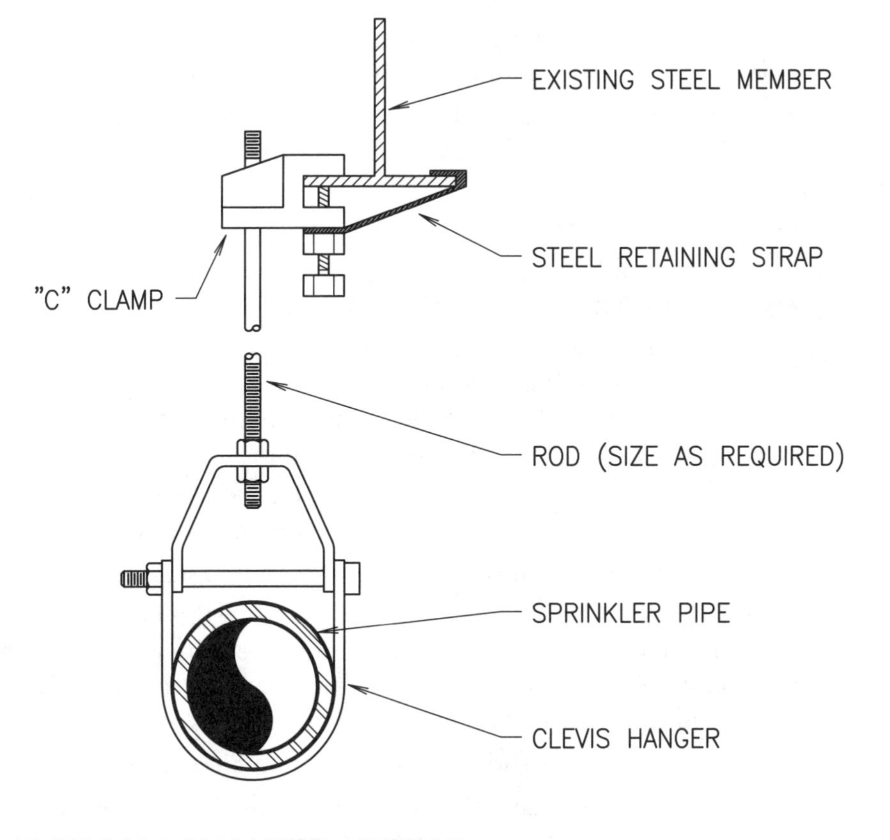

Figure 8-42. Typical hanger detail.

Summary

- Split air-conditioning systems consist of a condensing unit that is located outdoors and an air handler located indoors.
- There are many factors to consider when positioning the indoor and outdoor units in a split air-conditioning system. These factors typically affect system efficiency and serviceability.
- When selecting the condensing unit location, outside air must be able to mix freely and not recirculate back through the condenser coil. Other location considerations include local building codes, serviceability, sound transmission, electrical power placement, and proximity to the condensing unit.
- The condensing unit must always be securely set in place so it remains level and not directly placed on the ground.
- The location of an air handler is determined by the distance to the electric power supply, serviceability, ease of condensate removal, noise level, and return air.
- Air handlers are located as close to the conditioned space as possible.
- Duct systems are constructed of low-resistance sections to maintain proper airflow rates.
- Ductwork sections can be joined in a number of different ways, including slips and drives, bar slips, and flanged connectors.
- Ductwork and piping arrangements must be properly supported. They should be properly sealed using approved products, such as mastic.
- A duct pressurization test should be performed to ensure an installed duct system meets code requirements.
- Duct systems can be fabricated with acoustical lining to minimize noise transmission throughout the structure.
- Duct systems can be wrapped with insulation materials to help prevent condensation from forming on the duct surfaces.
- Condensate can be directed to floor or other drain locations, or can be pumped from the structure. Condensate traps may or may not be required at the outlet of the air handler, depending on the configuration of the air handler, blower, and drain line.
- Auxiliary drain pans and safety float switches reduce the possibility of water damage when air handlers are located above an occupied space.
- Condensate drain lines must be properly sized and pitched to ensure proper operation.
- Refrigerant lines must be installed correctly to ensure proper system operation. Refrigerant traps on the suction line are required when the air handler is located below the condensing unit.
- Refrigerant piping runs should be as short and direct as possible.
- All brazed connections must be completed with a dry-nitrogen purge to prevent oxidation within the refrigerant lines.
- Follow all project guidelines and install service valves as needed for future equipment service.

Chapter Assessment

Name ______________________ Date ____________ Class ______________

Know and Understand

_______ 1. Which of the following will have the least effect on system performance and efficiency?

A. The location of the indoor unit with respect to the power distribution panel

B. The distance between the indoor and outdoor units

C. An improperly sized air distribution system

D. An improperly charged system

_______ 2. Which of the following will result in an increase in system efficiency?

A. An overcharged system

B. A long return duct

C. A short refrigerant line set

D. An oversized system

_______ 3. If it is determined, as indicated by the project drawings, that the refrigerant lines are longer than the manufacturer allows, the HVACR contractor should _____.

A. refuse to install the system

B. consult with the project manager to find a solution

C. install the system as per the project plans

D. relocate the unit so the line set will be shorter

_______ 4. If the project plans do not call for a condensate line trap at the outlet of the air handler's drain line port, the HVACR contractor should _____.

A. not install a condensate line trap

B. install a condensate line trap anyway

C. ask the architect if a condensate line trap is needed

D. check the unit's installation literature to determine if a condensate line trap is required

_______ 5. Which of the following is a major concern if a natural return is used on a system that has the air handler installed in a utility closet?

A. Excessive pressure drop in the return duct

B. Fumes from the items stored in the utility closet

C. Decreased air velocity through the system

D. Increased potential for duct damage

_______ 6. What should the HVACR contractor do if the project manager verbally instructs the contractor to install a system component in a manner that differs from the project plan's instructions?

A. Refuse to follow the project manager's verbal instructions.

B. Perform the task as instructed, but first get the change order in writing, signed by both the architects and the project manager.

C. Perform the task as instructed, but first get written approval from the project owner.

D. Perform the task based solely on the verbal instructions.

_______ 7. Which of the following can have a negative effect on future system maintenance?

A. The distance between the attic-mounted air handler and the access point to the attic
B. The length of the refrigerant line set
C. The length of the return air duct
D. The installation of canvas connectors on the duct system

_______ 8. Which of the following is the point of contact if a property line concern exists?

A. Property line office
B. Boundary setback office
C. Local planning office
D. Neighboring property office

_______ 9. Which of the following steps will reduce air handler noise transmission through the return air grille?

A. Installing an air filter with a higher MERV rating
B. Incorporating a large loop in the return air duct
C. Shortening the return air duct
D. Replacing any canvas connectors from the return duct/air handler connection with rigid duct material

_______ 10. The non-hardening material used to seal joints and seams in a duct system is _____.

A. mylar
B. maylar
C. mastic
D. mystic

_______ 11. A project plan shows a 20′ horizontal run of condensate line connected to the air handler's condensate line port. No pitch is indicated on the plan. The HVACR contractor should _____.

A. install the line with no pitch as indicated on the plan
B. install the line with a 5″ pitch from end-to-end, sloping away from the unit
C. install the line with a 5″ pitch from end-to-end, sloping toward the unit
D. refuse to install the condensate line as indicated on the plan

_______ 12. Even if not indicated in the project plans, the HVACR contractor should plan to insulate the _____.

A. condensate line for at least the first 10′ from the air handler's condensate drain port
B. entire length of the condensate line
C. liquid refrigerant line for at least the first 10′ from the air handler's liquid line port
D. entire length of the liquid refrigerant line

_______ 13. Which of the following is *not* a best practice for installing refrigerant piping?

A. Refrain from using an excessive number of fittings.
B. Use a suction line that is not insulated.
C. Refrain from using short radius elbows.
D. Pitch the suction line back to the compressor.

_______ 14. Even if not specified in the project plans, all auxiliary drain pans should be equipped with a _____.

A. self-leveling feature
B. drain connection
C. normally open safety float switch
D. normally closed safety float switch

Name ______________________________ Date ____________ Class ____________________

_______ 15. What material should be used for a condensate drain line?

A. Schedule 40 PVC should always be used.

B. DWV copper should always be used.

C. Galvanized steel pipe should always be used.

D. Whatever material is called for in the project plans should be used, as long as the material is code-compliant.

_______ 16. A clevis hanger is best described as a _____.

A. two-piece hanger used to support ductwork

B. two-piece hanger used to support piping

C. three-piece hanger used to support ductwork

D. three-piece hanger used to support piping

_______ 17. Which of the following methods is acceptable to prevent C-clamp slippage off a steel beam?

A. Overtighten the clamp to the steel beam.

B. Use a threaded rod restraining strap.

C. Use a steel retaining strap.

D. C-clamps should not be used on steel beams.

_______ 18. When securing a threaded rod to a wood joist, _____.

A. the threaded hanger rod should be threaded into the wood joist

B. a steel U-support should first be fastened to the wood joist with lag bolts

C. the threaded rod should penetrate completely through the wood joist

D. a clevis pin should be used to secure the threaded rod to the wood joist

_______ 19. A condensate trap will most likely be required on an upflow vertical air handler if the _____.

A. blower is located below the evaporator coil

B. evaporator coil is located above the blower

C. blower is located above the evaporator coil

D. Both A and B.

_______ 20. Which of the following is most likely to be required for a hot water piping arrangement?

A. Flexible connector

B. Expansion loop

C. Condensate trap

D. C-clamp

_______ 21. Which of the following pipes is least likely to need insulation?

A. Condenser water piping

B. Chilled water supply piping

C. Hot water supply piping

D. Compressor suction line piping

Apply and Analyze

1. Referring to the following figure, why is the suction line the only line running between the condensing unit and air handler that is trapped?

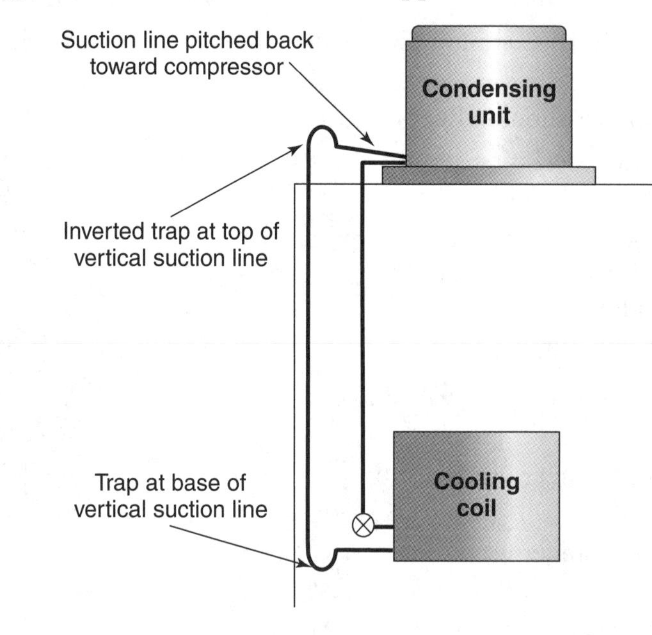

Lizardos Engineering Assoc., P.C.

____

____

____

____

Name ______________________ Date ______________ Class ______________________

2. Referring to the following figure, what additional piping component(s) can be added to the piping circuit that would make it easier for the domestic water heater to be removed or replaced?

BA
UN
RED
PG
TH
BA
DWH
D
CV
D
UN RED
UN
RED
AI
CP/DWHD
CONTROL VALVE
POSITION
PG
BA
UN RED
PV
BY PLUMBING
CONTRACTOR
P
DW1
DI
CP/BR
PUMP P/DW1
START/STOP
MS
DO
CP/BR
PUMP P/DW1
START/STOP
6"HHWS
2"HHWS
BA PV CH PG FPC
RED
RED FPC PG
DI
STR
2"HHWS
BA

Lizardos Engineering Assoc., P.C.

____

____

____

____

3. Referring to the following figure, does the ductwork shown need to be insulated? Provide examples of both a heating system and a cooling system to support your response.

____

____

____

____

4. The formula for airflow through a duct is cfm = velocity × area. Cfm is the amount of airflow expressed in cubic feet per minute, the velocity of the air is expressed in feet per minute, and the area, which is the cross-sectional area of the duct, is expressed in square feet. Consider an unlined section of duct that has cross-sectional dimensions of 12″ × 20″. If the velocity of the air is 700′ per minute, how many cfm of air are moving through the duct?

5. Using the information from question 4, recalculate the cfm through the duct if the duct section is lined with 1′ acoustical lining. What percentage of airflow has been gained/lost as a result of adding the lining to the duct section?

Name ______________________________ Date ____________ Class ____________________

Critical Thinking

1. Referring to the following figure, why is it important to have access doors installed on both sides of an in-duct heat transfer surface, or coil, such as the heating coil shown in the figure?

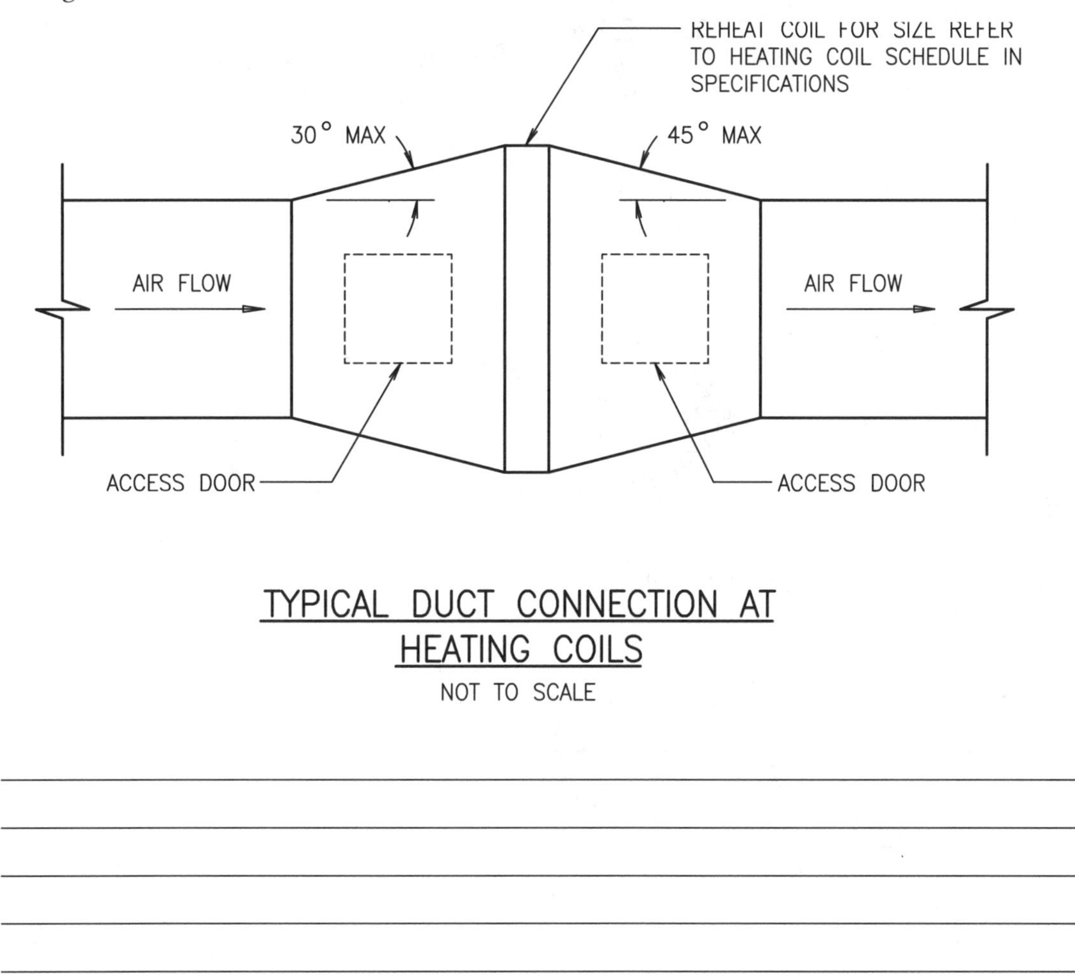

____

____

____

____

____

____

2. Referring to the following figure, why is it that piping runs simply rest in pipe hangers such as the clevis hanger depicted, as opposed to being physically connected to the hangers?

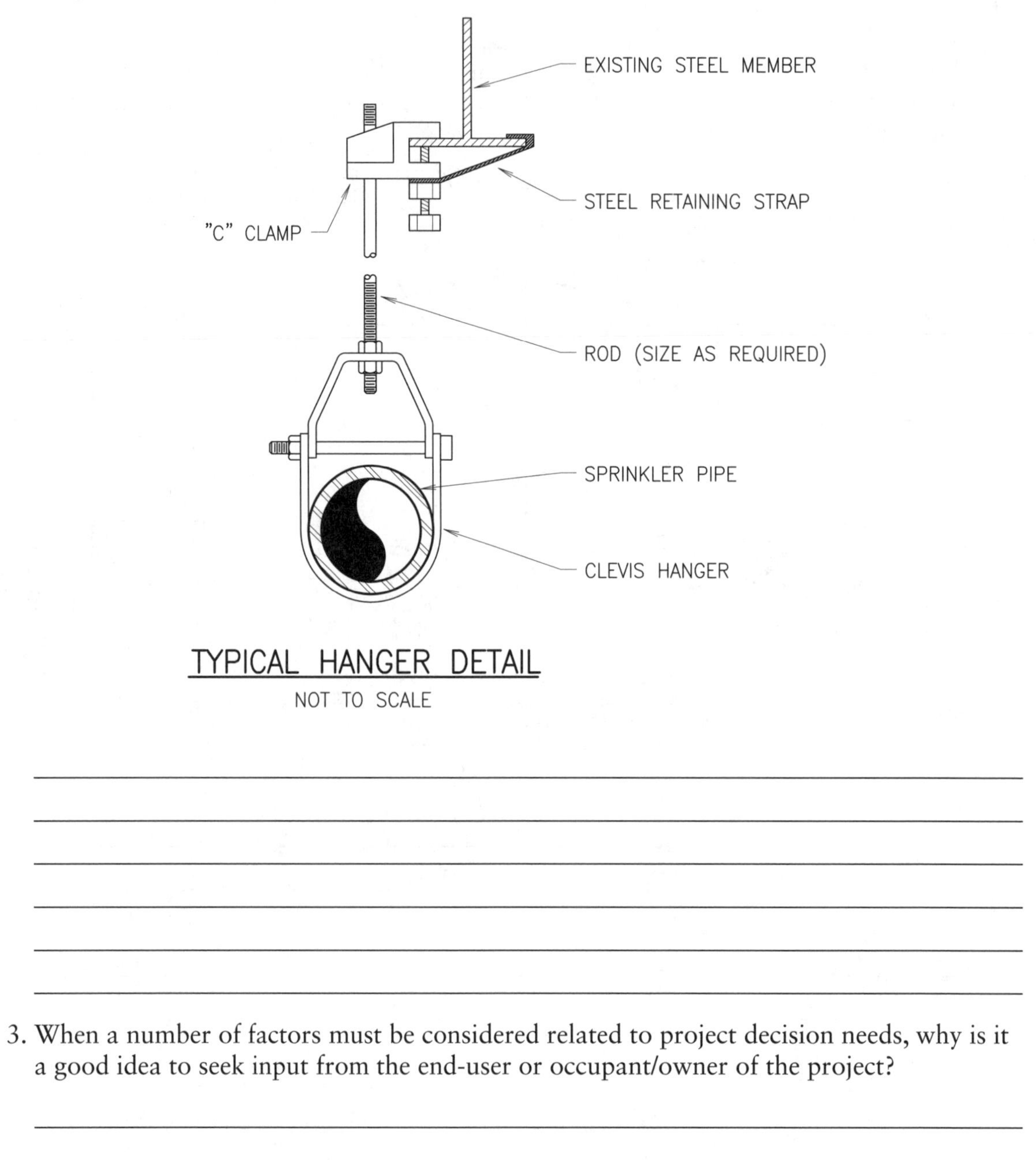

____

____

____

____

____

____

3. When a number of factors must be considered related to project decision needs, why is it a good idea to seek input from the end-user or occupant/owner of the project?

____

____

____

____

Large Prints Activity

Name ____________________ Date __________ Class ____________

Practice Using Large Prints

Refer to Print-25 in the Large Prints supplement to answer the following question. Note: The large prints are created at 50% of their original size. When extracting information from the print using an architect's scale, use a scale that is one-half of the one indicated on the drawing.

1. You plan to install a condensate line on AHU-3. The condensate line is to be connected to the right side of the unit and will run due east until the line penetrates the building. If the required slope of the condensate line is 1/4″ per foot of run, how much lower will the termination point of the condensate line be than its origination point?

____

____

____

____

9 Electrical Wiring Diagram Basics

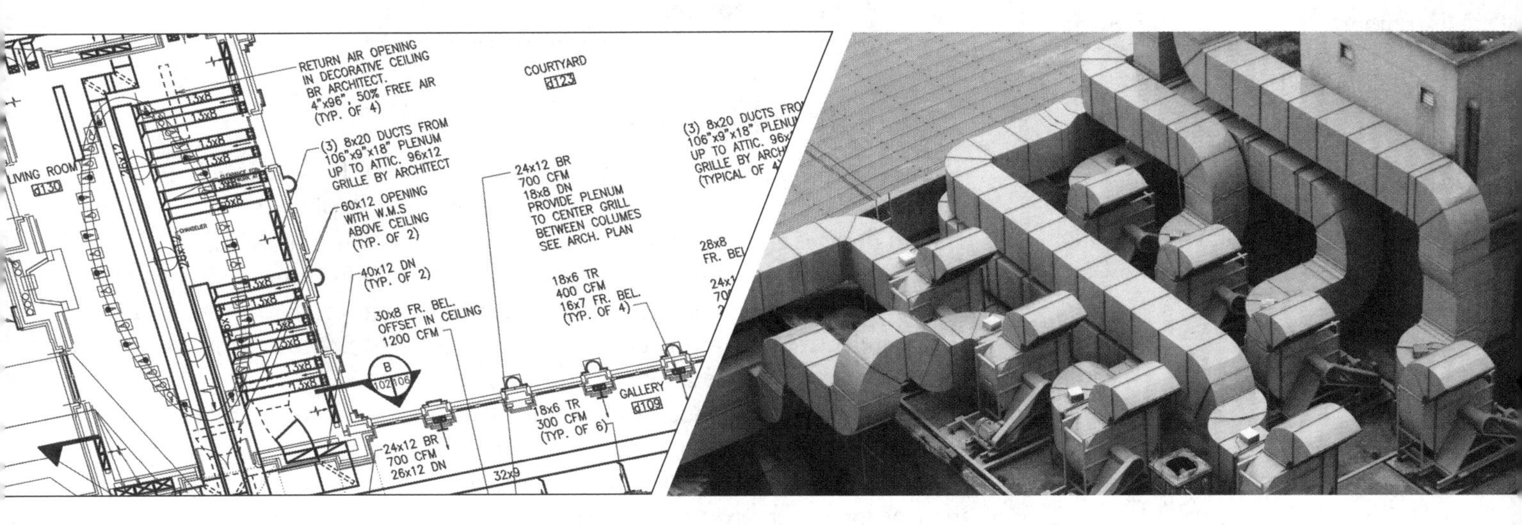

Chapter Outline

Learning Objectives

After studying this chapter, you will be able to:

- Describe the types of electrical devices found in a wiring diagram.
- Explain the purpose of electrical wiring diagrams.
- Identify various electrical symbols used in wiring diagrams.
- Discuss the different types of electrical wiring diagrams and their advantages and disadvantages.
- Convert a schematic wiring diagram to a ladder diagram.
- Determine the sequence of operations of an appliance by interpreting the appliance's wiring diagram.
- Interpret the wiring diagram of a window-type or through-the-wall air-conditioning unit.
- Interpret the wiring diagram of a split air-conditioning system with a low-voltage control circuit.
- Interpret the wiring diagram of a residential split heating and cooling system.

Technical Terms

contact
electrical wiring diagram
ladder diagram
normally closed (NC)
normally open (NO)
pictorial wiring diagram
pole
relay
schematic diagram
sequence of operations
solenoid
switch
thermostat
throw

Introduction

Project plans provide a road map, or set of instructions, that contractors follow to execute the job as envisioned by an architect or project owner. This includes being able to understand and use electrical wiring diagrams to complete a project. Electrical diagrams are different from mechanical drawings, which show the location of mechanical equipment in an HVACR or plumbing system. Instead, electrical diagrams indicate the path of the flow of electrical current through wiring and other system components, such as sensors and control devices.

Properly installing and connecting electrical components ensures that the system functions as intended. An HVACR contractor must understand system wiring and know how to read various electrical wiring diagrams. This chapter introduces different types of wiring diagrams and how to understand them. You will then learn how to apply this knowledge to effectively and efficiently troubleshoot various pieces of HVACR equipment.

9.1 Electrical Devices

Many types of electrical devices used in an HVACR system are noted in electrical wiring diagrams. Electrical circuits are employed to control the system devices that regulate or maintain the operation of an HVACR system. Some circuits respond to sensed external conditions, such as room temperature, to ensure the system operates as intended. Other circuits control electrical flow in response to internal system conditions, such as the unit's operating temperatures and pressures.

Some electrical devices are operational devices that open and close their contacts during the normal operation of the system. Other devices are classified as safety devices that remain closed and allow current to flow through them as long as the system is operating within its design parameters. If unsafe conditions are present, safety devices respond by de-energizing the entire system, or parts thereof. You must understand the function of electrical components to determine how they operate within an HVACR system.

A ***switch*** is a device that opens or closes one or more connections in an electrical circuit. Simple switches have only two positions: closed, where an electrical connection is completed; and open, where an electrical connection is interrupted. Switches are used for many HVACR applications and can respond to a number of different conditions, including pressure, temperature, humidity, water level, time, light, motion, and fluid flow.

Switches can be configured in many different ways, but are commonly identified by the number of poles and throws with which they have been fabricated. The term ***pole*** indicates how many power lines can be open, closed, or redirected at the same time. For example, a single-pole switch can open or close one single power line, while a two-pole switch can open or close two lines simultaneously. The term ***throw*** indicates how many possible circuits or current paths the switch controls. A single-throw switch controls the current flow through a single path. A double-throw switch directs current flow through either one of two possible current paths.

A switch opens or closes a circuit by connecting or disconnecting contacts. A ***contact*** is the part that is touched or manipulated to complete an electrical circuit. Sets of contacts are identified as normally open (NO) or normally closed (NC). ***Normally open (NO)*** indicates contacts that are open in their de-energized state. ***Normally closed (NC)*** indicates contacts that are closed in their de-energized state.

A ***relay*** is an electrical switch that is controlled by an external electrical signal. Its electrical contacts are operated by electromagnets. When an external electrical signal is applied, the relay's coil becomes energized and a magnetic field is generated. This magnetic field moves an armature to change the contact's positions. Relay contacts are either NO or NC.

A ***solenoid*** is an electromagnetic device that converts electric current to mechanical movement by producing a magnetic field. This device is commonly installed into a valve body and used to control the direction of flow of refrigerant, gas, or water by opening, closing, or repositioning the valve.

9.2 Electrical Symbols and Wiring Diagrams

Many symbols used in wiring diagrams represent specific electrical components and devices in a circuit. These symbols may vary based on who creates the prints, but they typically follow a standard convention. A separate reference document is sometimes used in completed diagrams that identifies the symbols employed in the diagram. Common electrical symbols are shown in **Figure 9-1.** A list of abbreviations is often provided with many electrical wiring diagrams and is shown in **Figure 9-2.**

An ***electrical wiring diagram*** is a representation of an electrical circuit and shows how electrical loads, switches, and other devices are connected in an appliance or system. These diagrams provide information that can be used by both installers and service personnel to install and troubleshoot electrical systems. There are three common types of electrical wiring diagrams: pictorial diagrams, schematic diagrams, and ladder diagrams. Each is used for a specific purpose. Depending on the type of wiring diagram being used, the information it contains may vary.

Electrical Symbols		
Device	**Type**	**Symbol**
Contacts	Normally open	
	Normally closed	
	Timed-controlled	
Relays	Relay coil	
	Motor starter	MS
Sensors	Light-dependent resistor	
	Thermistor	
	Electric sensor	

Goodheart-Willcox Publisher

Figure 9-1. Commonly used electrical symbols in wiring.

(*continued on next page*)

Switches	Single-pole single-throw (SPST)	
	Single-pole double-throw (SPDT)	
	Double-pole single-throw (DPST)	
	Double-pole double-throw (DPDT)	
	Limit switch (normally open)	
	Limit switch (normally closed)	
	Float switch (open on level rise)	
	Float switch (closed on level rise)	
	Flow switch	
	Manual switch (normally open)	
	Manual switch (normally closed)	
	Sail switch	
	High-pressure switch	
	Low-pressure switch	
	Heating thermostat	
	Cooling thermostat	
Transformers	Control transformer	
Overloads	Overload	

Figure 9-1. ***(continued)***

Goodheart-Willcox Publisher

Abbreviations

CAP	Capacitor	FLO	Flow switch	NO	Normally open
CB	Circuit breaker	FLS	Fan limit switch	ODT	Outdoor thermostat
CC	Compressor contactor	FLT	Float switch	OL	Overload
CC1, CC2	Contactor contacts 1, 2	GND	Ground	RC	Run capacitor
CCH	Crankcase heater	GV	Gas valve	ROS	Roll-out switch
CFM	Condenser fan motor	HPS	High pressure switch	RV	Reversing valve (coil)
COMP	Compressor	HR	Heating relay	SC	Start capacitor
CR	Cooling relay	HSI	Hot surface igniter	SAIL	Sail switch
CT	Control transformer	IFM	Indoor fan motor	SEQ	Sequencer
DFT	Defrost timer	IFR	Indoor fan relay	SS	Spill switch
DIM	Draft inducer motor	L1, L2	Line 1, line 2	T-STAT	Thermostat
EDH	Electric duct heater	LIM	Limit switch	T1, T2	Terminal 1, terminal 2
F	Fuse	LPS	Low pressure switch	TM	Timer motor
FLK	Fusible link	NC	Normally closed	X-FRM	Transformer

Goodheart-Willcox Publisher

Figure 9-2. Common abbreviations found on electrical wiring diagrams.

9.2.1 Pictorial Wiring Diagrams

A ***pictorial wiring diagram*** is an illustration of a circuit that shows the location of electrical components and all electrical connections. A technician uses this diagram to see each wire connection between devices. Pictorial diagrams can contain renderings that resemble the components in the actual circuit. See **Figure 9-3**. These diagrams often include details, such as the device ratings, wire gage, wire colors, number of phases, voltage rating, and amperage rating.

Pictorial diagrams look similar to schematic diagrams, but show more exact physical locations of the components in the device. See **Figure 9-4**.

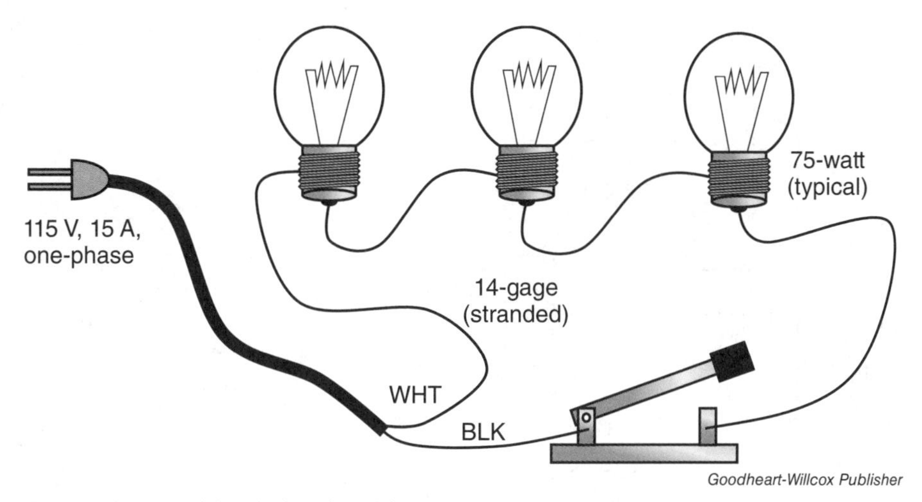

Figure 9-3. A traditional pictorial wiring diagram.

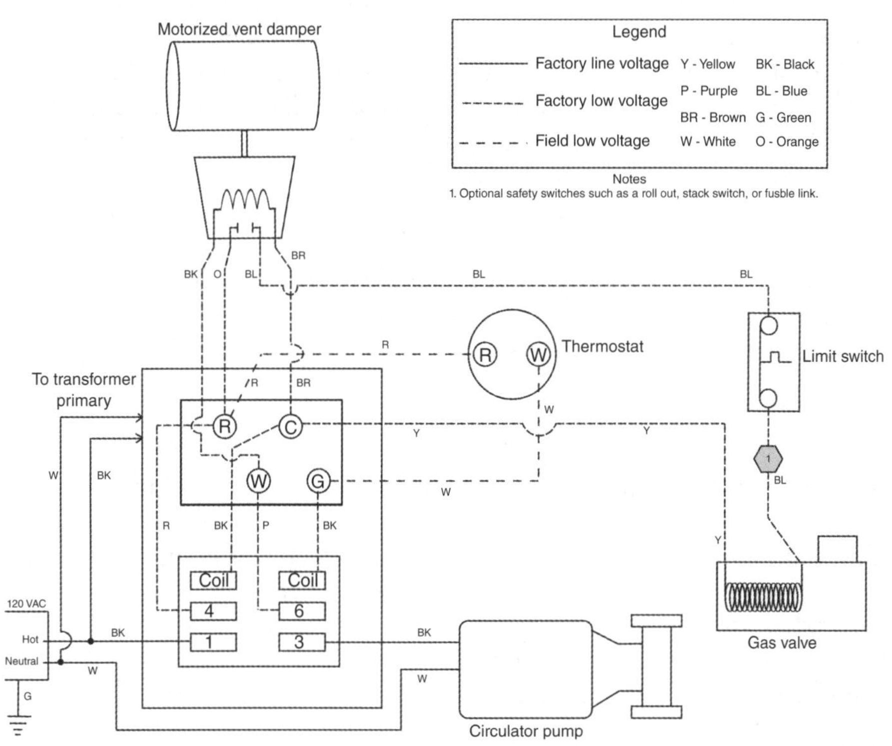

Figure 9-4. Another version of a pictorial wiring diagram.

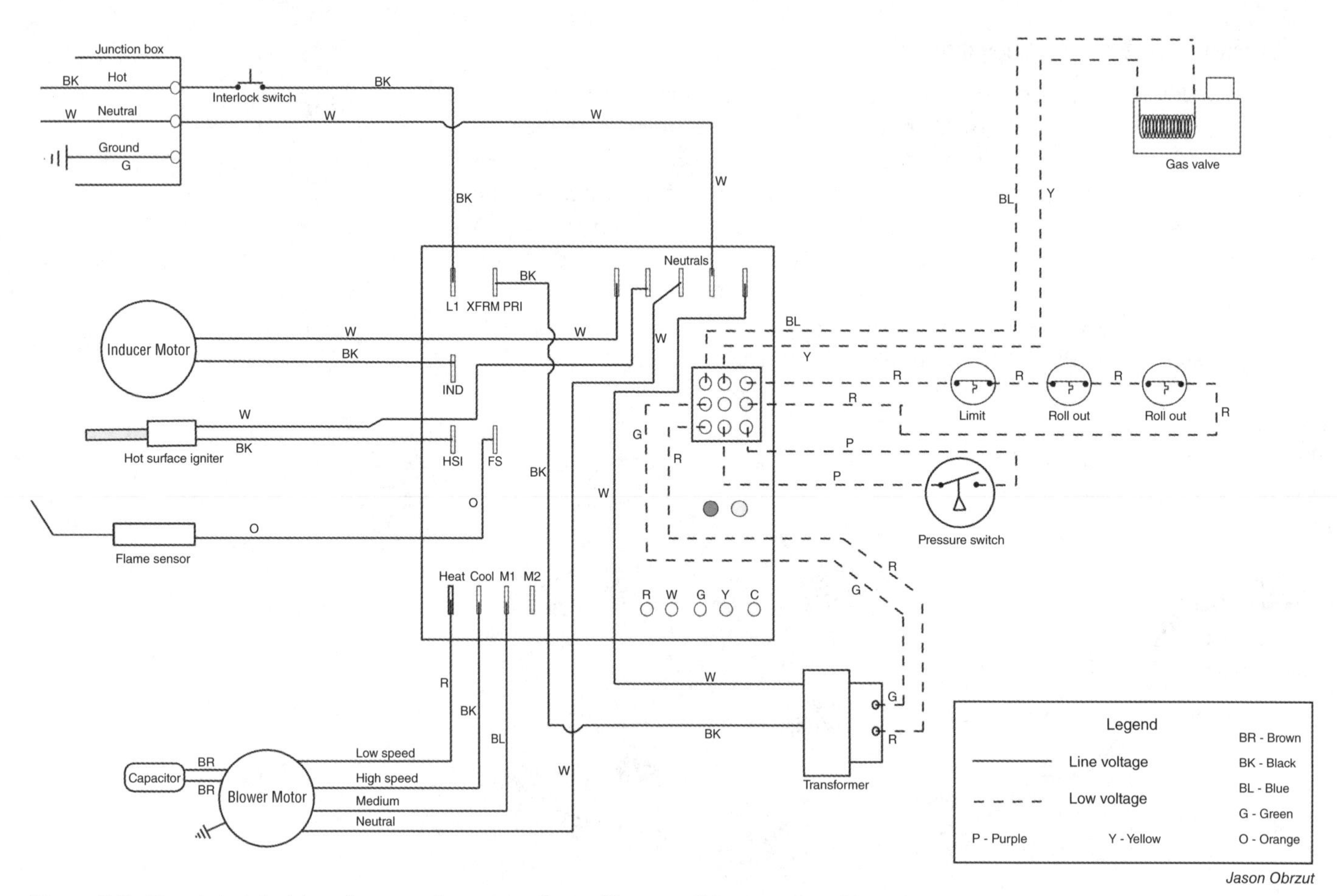

Figure 9-5. The pictorial wiring diagram of a system that utilizes a solid-state control board.

This diagram provides a combination of easily recognizable components and some features commonly found on schematic diagrams. The pictorial diagram in **Figure 9-5** includes a solid-state control board. This type of pictorial diagram is useful when a service technician needs to know the exact location of a particular terminal on the board or where one of the system components, such as the inducer motor, is connected.

Another variation of the pictorial diagram is shown in **Figure 9-6**. This is a diagram of a solid-state HVACR system control board that includes multiple relays, chips, and modules. Each item in the diagram is coded and can be easily located because the positions of the components on the diagram are the same as on the actual board.

9.2.2 Schematic Wiring Diagrams

A ***schematic diagram*** is a detailed illustration of the wiring in a piece of equipment or device. Each wire in the circuit appears on the diagram, along with its color and associated electrical connections. The circuit wiring of the system is represented by lines, and its components and associated terminals are identified by symbols. See **Figure 9-7**. Other information may be included in the diagram, such as the wire gage, circuit voltage, and factory- and field-installed wiring.

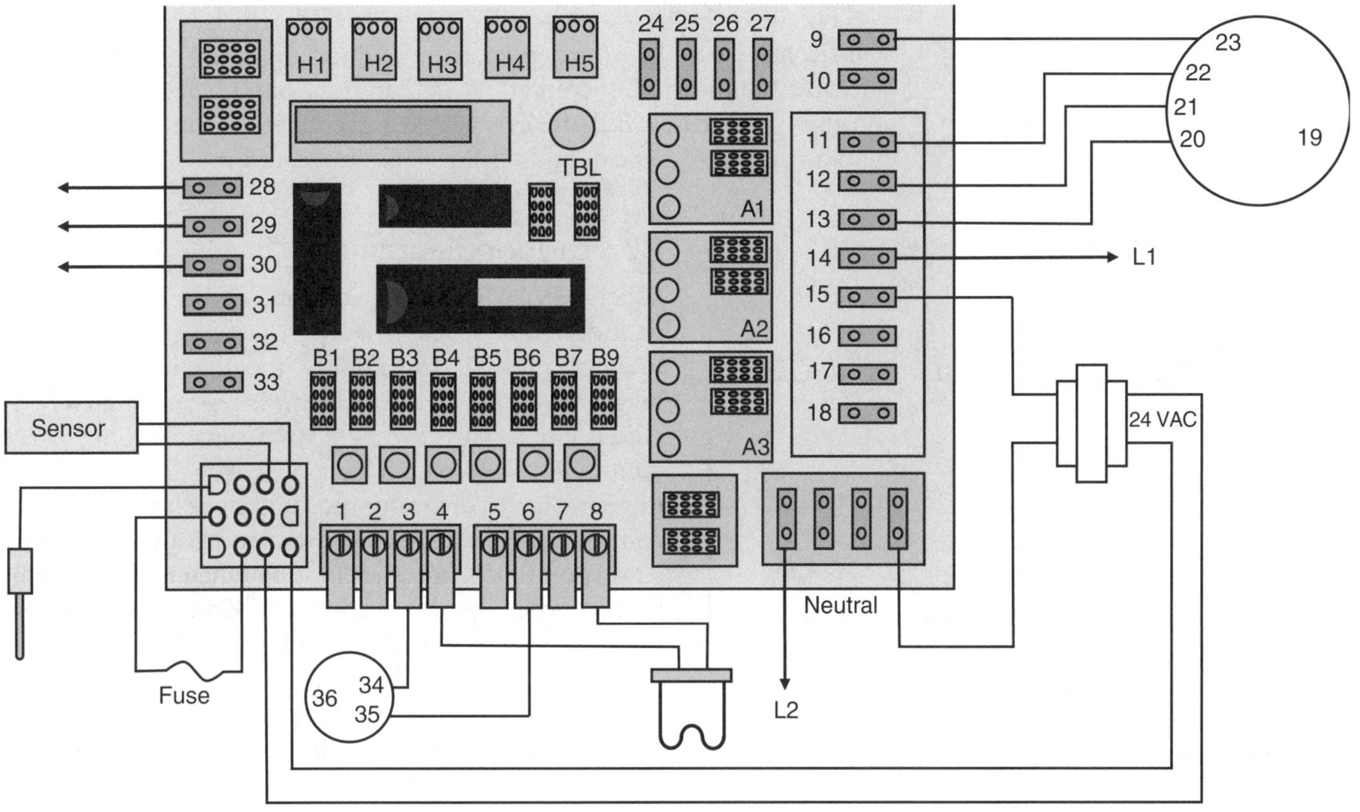

Goodheart-Willcox Publisher

Figure 9-6. The pictorial wiring diagram of a control board with multiple relays and control modules.

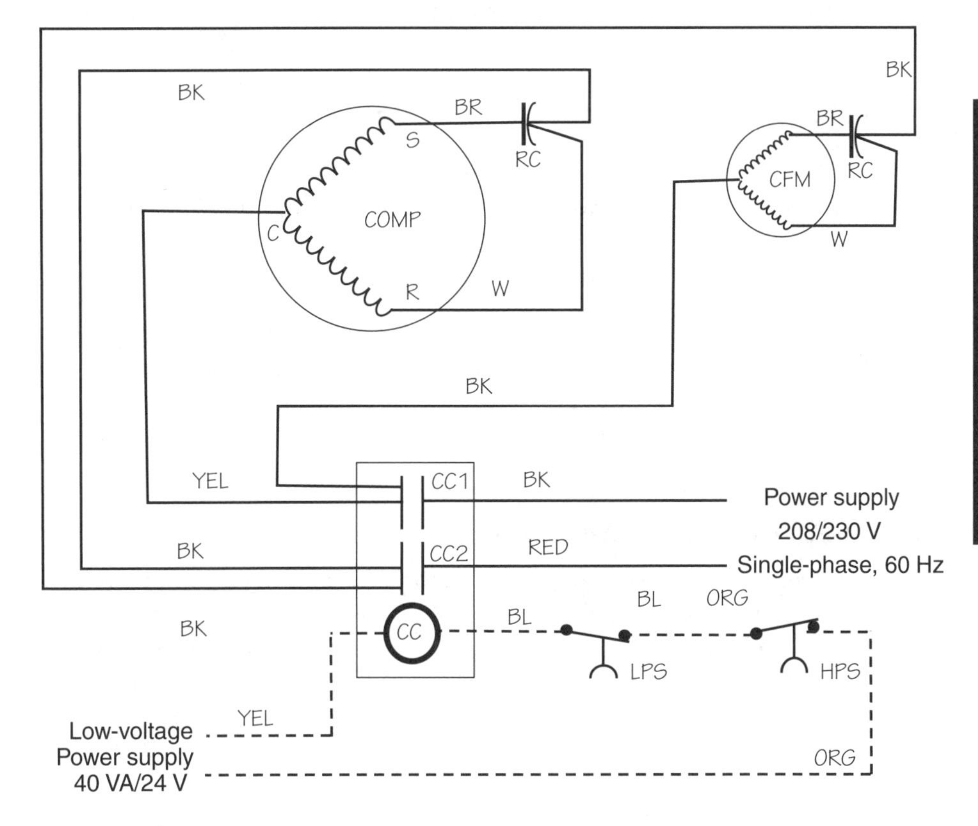

Legend and Abbreviations

C	Compressor common
CC	Contactor coil
CC1, 2	Contactor contacts
CFM	Condenser fan motor
COMP	Compressor
HPS	High pressure switch
LPS	Low pressure switch
R	Compressor run
RC	Run capacitor
S	Compressor start
————	Line voltage
- - - - - -	Low voltage

Goodheart-Willcox Publisher

Figure 9-7. A typical schematic wiring diagram.

The wires are also color-coded in the circuit, which is helpful to the field technician. It is much easier for a technician to try to locate a red wire in a bundle of multicolored wires than trace out a black wire in a sea of other black wires. Because every wire is accounted for in the schematic diagram, a technician can find wiring errors more easily.

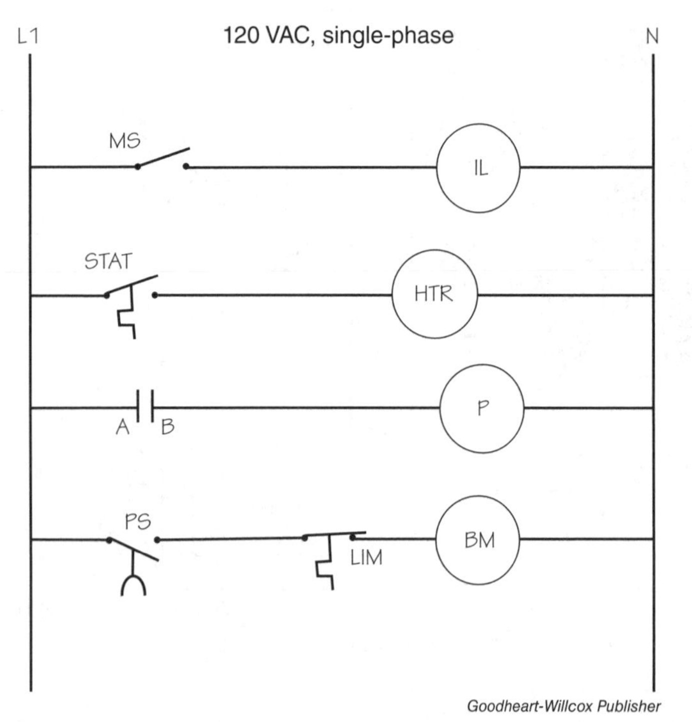

Figure 9-8. An example of a ladder diagram.

9.2.3 Ladder Diagrams

When HVACR systems become more complex and include many wires, it is often difficult to read schematic diagrams. To simplify the diagram, a ladder diagram is used. ***Ladder diagrams*** represent each individual circuit in the system on a separate line. These diagrams allow technicians to easily locate a component on the diagram and determine all the system control or switching devices that control its operation. This is useful for troubleshooting when a system component is not operating correctly. Most manufacturers provide both a schematic and a ladder diagram for equipment.

An example of a ladder diagram is shown in **Figure 9-8**. The power supply, in this case L1 and N, is represented as the vertical rails of the ladder. The control devices are located on the horizontal rungs on the ladder, as are the power-consuming devices. Each power-consuming device is typically located on its own line or rung. As more circuits are added to the system, more rungs are added to the ladder diagram.

In the ladder diagram shown, several observations can be made:

- The indicator light, IL, is controlled by only the manual switch, MS.
- The heater, HTR, is controlled by only the heating thermostat, STAT.
- The pump, P, is controlled by a set of normally open contacts, A-B.
- The blower motor, BM, is controlled by both the pressure switch, PS, and the limit switch, LIM. These switches are wired in series with each other. For the blower motor to be energized, both the pressure switch and the limit switch must be in the closed position.

9.2.4 Sequence of Operations

Ladder diagrams are used to troubleshoot a system device because they allow the technician to easily determine the system's sequence of operations. The ***sequence of operations*** is the order of events that occur to activate a device. Understanding the sequence of operations for a system helps a technician effectively evaluate and troubleshoot a system.

Procedure Sequence of Operations

Consider the ladder diagram shown in **Figure 9-9**. This ladder diagram is a representation of the schematic diagram shown in **Figure 9-4**. This system has a relay with its coil marked COIL and two sets of normally open contacts, 1-3 and 4-6. These two sets of contacts close when the relay coil is energized. This system also has a set of normally open vent damper contacts, VD, that closes once the vent damper, controlled by the vent damper motor, VDM, has fully opened.

Determining the sequence of operations for this system can be broken down into a series of individual, logically organized steps:

1. Identify the components with no switches in series with them that are energized all the time. For instance, the primary size of the transformer is powered all the time, so 24 V is always supplied to the control circuit. Also, line voltage power is always supplied to the top portion of the circuit from L1 and L2, **Figure 9-10A**.
2. Identify the action that starts system operation. The action that typically starts the system is often an external event, such as the flipping of a switch or the closing of the thermostat. In this circuit, the action that starts the system is the thermostat, identified as STAT.

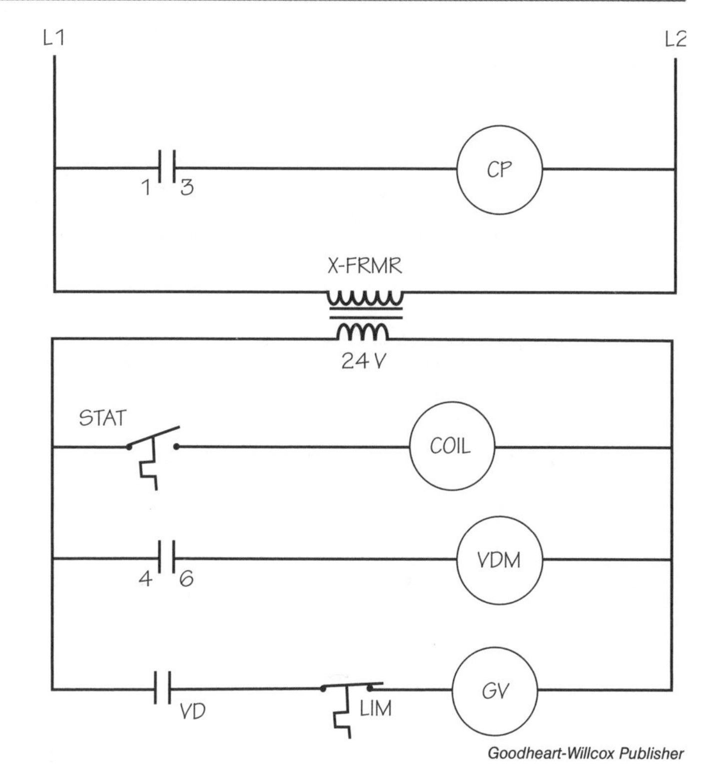

Goodheart-Willcox Publisher

Figure 9-9. A ladder diagram that indicates 24 V is always being supplied to the control circuit.

BETWEEN THE LINES

Determining the Start Action

The action that starts a system can be easily found by the process of elimination. For example, the circulator pump, CP, circuit at the top of the diagram cannot initiate system operation because that circuit is waiting for the closing of contacts 1-3 for the circulator pump to turn on. Similarly, the VDM and GV circuits do not become energized until other system events occur.

3. Once the STAT closes, the relay coil, COIL, becomes energized, **Figure 9-10B**.
4. The energizing of the COIL causes the 1-3 and 4-6 sets of contacts to close.
5. The closed 1-3 contacts energize the circulator pump, CP, and the closed 4-6 contacts energize the vent damper motor, VDM, **Figure 9-10C**.
6. VDM operation opens the vent damper and closes the vent damper contacts, VD.
7. As long as the limit switch, LIM, is closed, the gas valve, GV, becomes energized and opens, starting the heating cycle, **Figure 9-10D**.
8. Under normal system operation, the system remains in heating mode until the space thermostat no longer calls for heat. The heat cycle can also terminate in the event there is an unsafe temperature condition in the appliance that causes the limit switch to open, de-energizing the heat source.

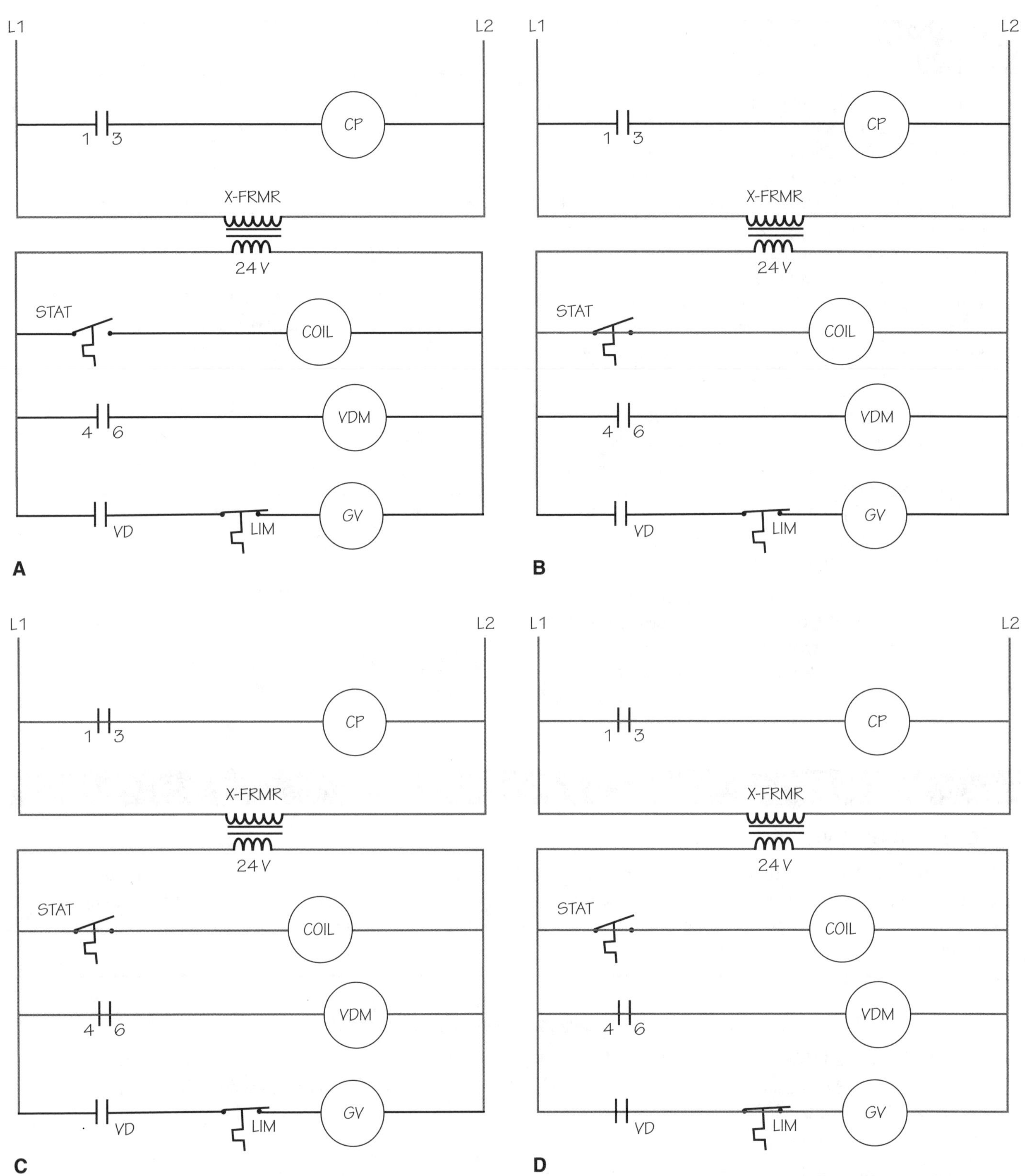

Figure 9-10. A—The transformer supplies power to the low-voltage control circuit. B—The relay coil is energized when the space thermostat closes its contacts. C—When the relay coil is energized, both sets of normally open contacts on the relay close. D—Once vent damper motor operation has opened, the VD contacts close and the gas valve is energized.

9.2.5 Converting Schematic Diagrams to Ladder Diagrams

Not all equipment manufacturers provide ladder diagrams with their equipment. However, a schematic diagram can be easily converted to a ladder diagram for troubleshooting purposes.

Procedure Converting Schematic Diagrams to Ladder Diagrams

One of the easiest ways to accomplish this is to use two different-color highlighters to identify the power supply to the circuit. Consider the schematic diagram in **Figure 9-11**.

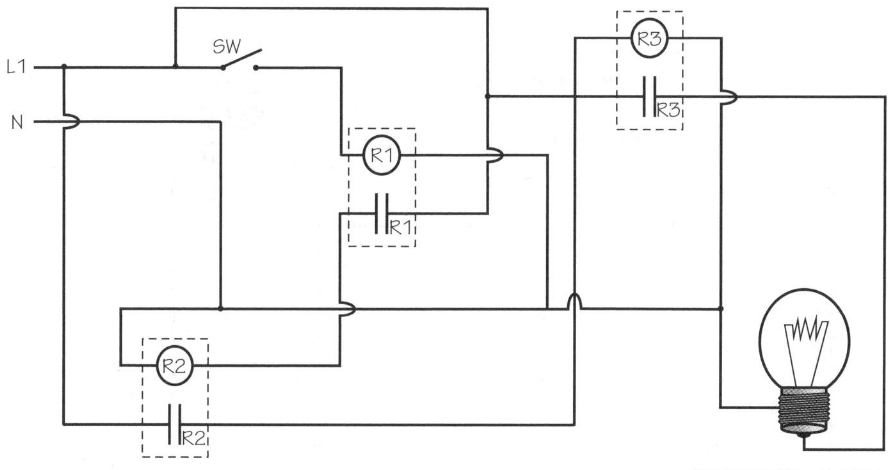

Goodheart-Willcox Publisher

Figure 9-11. An example of a schematic wiring diagram to be converted into a ladder diagram.

1. Color all the lines connected to L1 one color, **Figure 9-12A**. Only wires that are electrically the same L1 should be colored. Once an electrical component, such as a switch or load, has been reached, the line stops.
2. Color all the lines connected to N a different color, **Figure 9-12B**. The red and green color-coded lines now represent the rails of the ladder, with the red line representing Line 1 on the left, and green representing N on the right, **Figure 9-12C**.
3. Add the first remaining, uncolored line, SW and R1 coil as the first rung of the ladder diagram. Each line segment that begins at one color and ends at the other, is one rung on the ladder, **Figure 9-12D**.
4. Add the line with R1 contacts and an R2 coil to the second rung of the ladder diagram.
5. Add the line with R2 contacts and an R3 coil to the third rung of the ladder diagram.
6. Add the line with R3 contacts and the light bulb to the final rung of the ladder diagram.

By adding each of these rungs to the diagram, the completed ladder diagram is created. From this completed diagram, it can be seen that the switch, when closed, energizes the coil of relay 1. This causes the contacts of relay 1 to close, energizing the coil of relay 2. This causes the contacts of relay 2 to close, energizing the coil of relay 3. This causes the contacts of relay 3 to close, energizing the light bulb. Although the process of converting a schematic diagram to a ladder diagram takes practice, the result is worth the effort. Note that the order of rungs on the ladder diagram is not important.

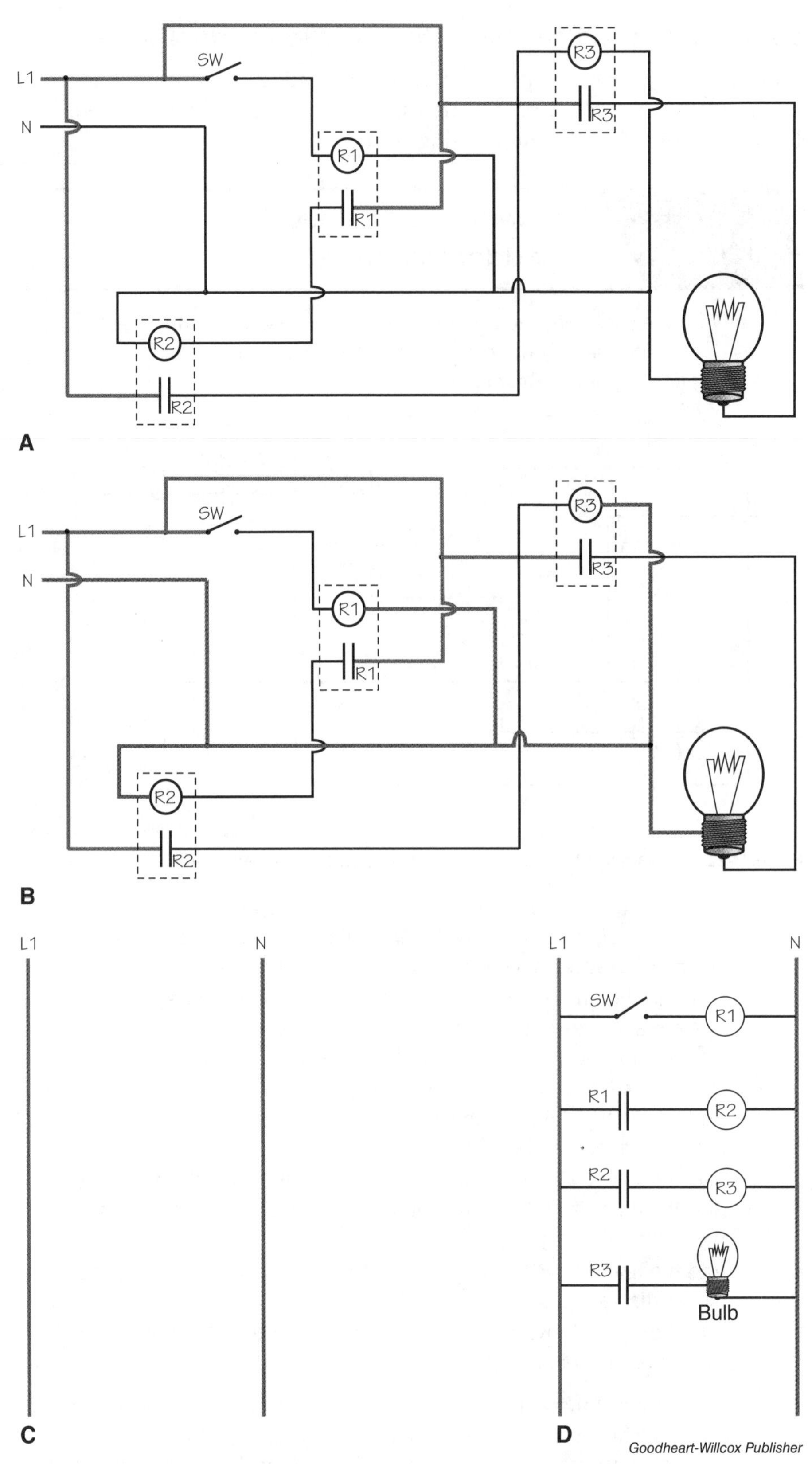

Goodheart-Willcox Publisher

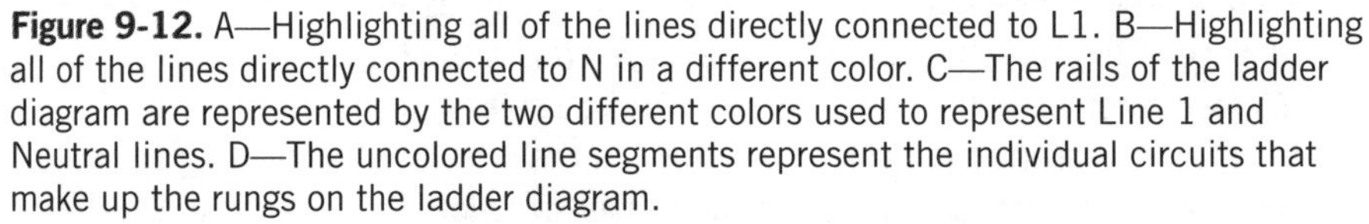

Figure 9-12. A—Highlighting all of the lines directly connected to L1. B—Highlighting all of the lines directly connected to N in a different color. C—The rails of the ladder diagram are represented by the two different colors used to represent Line 1 and Neutral lines. D—The uncolored line segments represent the individual circuits that make up the rungs on the ladder diagram.

9.3 Basic Window Air-Conditioning Unit Wiring Diagrams

The window-type unit is a plug-in appliance mounted in the window. This unit is similar in electrical wiring configuration to through-the-wall units, but does not have solid sides and is not installed through the wall. Window units are considered more of a temporary installation, while through-the-wall unit installations require structural modifications. The following factors make electrical diagrams for window and through-the-wall units easier to understand:

- There is typically only one fan motor on window and through-the-wall units. The fan motor has two shafts—one to turn the condenser fan blade and one to turn the evaporator blower.
- The thermostat and other associated controls are all line voltage devices. There is no control transformer.
- No field-installed wiring is required.
- The unit has few control devices. This makes the circuits relatively easy to trace out and evaluate.

The wiring diagram for a typical window unit is shown in **Figure 9-13.** A legend is provided, denoting the electrical connections made in each mode of system operation. Thus, these diagrams are easy to follow even though the internal workings of the switch are not shown. For example, if the selector switch is set to the HIGH FAN position, contacts L1-1 will close on the switch. This action causes the fan to operate at high speed, as indicated by the current path highlighted in **Figure 9-14A**. If the selector switch is set to the HIGH COOL position, contacts L1-1-C will close on the switch, which causes the high-speed fan and the compressor to operate, **Figure 9-14B**.

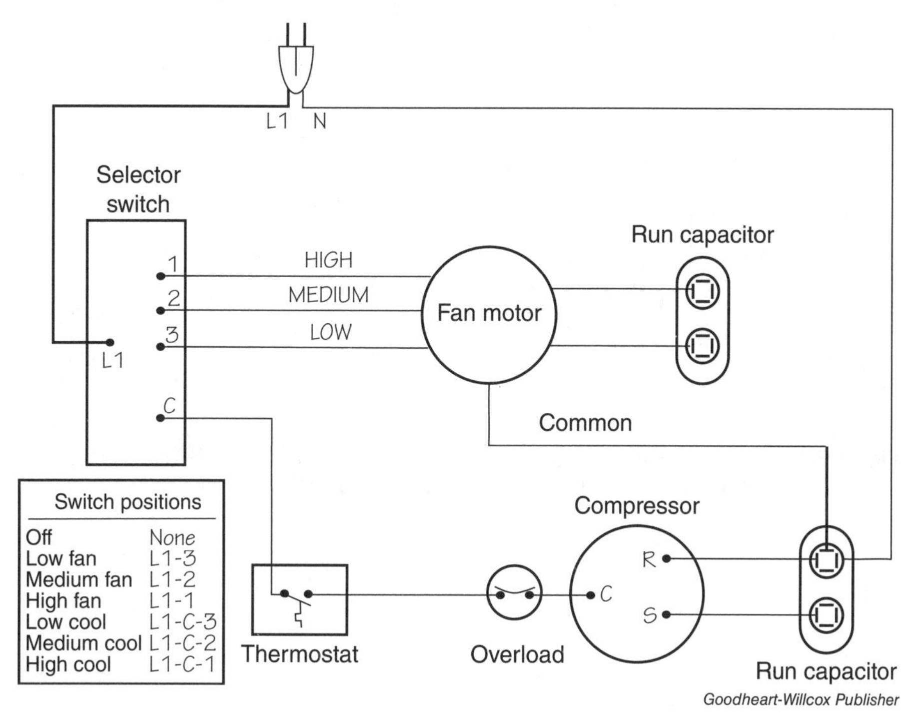

Figure 9-13. The schematic wiring diagram of a typical window or through-the-wall air-conditioning unit.

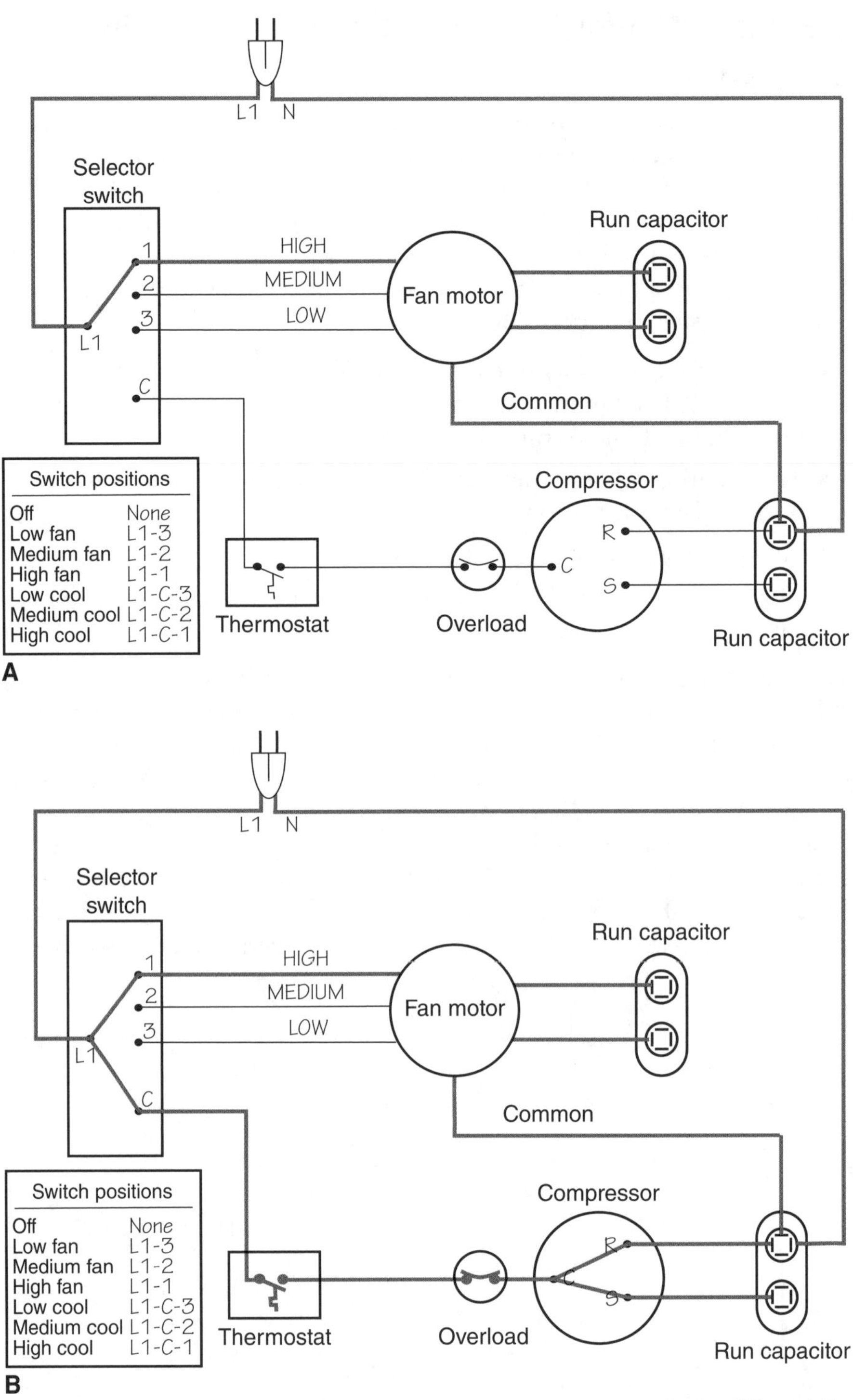

Figure 9-14. A—A window air-conditioning unit operating in the HIGH FAN mode. B—A window air-conditioning unit operating in the HIGH COOL mode.

A ladder diagram, as shown in **Figure 9-15**, can be used to troubleshoot the unit. Assume the unit was set to operate in the HIGH COOL mode, and the compressor does not operate but the fan does. The problem can be isolated to the bottom rung of the ladder diagram, which indicates one of the following issues:

- The contacts L1-C are not closing.
- The thermostat might be open.
- The overload might be open.
- The compressor or its starting components might be defective.
- There is a problem with the interconnecting wiring in the compressor circuit.

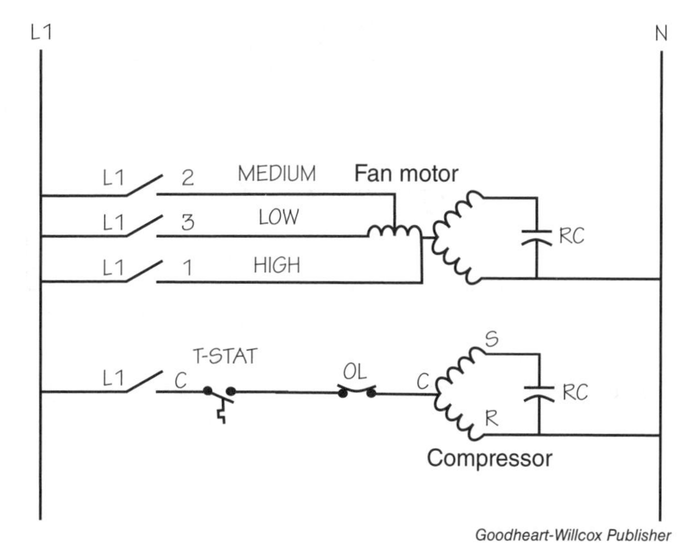

Figure 9-15. A ladder diagram for a typical window or through-the-wall air-conditioning unit.

9.4 Basic Split Air-Conditioning System Wiring Diagrams

Split air-conditioning systems are comprised of two pieces of equipment called the indoor and outdoor units. This is unlike air-conditioning systems where all four major system components are enclosed in one cabinet. Each of the sections of a split air-conditioning system has its own power supply and control devices. These control devices are powered through the space thermostat. The space thermostat is also an integral part of an air-conditioning system. The wiring diagram of a residential split air-conditioning system is shown in **Figure 9-16**. Each of the components that make up the system, the thermostat, indoor unit, and outdoor unit, and their wiring diagrams will be discussed in the following sections.

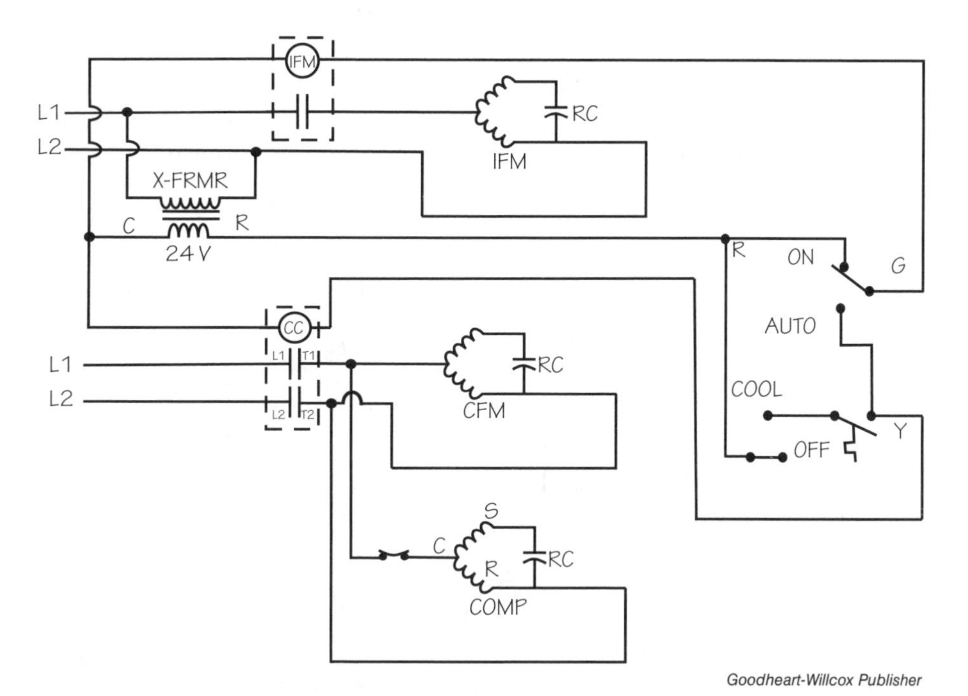

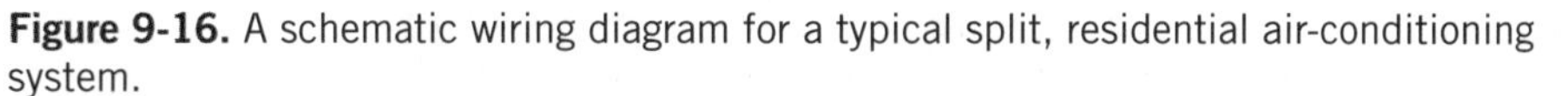

Figure 9-16. A schematic wiring diagram for a typical split, residential air-conditioning system.

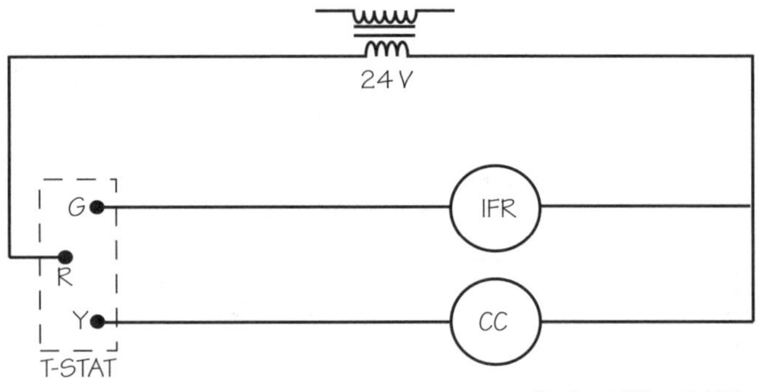

Figure 9-17. Typical thermostat terminals on a simple cooling-only air-conditioning system.

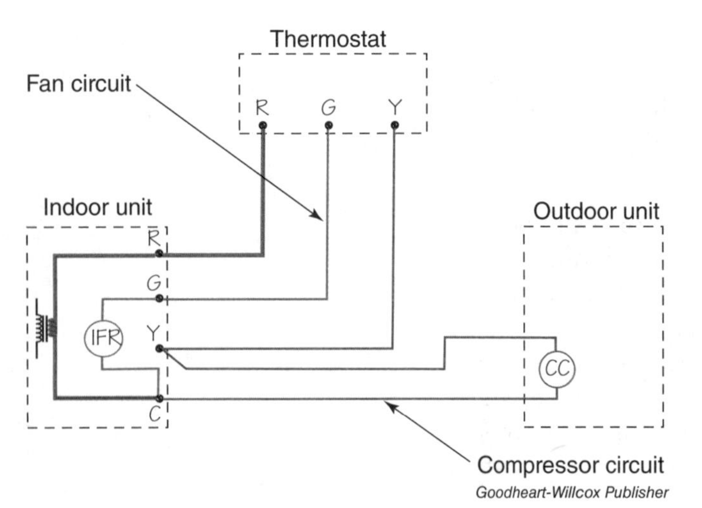

Figure 9-18. Low-voltage control wiring for a simple, cooling-only split air-conditioning system.

9.4.1 Thermostat

The ***thermostat*** is the operational control interface between the air-conditioning system and the system operator. For most applications, the space thermostat is a low-voltage device, which controls the operation of the main system components. The operator sets the thermostat to the desired mode of operation and the desired temperature set point. The system then, if properly wired and configured, operates based on these settings.

On a simple cooling-only system, the thermostat has three terminals: R, G, and Y. The R terminal is powered by the transformer, the G terminal energizes and de-energizes the indoor fan relay coil, IFR, and the Y terminal energizes and de-energizes the compressor contactor coil, CC, **Figure 9-17**. A typical wiring diagram for the low-voltage control circuit on a split air-conditioning system is provided in **Figure 9-18**.

The main selector switch has two mode settings: COOL and OFF. The operator can also select between two fan settings: ON and AUTO. When in the ON position, the evaporator blower motor remains on regardless of the position of the main selector switch. In the AUTO position, the evaporator blower motor cycles on and off with the compressor. The thermostat must also be set to the desired space temperature. When the system is set to COOL, the system cycles on and off to maintain the proper temperature setting on the thermostat. When the indoor blower is in cooling mode, the G terminal on the thermostat is powered, and the IFR coil in the indoor unit is energized. Various thermostat positions are shown in **Figure 9-19**.

In a heating-cooling system, the thermostat has an additional terminal that controls the heating mode of operation, called the W terminal.

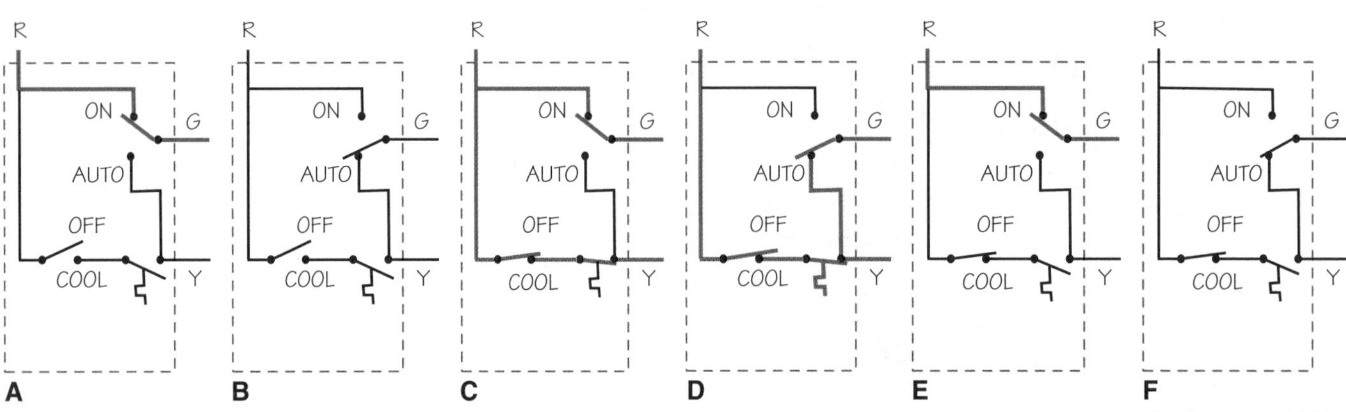

Figure 9-19. Various internal thermostat connections. A—System: OFF, Fan: ON. B—System: OFF, Fan: AUTO. C—System: COOL, Fan: ON, thermostat calling for cooling. D—System: COOL, Fan: AUTO, thermostat calling for cooling. E—System: COOL, Fan: ON, thermostat not calling for cooling. F—System: COOL, Fan: AUTO, thermostat not calling for cooling.

This terminal controls the coil of the heat relay, HR, **Figure 9-20**. The thermostat's internal temperature-sensing switch, in series with the W terminal, closes on a drop in temperature, initiating the heating cycle, **Figure 9-21A**. The thermostat's temperature-sensing switch, wired in series with the Y terminal, closes upon a rise in temperature, thus initiating the cooling cycle.

Like a cooling-only thermostat, the fan switch on a heating-cooling thermostat can be set to either the ON or AUTO position. When in the ON position, the indoor blower remains on regardless of the position of the contacts between terminals R and W, **Figure 9-21B**. When in the AUTO position, the indoor blower motor circuit is isolated from the thermostat because the blower is controlled by the temperature in the appliance's heat exchanger, **Figure 9-21C**.

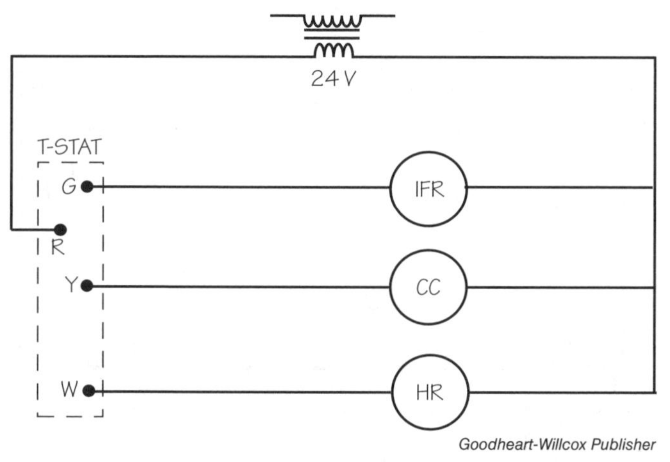

Figure 9-20. The thermostat of a heating-cooling system uses the W terminal to control the heating mode of operation.

9.4.2 Indoor Fan Controls on a Ducted Heating and Cooling System

On a ducted heating and cooling system, there is one indoor blower located in the air handler. The air handler is responsible for delivering air in both the heating and cooling modes of system operation. Most often, the blower operates at high speed in the cooling mode, and low speed in the heating mode. The indoor blower should be operational in the heating mode only when the heat exchanger is warm. This is to prevent the occupants of the conditioned space from experiencing cold drafts from the duct system at the start and end of the heating cycle.

The strategy commonly used to control the motor under this set of circumstances is shown in **Figure 9-22**. In this wiring diagram, the speed of the motor is determined by an indoor fan motor relay, IFR. The high-speed motor lead is wired through a normally open set of contacts on the relay, and the low-speed motor lead is wired through a set of normally closed contacts on the same relay.

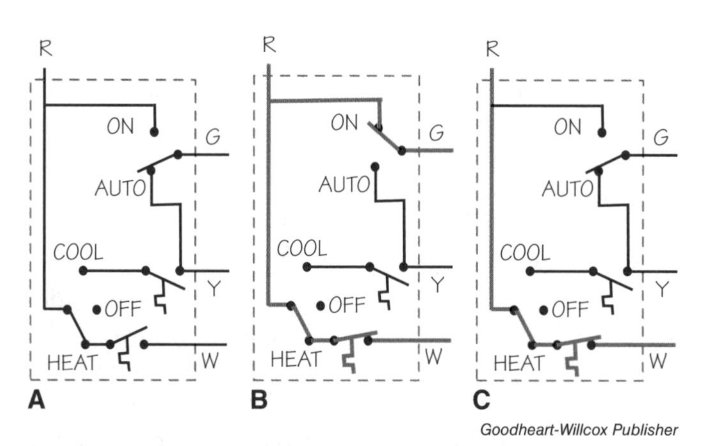

Figure 9-21. A—Simplified wiring diagram of a heating and cooling thermostat. B—Fan switch in the ON position. C—The thermostat does not control the indoor fan motor when the system is in the heating mode and the fan switch is set to AUTO.

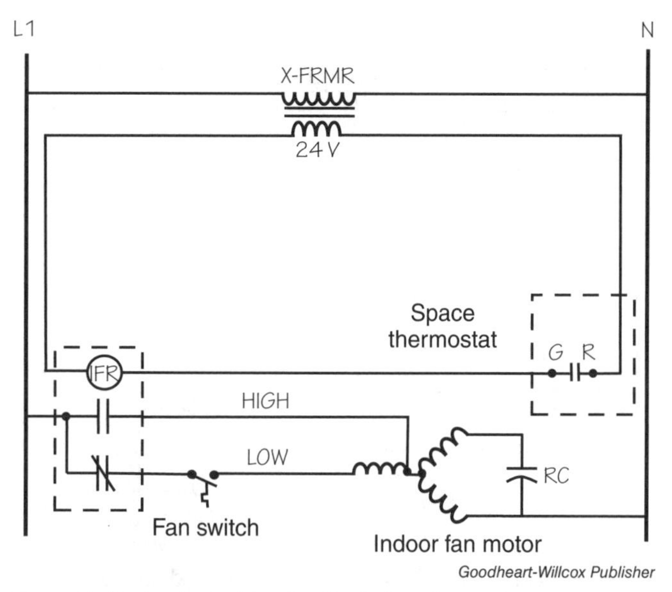

Figure 9-22. Control wiring for the indoor fan motor on a heating and cooling system.

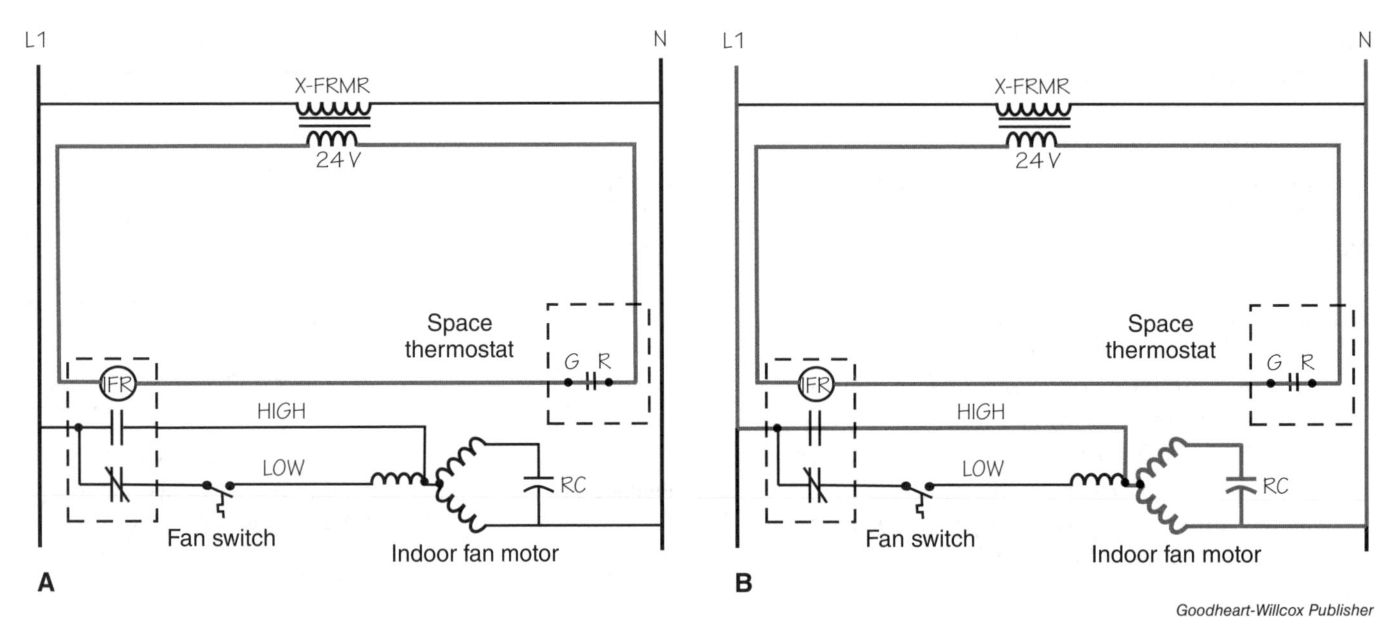

Figure 9-23. A—When the G terminal on the thermostat is energized, the IFR relay coil is energized. B—The normally open set of contacts on the IFR relay close to allow the fan motor to operate at high speed.

This wiring strategy prevents the two motor speed leads from becoming energized at the same time. The coil of the IFR relay is controlled by the G terminal on the space thermostat. If the fan switch on the thermostat is set to the ON position, the R-G contacts on the thermostat close, and the IFR coil becomes energized, **Figure 9-23A**. This causes the normally open contacts on the relay to close, the normally closed contacts on the relay to open, and the motor to operate at high speed, **Figure 9-23B**. The indoor fan motor will also operate at high speed if the fan switch is set to the AUTO position and the thermostat is set to COOL, as long as the temperature in the conditioned space is above the thermostat's set point.

If the R-G contacts on the thermostat are not closed, the normally closed set of contacts on the IFR relay remain closed and pass power to the low-speed motor lead if the fan switch contacts are closed. The fan switch is positioned in the air handler near the heating system's heat exchanger, so these contacts only close when the heat exchanger is hot.

In operation, on a call for heat, the heat source energizes, but the heat exchanger is still too cool to allow the indoor blower to operate. After some time, the temperature in the heat exchanger rises to a level that causes the fan switch to close. This allows the blower to cycle on in low speed, **Figure 9-24**. When there is no longer a call for heating, the heat source de-energizes. Because the heat exchanger is still warm, the fan continues to operate and dissipate heat from the air handler. Once the heat exchanger has cooled to the set point on the fan switch, the fan switch's contacts open and the fan cycles off. If the equipment owner turns the fan switch on the thermostat to the ON position during the heating cycle, the R-G contacts on the thermostat close, and the blower operates at high speed.

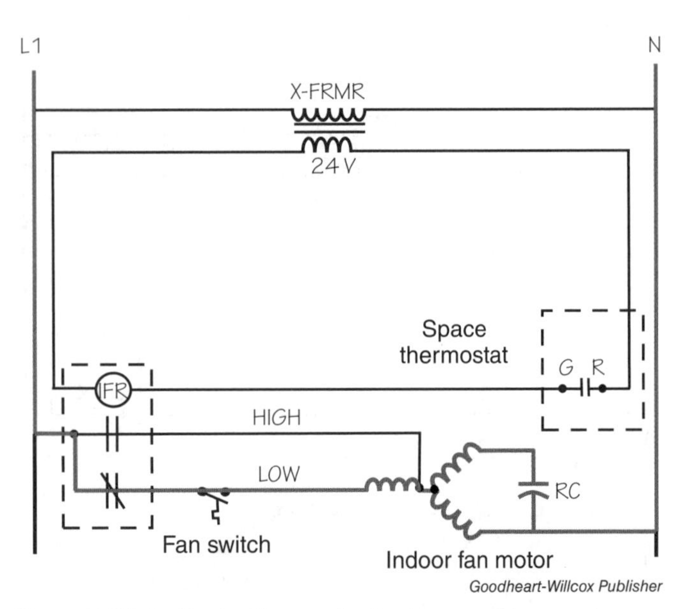

Figure 9-24. In the heating mode, the increased temperature of the heat exchanger closes the fan switch contacts and allows the indoor fan motor to operate at low speed.

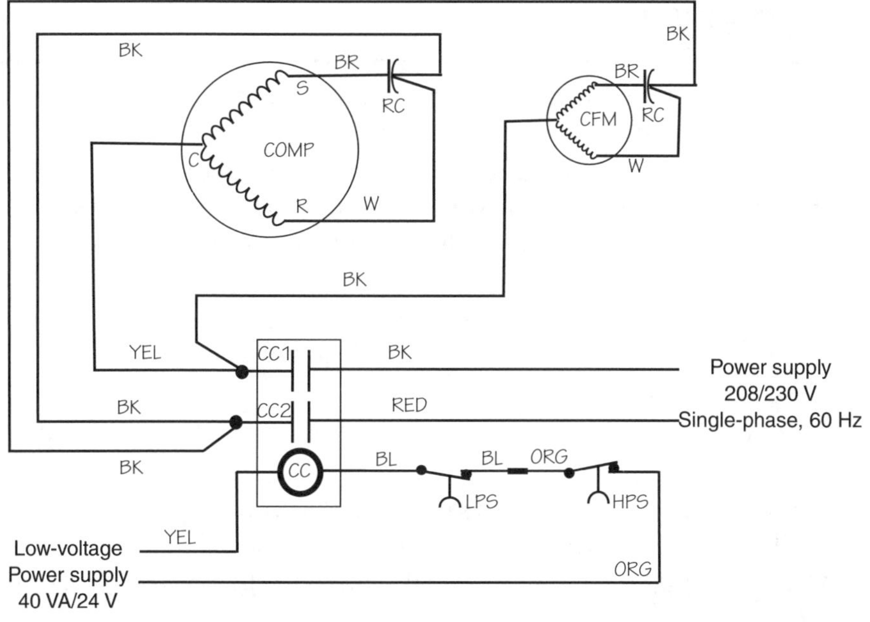

Goodheart-Willcox Publisher

Figure 9-25. Schematic wiring diagram of the outdoor unit of a split air-conditioning system.

9.4.3 Outdoor Controls on a Ducted Heating and Cooling System

The schematic wiring diagram for the outdoor unit of a split air-conditioning system is shown in **Figure 9-25**. Converting this schematic to a ladder diagram, **Figure 9-26**, shows that the system compressor and condenser fan motor cycle on and off together, as long as the CC1 and CC2 contacts on the contactor are closed.

Most air-conditioning systems are protected by both high-pressure and low-pressure switches. These controls are wired in series with the contactor coil and prevent the compressor from operating if the system's operating pressure gets either too high or too low.

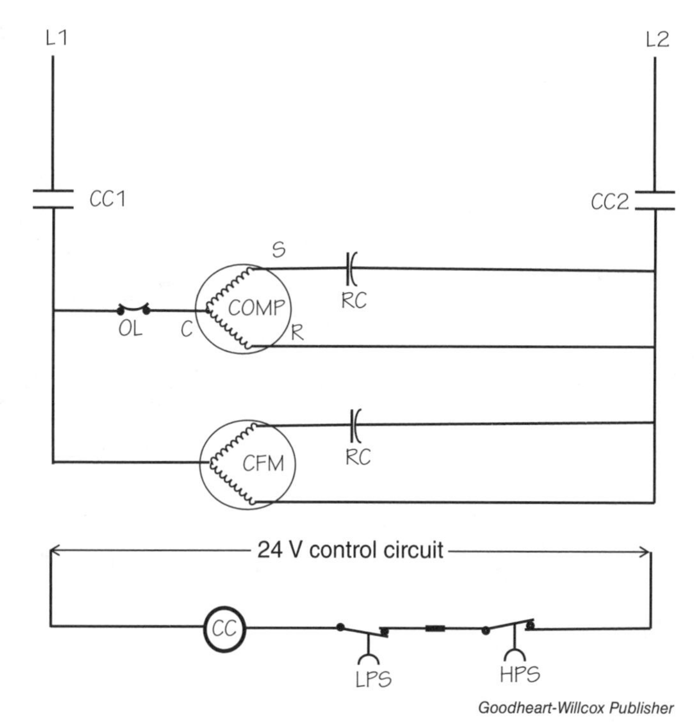

Goodheart-Willcox Publisher

Figure 9-26. Ladder wiring diagram of the outdoor unit of a split air-conditioning system.

Summary

- A switch is used to open or close a connection in an electrical circuit. It has two states: open and closed.
- The operation of a switch is identified by its number of poles and throws. The number of poles indicates how many power lines are controlled, while the throw indicates how many possible current paths there are.
- A set of contacts is the part of a switch that closes to complete an electrical circuit. Sets of contacts are identified as normally open (NO) or normally closed (NC).
- A relay is an electrical switch controlled by an external electrical signal.
- A solenoid is an electromagnetic device that converts current to mechanical movement by producing a magnetic field.
- Wiring diagrams serve as electrical road maps for electric current to follow.
- Electrical symbols and abbreviations are not uniform throughout the industry, so diagram legends must be referred to when reading wiring diagrams.
- Be sure to always follow the manufacturer's literature when wiring, or evaluating the wiring, of a system.
- Pictorial diagrams often provide a visual representation of the actual appearance of system and circuit components.
- Schematic diagrams provide circuit details, including all wires, wiring connections, and color-coding.
- The ladder, or line, diagram is especially useful for system troubleshooting.
- The system's sequence of operations can often be determined by evaluating the system's wiring diagrams.
- Wiring diagrams often include both line and low-voltage wiring circuits. Some systems utilize both line voltage and low-voltage control devices.
- Controls devices, when wired in series with the load, must be able to withstand the current draw of the load.

Chapter Assessment

Name ______________________ Date ____________ Class ______________

Know and Understand

_______ 1. A _____ relay indicates contacts that are closed in an energized state.

A. pole
B. throw
C. normally open
D. normally closed

_______ 2. If a switch is wired in series with a power-consuming device, the contacts on the switch must be _____.

A. constructed of a material that is the same as the material used as the conductor
B. constructed of a material that is different from the material used as the conductor
C. rated at an amperage that is lower than the expected amperage draw of the device being controlled
D. rated at an amperage that is higher than the expected amperage draw of the device being controlled

_______ 3. Which of the following wiring diagram types typically includes information such as all wire connection points, wire gage, and wire color, as well as information regarding factory and field-installed wiring?

A. Pictorial wiring diagram
B. Line diagram
C. Schematic wiring diagram
D. Ladder diagram

_______ 4. The thermostat terminal that controls the operation of the indoor fan motor on a cooling-only, split air-conditioning system is the _____.

A. R terminal
B. G terminal
C. W terminal
D. Y terminal

_______ 5. The thermostat terminal that controls the operation of the compressor on a cooling-only, split air-conditioning system is the _____.

A. R terminal
B. G terminal
C. W terminal
D. Y terminal

Apply and Analyze

_______ 1. Referring to the wiring diagram below, which of the following best describes the contacts CC1 and CC2?

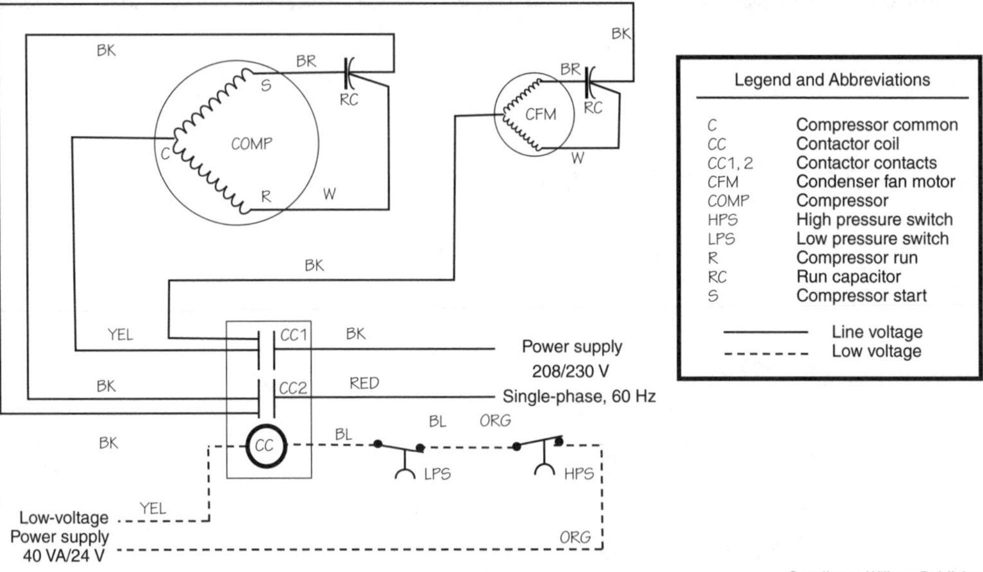

Goodheart-Willcox Publisher

A. CC1 and CC2 are both normally closed sets of contacts.
B. CC1 and CC2 are both normally open sets of contacts.
C. CC1 is a normally closed set of contacts, and CC2 is a normally open set of contacts.
D. CC1 is a normally open set of contacts, and CC2 is a normally closed set of contacts.

_______ 2. The IFR in in the following figure is best described as what type of switching device?

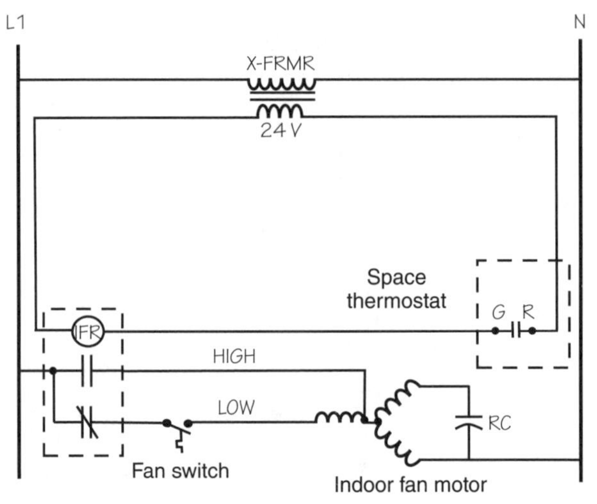

Goodheart-Willcox Publisher

A. Single-pole single-throw
B. Single-pole double-throw
C. Double-pole single-throw
D. Double-pole double-throw

Name ______________________________ Date ____________ Class ____________________

_______ 3. Referring to the R3 relay below, which of the following best describes the switching mechanism in the relay?

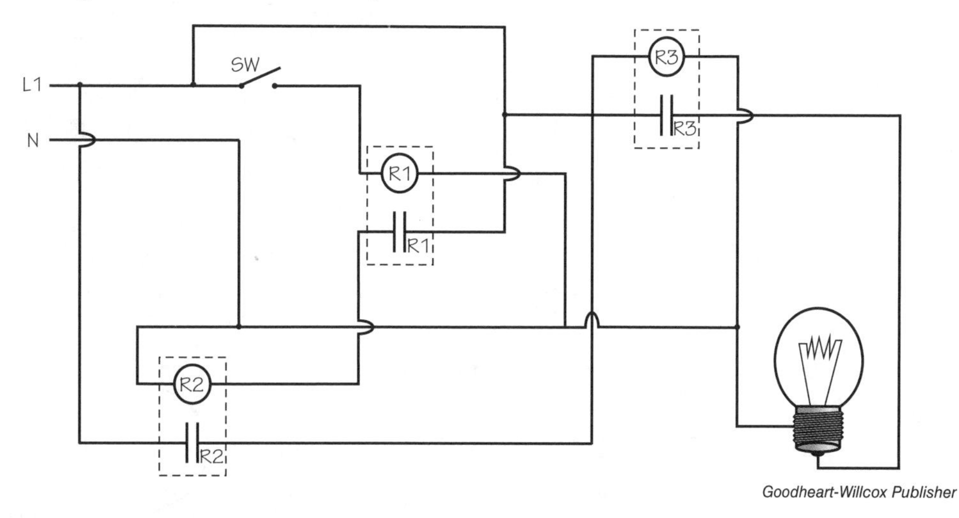

Goodheart-Willcox Publisher

A. Single-pole single-throw
B. Single-pole double-throw
C. Double-pole single-throw
D. Double-pole double-throw

_______ 4. Referring to the wiring diagram below, which of the following components initiates system operation?

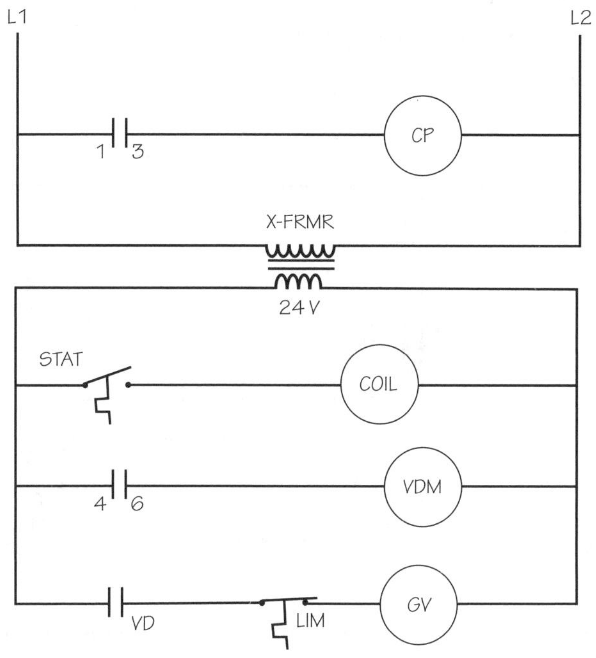

Goodheart-Willcox Publisher

A. Contacts 1-3
B. Thermostat
C. Circulator pump
D. VD contacts

_______ 5. Which of the following will prevent the compressor in the following system from operating?

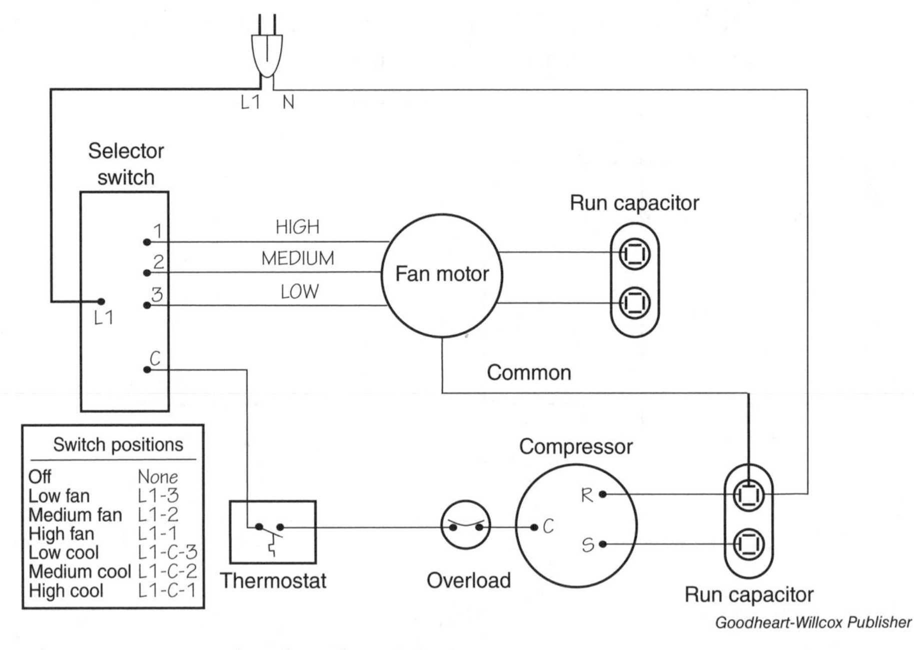

Goodheart-Willcox Publisher

A. The thermostat is in the closed position.
B. The L1-C contacts on the selector switch are open.
C. The L1-2 contacts on the selector switch are open.
D. The overload is in the closed position.

Critical Thinking

Refer to the figure below to answer question 1.

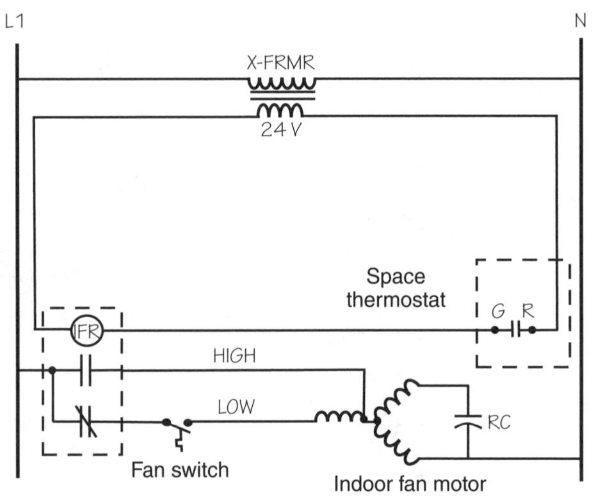

Goodheart-Willcox Publisher

1. Why is it that both the high and low motor speeds cannot be energized at the same time?

____

____

____

____

Name ______________________ Date __________ Class ______________

Refer to the figure below to answer question 2.

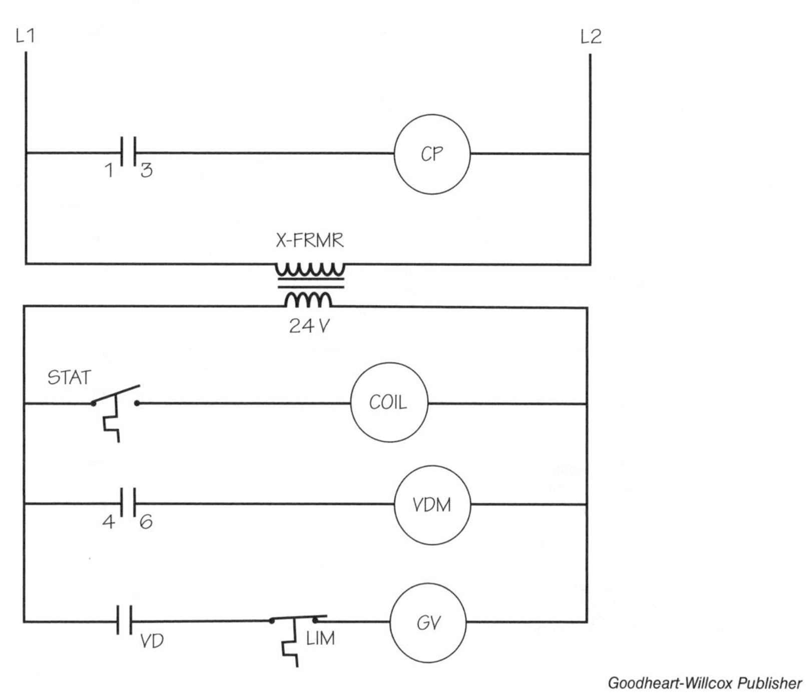

Goodheart-Willcox Publisher

2. Referring to the following figure, what are three possible causes for the appliance to appear to be completely off?

____

____

____

____

Refer to the figure below to answer question 3.

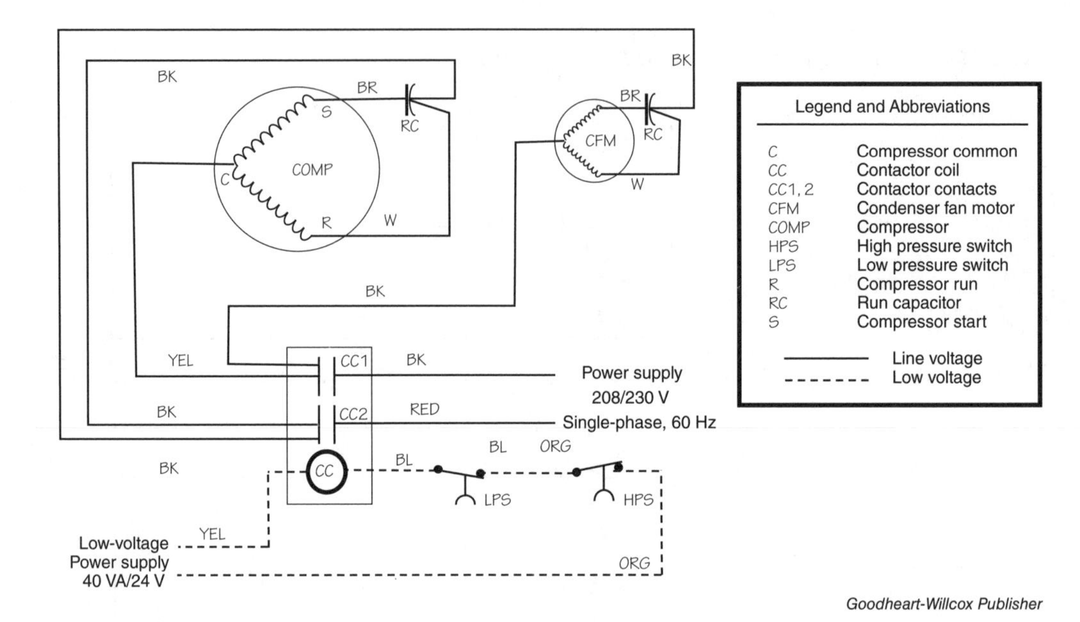

Goodheart-Willcox Publisher

3. Create a complete list of conditions that must be satisfied in order for the CC1 and CC2 contacts to be closed in the following figure.

____

____

____

____

10 Electrical Wiring Diagrams for Furnaces, Hot Water Heating, and Heat Pump Systems

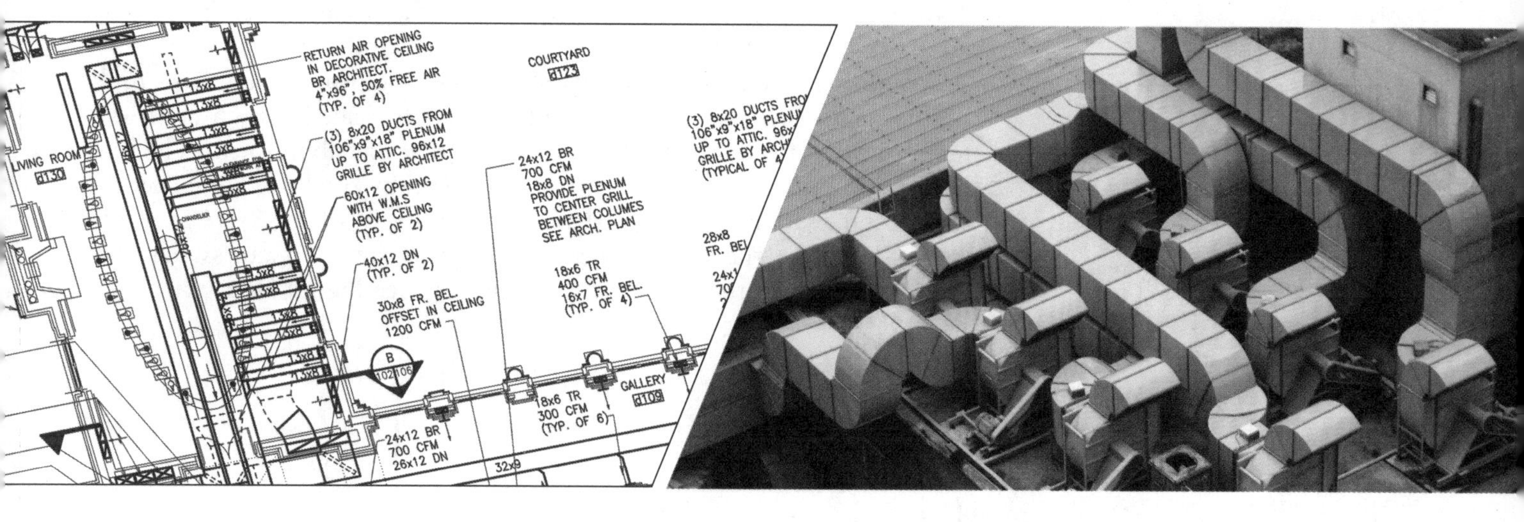

Chapter Outline

10.1 Furnace Wiring Diagrams

10.2 Hot Water Heating System Wiring Diagrams

10.2.1 Low-Mass Boiler on a Single-Zone Heating System
10.2.2 Low-Mass Boiler on a Multi-Zone Heating System
10.2.3 High-Mass Boiler on a Single-Zone Heating System
10.2.4 High-Mass Boiler on a Multi-Zone Heating System

10.3 Heat Pump System Wiring Diagrams

10.3.1 The Four-Way Reversing Valve
10.3.2 Holdback Thermostat
10.3.3 Heat Pump System Thermostat
10.3.4 Lockout Relay
10.3.5 Electric Strip Heaters

10.4 Heat Pump Modes of Operation

10.4.1 Cooling
10.4.2 First-Stage Heating
10.4.3 Second-Stage Heating with Warm Outside Ambient Temperature
10.4.4 Second-Stage Heating with Cold Outside Ambient Temperature
10.4.5 Defrost Mode
10.4.6 Emergency Heat Mode

Learning Objectives

After studying this chapter, you will be able to:

- Describe the function of a furnace and the components included in its control circuit.
- Compare a hot water heating system to a furnace.
- Interpret the wiring diagrams of various hot water heating systems.
- Explain the control strategies for a hot water heating system based on the type of boiler and the number of conditioned zones.
- Explain the purpose of a heat pump and how it is operated.
- Describe how a reversing valve is used to change the operation of a heat pump from heating and cooling.
- List the main system components controlled during normal air-conditioning and heat pump system operation.
- Discuss the function and operation of the holdback thermostat.
- Describe the function and operation of the lockout relay.
- Discuss the modes of heat pump operation.

Introduction

The intricacy of a wiring diagram is based on the complexity of the strategy used to control the operation of a system's components. Also, the functions performed in each mode of system operation affects the intricacy of wiring diagrams. The blower on a furnace equipped with a cooling coil, for example, operates differently in the heating and cooling mode. Hot water heating and heat pump systems are examples where system components perform different functions and are controlled differently in different modes of operation or when system configurations change. This chapter discusses some commonly encountered heating systems, along with their more complex wiring diagrams.

Technical Terms

defrost mode
heat pump
hot water heating system
limit switch
lockout relay
reversing valve
rollout switch
time-delay relay
vent switch

10.1 Furnace Wiring Diagrams

A furnace is a piece of HVACR equipment that provides comfort heating to an occupied space. Just as a central air-conditioning system delivers comfort cooling, a furnace uses a blower and duct system to supply heated air to the space being conditioned. A furnace has two main systems, one that produces heat and the other that distributes heat. The heat generation portion of the system is controlled by the W terminal on the thermostat, whereas the heat distribution portion of the system is controlled by either the G terminal on the thermostat or a fan swtich that senses the temperature of the heat exchanger. See **Figure 10-1**.

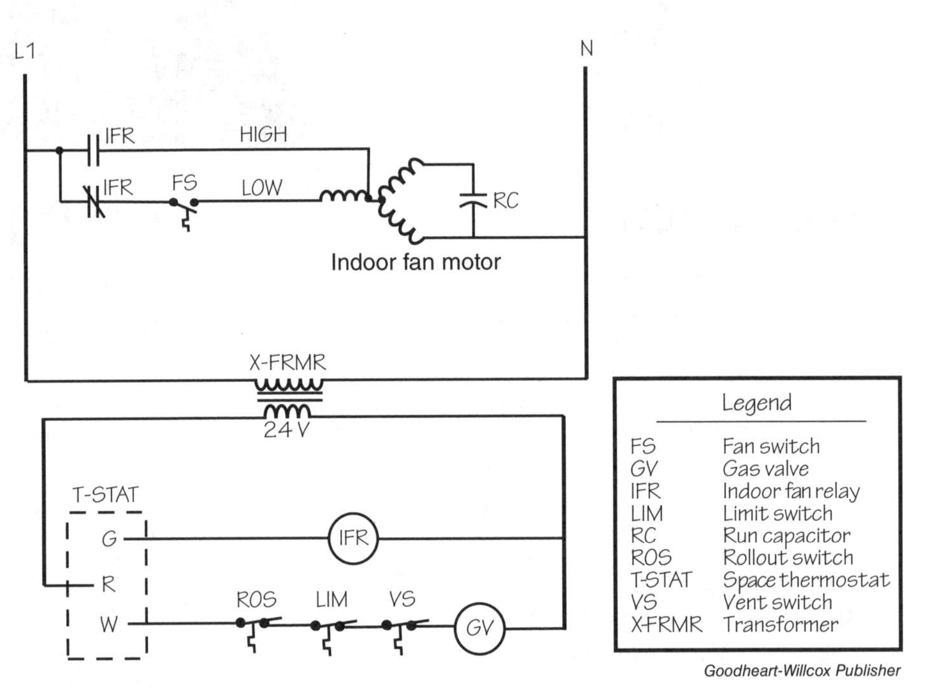

Figure 10-1. Ladder diagram of a simple gas-fired furnace.

The control circuit has several safety elements. This helps ensure the fuel does not flow into the appliance unless system conditions are safe. In addition to the thermostat contacts, three safety switches are shown in the diagram: the rollout switch, ROS; the limit switch, LIM; and the vent switch, VS. The ***rollout switch*** is a normally closed switch that opens its contacts if a gas flame rolls out of the appliance. The ***limit switch*** is also a normally closed switch. It opens its contacts if the temperature in the appliance gets too high. The ***vent switch*** prevents the gas valve from being energized unless adequate appliance venting is established.

In operation, the heating cycle is initiated when the thermostat is set to call for heat. The fan switch is set to the AUTO position. As long as the three safety devices are closed, power will pass to the gas valve and the heat generation process will start, **Figure 10-2**. The indoor blower motor does not operate because the G terminal on the thermostat is not energized (the normally open IFR relay contacts are open) and the heat exchanger is still cool (the FS contacts are open).

Once the heat exchanger warms to the set point on the fan switch, the FS contacts close and the indoor blower operates at low speed, **Figure 10-3**. If the equipment operator turns the fan switch to the ON position during the heating cycle, the G terminal on the thermostat energizes the IFR coil and the fan motor switches to high speed, **Figure 10-4**. The heating cycle continues as long as heat is called for and the safety devices remain in the closed position.

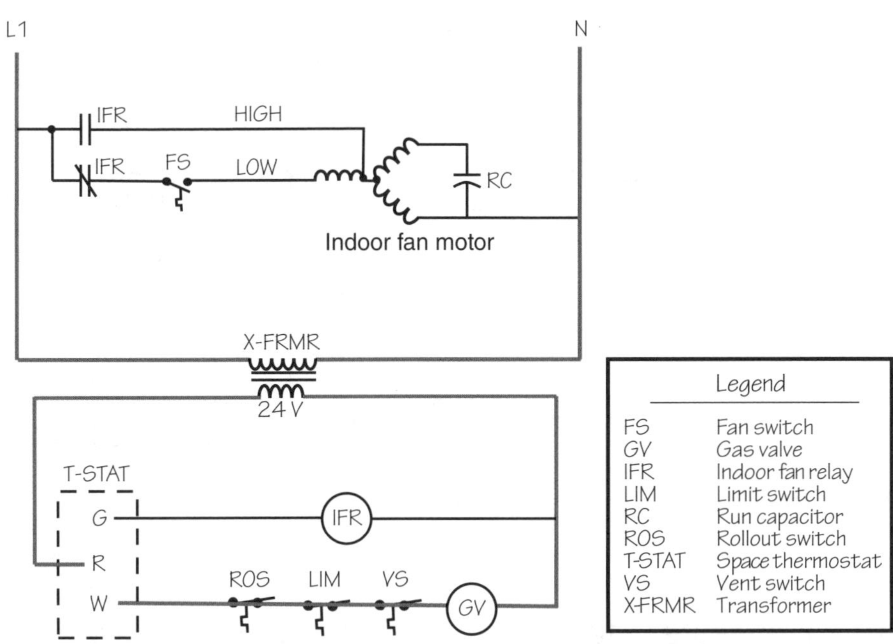

Figure 10-2. The gas valve is energized when the thermostat calls for heat.

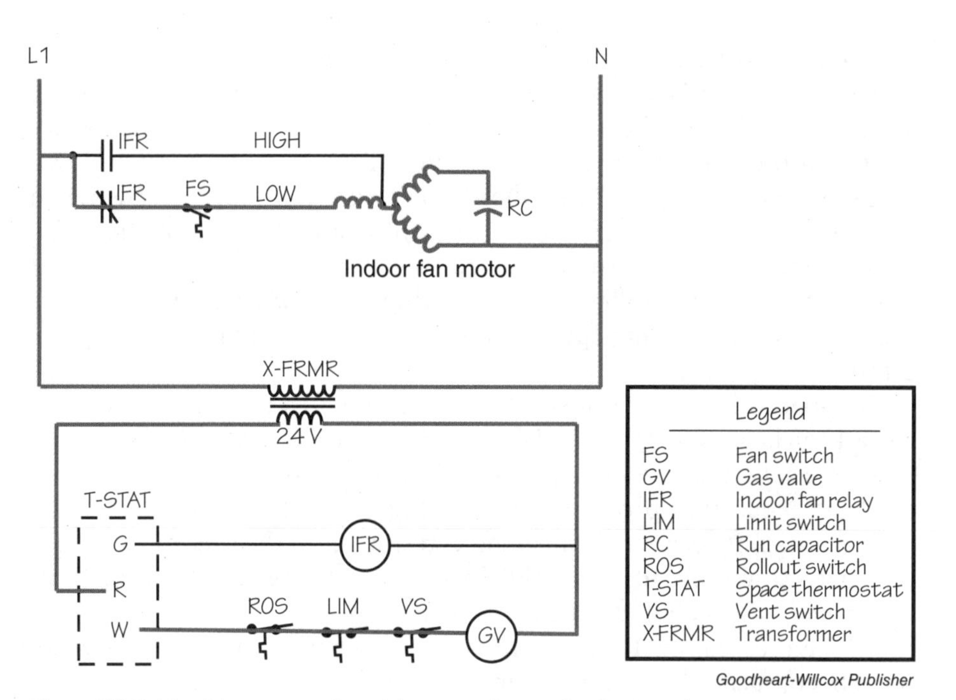

Figure 10-3. The blower operates at low speed once the heat exchanger warms up.

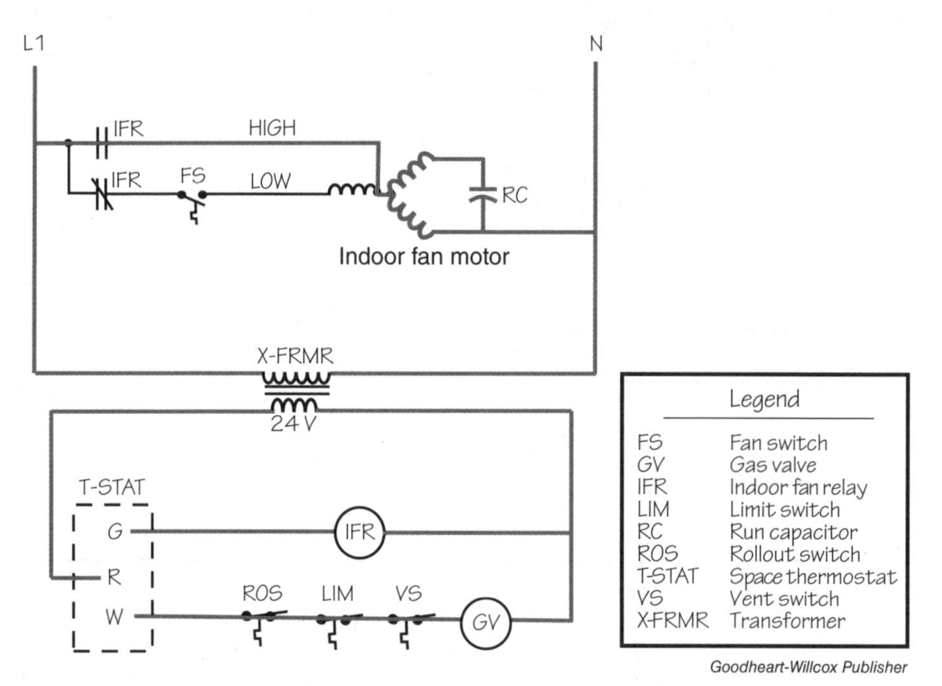

Figure 10-4. The blower operates at high speed if the fan switch is set to the ON position.

10.2 Hot Water Heating System Wiring Diagrams

Hot water heating systems are also used to provide heating to an occupied space. Unlike a furnace, hot water heating systems use boilers to heat the water, as well as a piping arrangement to carry the heated water to the occupied spaces. The heated water is distributed to the conditioned areas by one or more circulator pumps. A circulator pump is the system component responsible for moving water through a closed loop system.

The pump removes hot water from the boiler, where it passes through a piping system, onto the terminal units, where heat is transferred from the water to the heated space. The pump also creates the pressure difference required to move the water from the terminal units back to the boiler, where the water is heated again for reuse.

A heating system can distribute heat to one or more zones in a structure. A simplified piping diagram of a single-zone, hot water heating system is shown in **Figure 10-5**, while a piping diagram of a three-zone, hot water heating system is shown in **Figure 10-6**. The heat source must be controlled, as well as the circulator pumps that distribute the heated water. The type of boiler used in a particular application affects how the heat source and circulator pumps are controlled. Boilers, for example, can be classified as either high-mass or low-mass appliances and are controlled differently. Knowing how the boiler functions will help you properly wire its controls and associated system components.

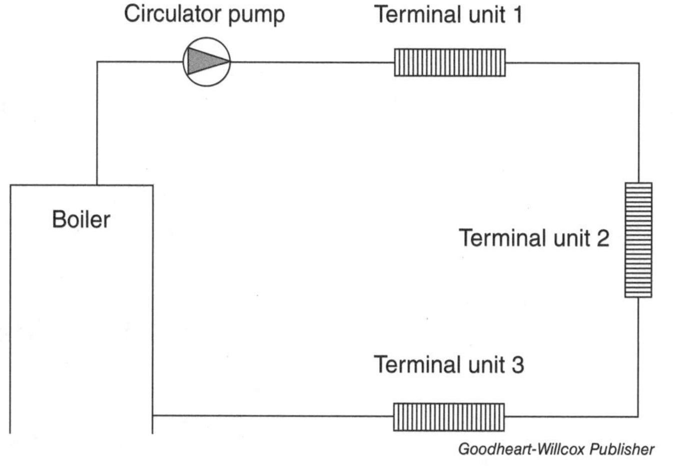

Figure 10-5. Simplified piping diagram of a single-zone hot water heating system.

High-mass boilers have a large thermal mass, and thus take a long time to heat up. Due to this, these appliances are typically wired so they maintain system water at the desired temperature all the time. One benefit of high-mass boilers is that, because of their large thermal mass, they maintain heat for a significant amount of time and take a long time to cool down. Low-mass boilers, on the other hand, do not have a large thermal mass and heat up quickly, so they are wired where the heat generation process starts when a call for heat is initiated. Because of their low thermal mass, low-mass boilers cool down quickly.

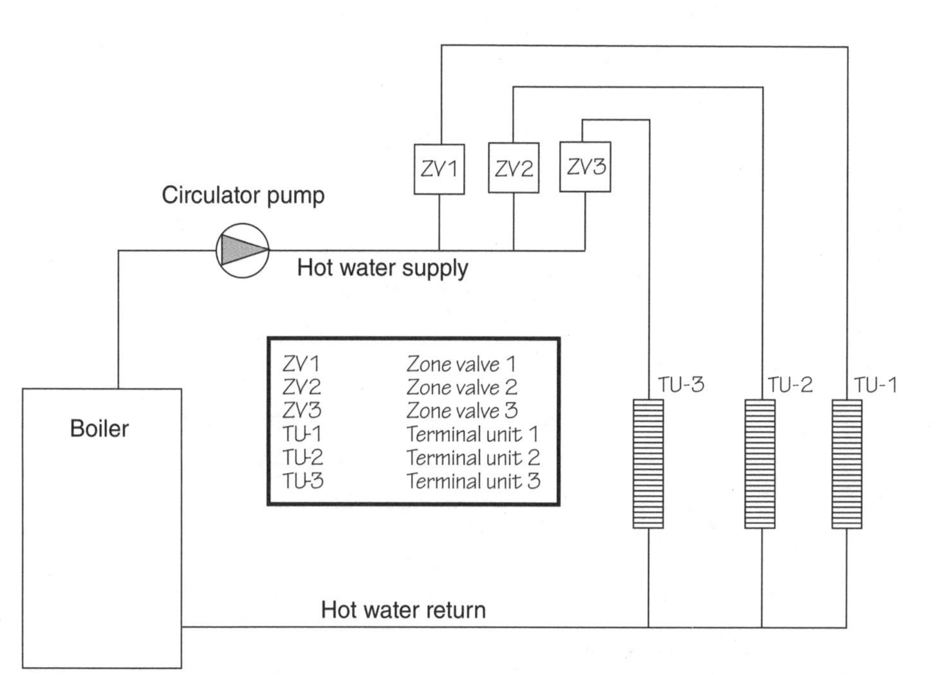

Figure 10-6. Simplified piping diagram of a three-zone hot water heating system.

Based on the mass and number of heating zones, there are different wiring configurations for a boiler. These include:

- Low-mass boiler on a single-zone heating system
- Low-mass boiler on a multi-zone heating system
- High-mass boiler on a single-zone heating system
- High-mass boiler on a multi-zone heating system

10.2.1 Low-Mass Boiler on a Single-Zone Heating System

The function of a low-mass boiler that services a single heating zone is simple and straightforward. There is only one circulator pump in the system. Because this boiler's heat source can be energized on a call for heat, the boiler can heat water to the desired temperature rapidly.

A sample ladder diagram is shown in **Figure 10-7**. The thermostat passes power to the circulator pump relay coil and the heating relay coil circuit, which are wired in parallel with each other. On a call for heat, the circulator pump starts and the heat source is energized, as long as the water temperature is below the set point on the limit switch. When the water reaches the cut-out temperature on the limit switch, the heat relay coil de-energizes. However, the circulator pump continues to operate as long as a call for heat remains.

10.2.2 Low-Mass Boiler on a Multi-Zone Heating System

A multi-zone system has more than one zone in the system, with zone valves for each zone. In this system, the zone valves are normally closed (in a de-energized state), but once energized, they move to the open position. Moving from the fully closed to fully open position can take up to one minute. When power is sent to the zone valve, the valve begins to open and the end switch contacts close. This signals that the zone valve is fully open and the circulator pump can turn on.

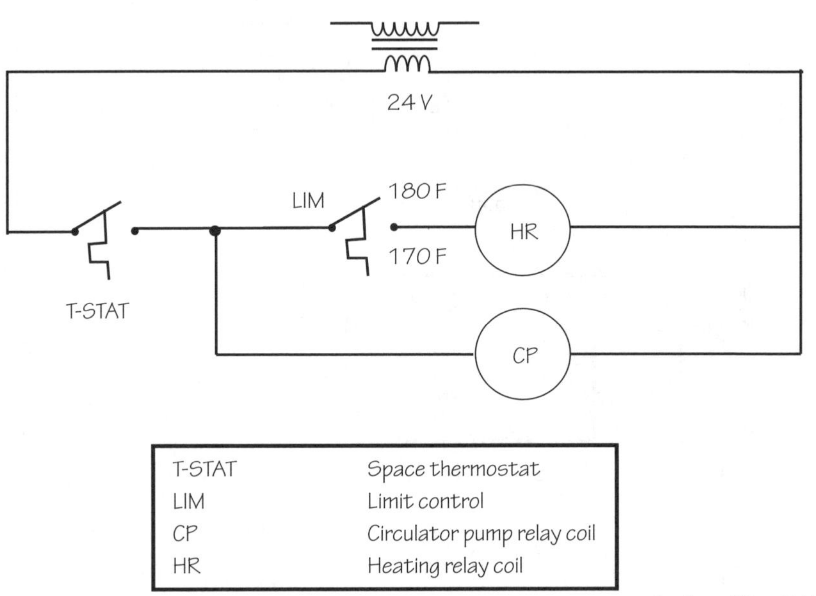

Goodheart-Willcox Publisher

Figure 10-7. Wiring diagram of the control circuit for a low-mass boiler on a single-zone, hot-water heating system.

BETWEEN THE LINES

Deadheading a Pump

Energizing the circulator pump at the same time as the zone valve can deadhead the pump and cause damage. Deadheading occurs when the pump's discharge is obstructed to the point where there is no flow. In addition, if a particular zone valve fails and is the only zone valve attempting to open, severe pump damage can result. This is why zone valves with end switches are often used, **Figure 10-8**.

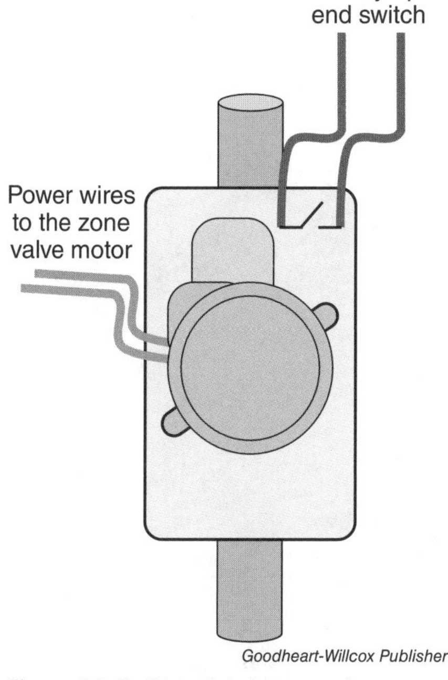

Goodheart-Willcox Publisher

Figure 10-8. Pictorial diagram of a zone valve with a normally open end switch.

A low-voltage wiring diagram for a low-mass boiler on a multi-zone, hot-water heating system is shown in **Figure 10-9**. This figure shows three heating zones, each with its own thermostat. The thermostat in each zone controls the operation of the zone valve in that particular zone and the coil of a heat relay. The heat relay controls the heat source, HS, for the system.

A multi-zone system typically has more than one heat relay, as shown in **Figure 10-10**. The figure's line voltage circuit illustrates three heat relays. Any of these three heat relays can cause the heat source to become energized if the limit switch is in the closed position. Three sets of parallel switches are also shown in the figure: ES1, ES2, and ES3. These sets of contacts are wired in parallel with each other, but are in series with the circulator pump. These are the end switches located in each of the three zone valves.

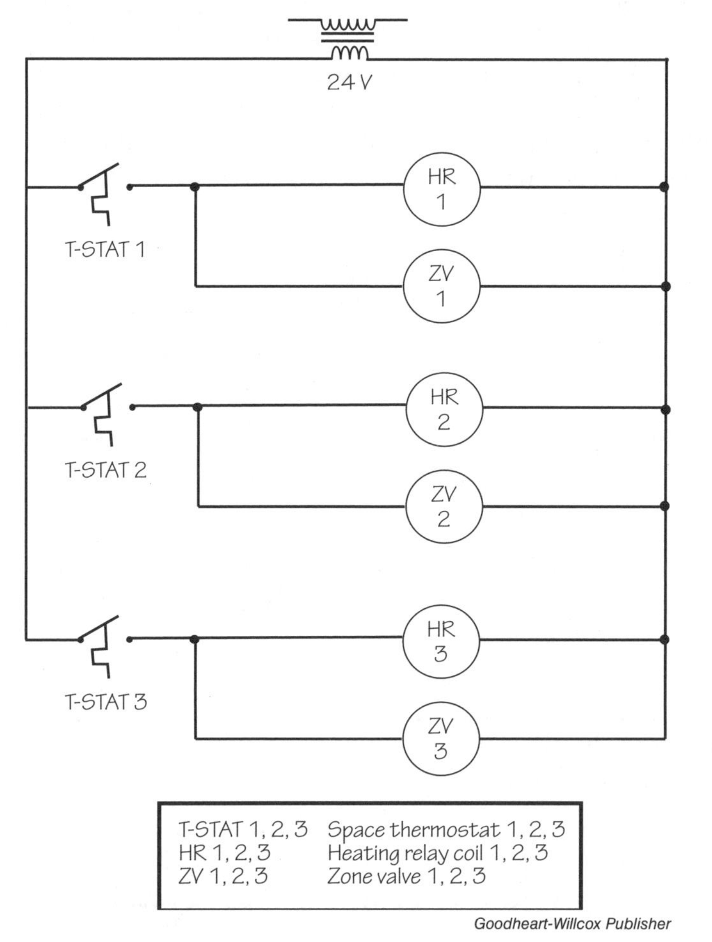

Goodheart-Willcox Publisher

Figure 10-9. Ladder diagram of the low-voltage control circuit for a low-mass boiler on a multi-zone system.

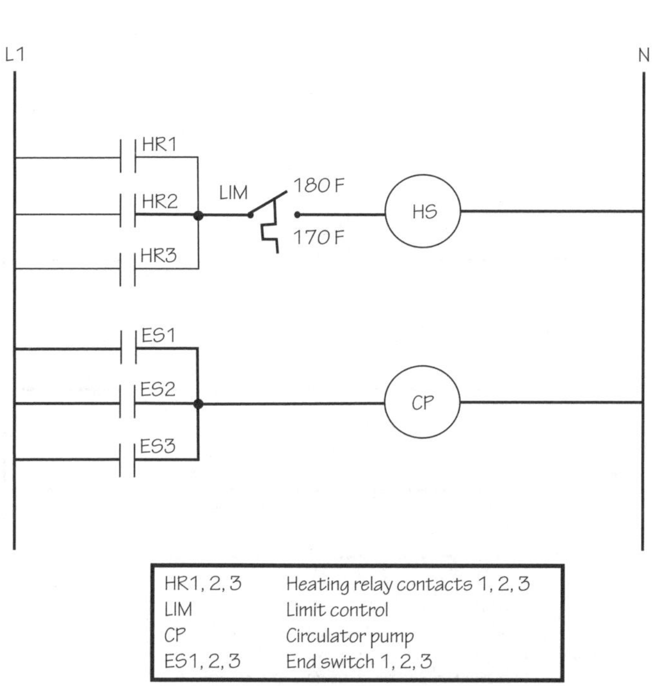

Goodheart-Willcox Publisher

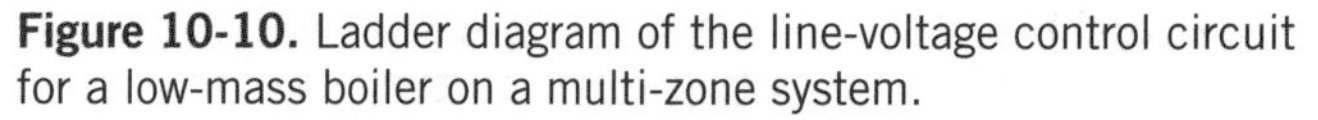

Figure 10-10. Ladder diagram of the line-voltage control circuit for a low-mass boiler on a multi-zone system.

If any of the zones initiates a call for heat, the heat relay coil in that zone becomes energized. The power then passes to the heat source relay, HS, and the zone valve in that zone starts to open. The end switch in that zone valve then closes, energizing the circulator pump. Note that if the circulator pump is already operating, closing an additional set of end switch contacts does not affect circulator pump operation.

10.2.3 High-Mass Boiler on a Single-Zone Heating System

Because of the lengthy time needed to heat up, high-mass boilers maintain water at the desired temperature at all times, making it available immediately when needed. Thus, the space thermostat is not a controlling factor for the heat source. Instead the heat source is controlled by the limit switch, which senses the temperature of the water in the boiler. Because this system is configured as a single zone with no zone valves, the thermostat can directly control the operation of the circulator pump. The limit switch opens and closes its contacts to maintain a water temperature in the boiler between 170°F and 180°F. Upon a call for heat, the circulator pump energizes to provide hot water to the occupied space. See **Figure 10-11**.

10.2.4 High-Mass Boiler on a Multi-Zone Heating System

Similar to a high-mass boiler on a single-zone system, a thermostat does not control the operation of the heat source on a multi-zone system. The three space thermostats control only the operation of the three zone valves, **Figure 10-12**. When a thermostat initiates a call for heat, the zone valve controlled by that thermostat begins to open. Once that zone valve has opened completely, the end switch on that valve closes and the circulator pump operates. The circulator pump control circuit resembles that shown for a low-mass boiler.

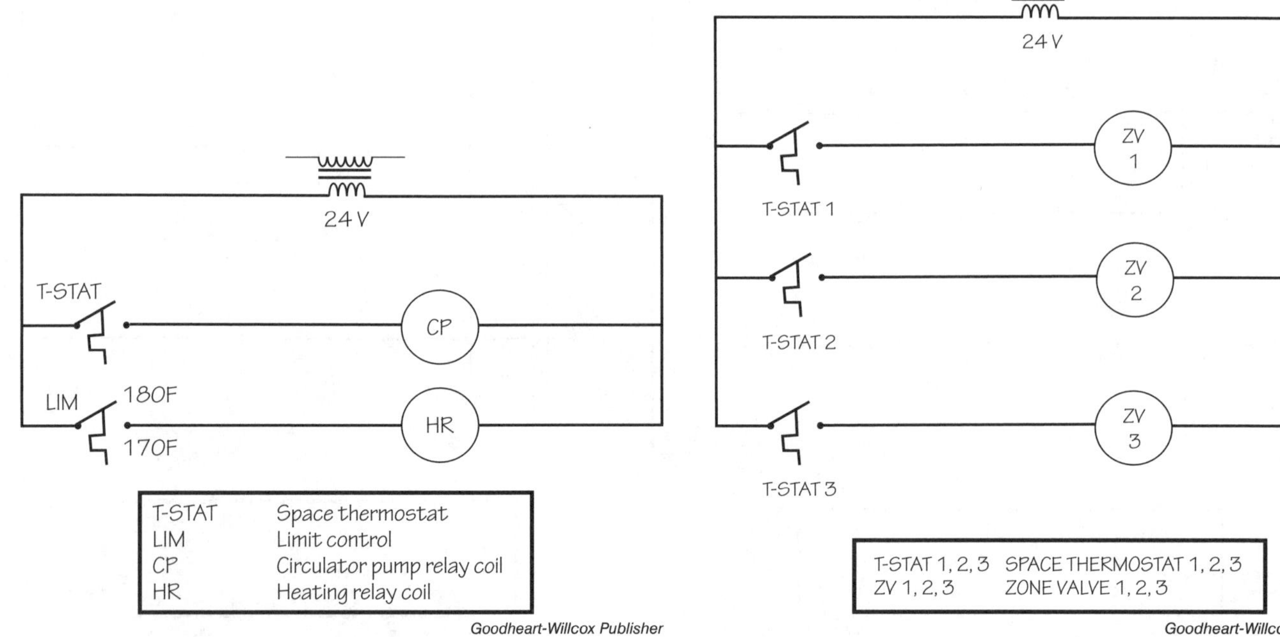

Figure 10-11. Ladder diagram of the low-voltage control circuit for a high-mass boiler on a single-zone system.

Figure 10-12. Ladder diagram of the low-voltage control circuit for a high-mass boiler on a multi-zone system.

10.3 Heat Pump System Wiring Diagrams

Heat pumps are unique heating-cooling systems because the same four major system components used to provide comfort cooling are the same components that provide comfort heating. A ***heat pump*** has the ability to add heat to, or remove heat from, a conditioned space by using mechanical energy. This occurs when heat is absorbed in one location and released in another, which is similar to a basic cooling system. Heat pumps change between cooling and heating modes of operation by changing the direction of refrigerant flow through the system, as well as the function of some of the system's components.

Heat pumps, like air-conditioning systems, have operational elements that must be properly controlled:

- The solenoid coil on the reversing valve
- The relays and controls for the supplementary and emergency heat
- Relays that control the outdoor fan operation
- Defrost timers and sensors

Proper wiring of a heat pump system affects overall performance. A service technician must have a solid understanding of how system components are connected in a specific circuit and their function. There is generally more wiring and circuitry in heat pumps than in conventional air-conditioning systems. See **Figure 10-13**. Some electrical components

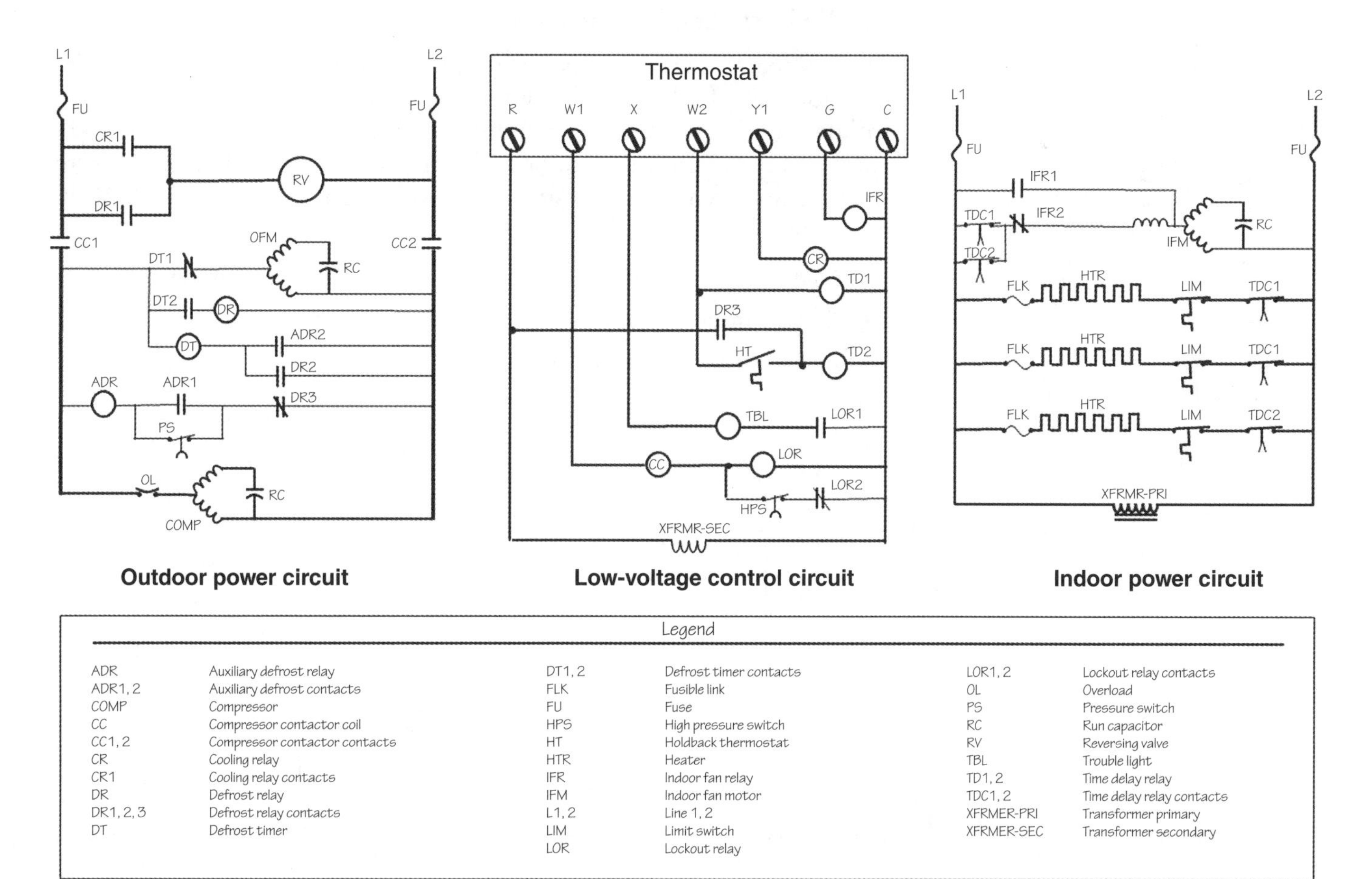

Figure 10-13. Complete wiring diagram of a typical air-source heat pump system.

found on heat pump systems that are not typically found in cooling-only systems are the reversing valve, the defrost control circuitry, electric strip heaters, lockout circuitry, and outdoor thermostats. In addition, the thermostat used on a heat pump system controls more circuits, and is, overall, more complicated than traditional, cooling-only or heating-cooling devices.

10.3.1 The Four-Way Reversing Valve

A ***reversing valve*** is used on a heat pump to change operation between heating and cooling. Most heat pumps are designed so the reversing valve is energized when the system is in the cooling or defrost mode. When a heat pump is in ***defrost mode***, supplementary heat is provided to increase the temperature of the air circulated within the occupied space. This helps reduce cold drafts from being felt in the space. Defrost mode is only initiated when the system is in heating mode.

Because the reversing valve is energized in both the cooling and defrost modes, different electrical circuits are used. One circuit energizes the reversing valve in cooling mode, whereas another energizes the reversing valve in defrost mode. This is shown in **Figure 10-14**. Contacts C1 are controlled by the cooling terminal on the thermostat, and contacts D1 are controlled by the defrost relay coil or circuit board.

10.3.2 Holdback Thermostat

The holdback thermostat is an outdoor thermostat located within the outdoor unit of a heat pump system. The holdback thermostat is a control device that determines the number of heating elements that are to be energized based on the outdoor ambient temperature. All or some of the heat pump system's electric strip heaters are energized under each of the following conditions:

- When the system is operating in the emergency-heat mode
- When supplementary, second-stage heat is needed
- When the system is in defrost mode

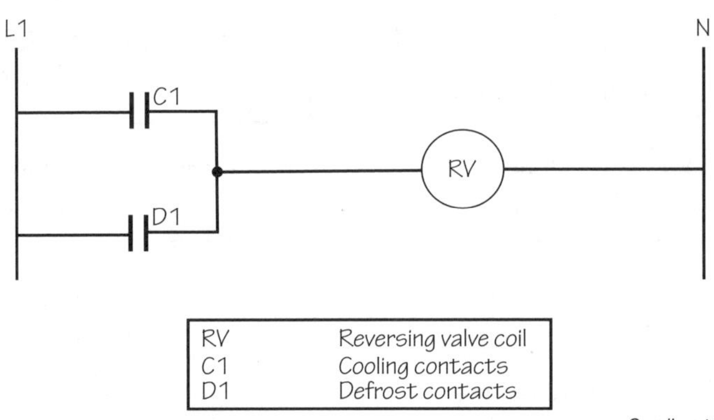

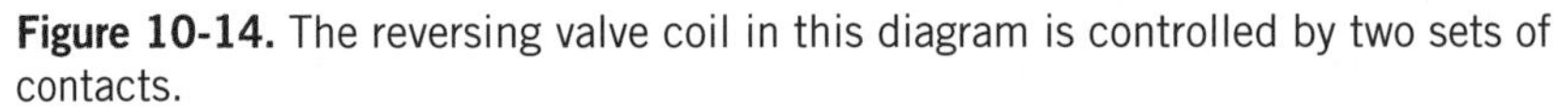

Figure 10-14. The reversing valve coil in this diagram is controlled by two sets of contacts.

The holdback thermostat closes its contacts as the outside temperature drops. With the holdback thermostat's contacts in the closed position, more electric strip heaters are energized when in emergency heat mode, second-stage heating mode, or defrost mode. When the outside ambient temperature is high, the number of strip heaters energized during these three modes of operation is reduced.

A simplified wiring diagram of a holdback thermostat circuit is provided in **Figure 10-15**. The D2 in the diagram represents a normally open set of contacts that are closed when the system goes into defrost mode. The holdback thermostat is labeled HT in the diagram. The components labeled TDR1 and TDR2 represent time-delay relay coils that control the electric strip heaters in the system.

When the heat pump system enters defrost mode when the holdback thermostat is open, this indicates a warm outside ambient temperature. Only TDR2 will be energized, and one heater will be producing heat, **Figure 10-16**. If, however, the outside ambient temperature is low when the system goes into defrost mode, the holdback thermostat closes, and both TDR1 and TDR2 are energized. This powers all three electric strip heaters, **Figure 10-17**.

10.3.3 Heat Pump System Thermostat

Because heat pump systems are designed to provide both comfort heating and comfort cooling, the control strategies employed are more intricate. For example, the space thermostat on a conventional heating-cooling system typically has four terminals.

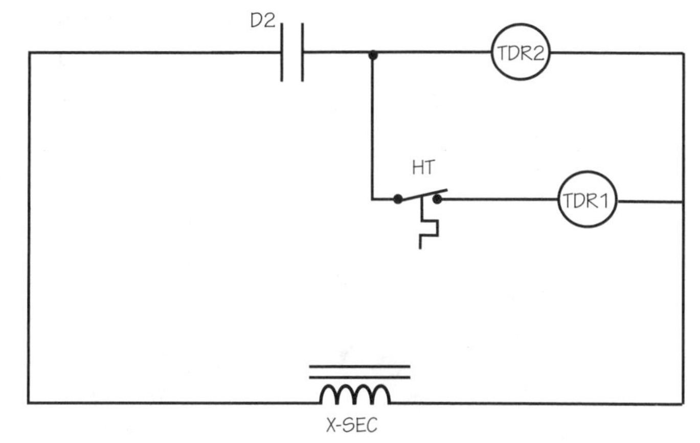

Goodheart-Willcox Publisher

Figure 10-15. Simplified wiring diagram of a holdback thermostat.

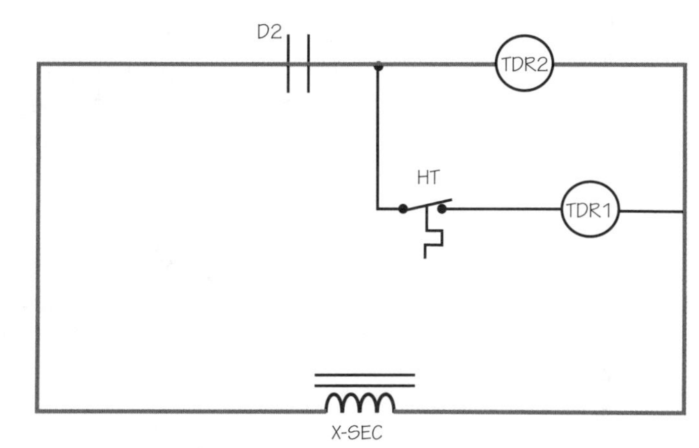

Goodheart-Willcox Publisher

Figure 10-16. Holdback thermostat circuit during a period of warm outside ambient air temperature.

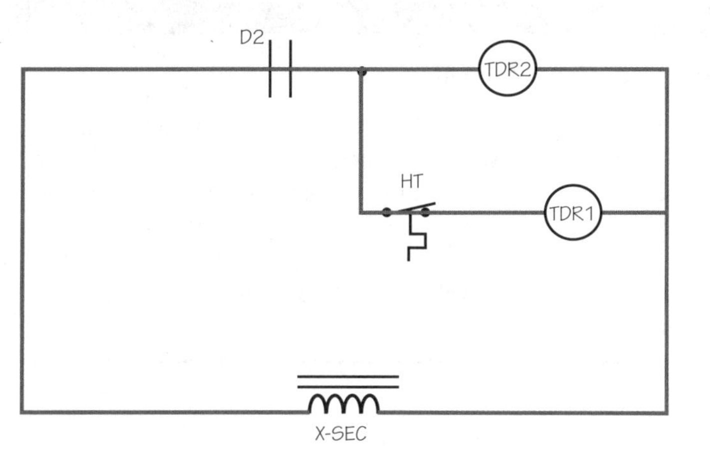

Goodheart-Willcox Publisher

Figure 10-17. Holdback thermostat circuit during a period of cold outside ambient air temperature.

These terminals are R (24-volt power), G (indoor fan motor), Y (cooling), and W (heating). A heat pump system thermostat has many more control terminals. See **Figure 10-18.** These terminals may include the following:

- **R.** Connected directly to the transformer. This terminal brings power to the thermostat.
- **W2.** Second-stage heat terminal.
- **X.** Trouble light circuit (when system goes into lockout).
- **L.** An additional set of contacts used in emergency heat mode.
- **B/O.** Reversing valve terminal. The B terminal is where the reversing valve is connected if the heat pump system is designed to fail in the cooling mode or if the reversing valve must be energized for the system to operate in the heating mode. The O terminal is where the reversing valve is connected if the heat pump system is designed to fail in the heating mode or if the reversing valve must be energized for the system to operate in the cooling mode.
- **Y.** Connected to a compressor contactor coil.
- **W1.** Connected to a compressor contactor coil. This terminal is primarily used if two separate compressors are employed for first-stage heating and cooling. If the same compressor is used for both first-stage heating and first-stage cooling, a jumper wire is installed between the Y and W1 terminals.
- **E.** Emergency heat terminal.
- **G.** Indoor fan motor relay.
- **C.** Common terminal used to return power to the transformer. It allows the thermostat to be powered continuously.

A technician may be required to understand how each terminal operates but not all its interconnections. The technician is responsible for testing the thermostat to ensure the correct sets of contacts are opening and closing at the proper times for adequate heat pump system operation.

BETWEEN THE LINES

Defective Thermostat

If the thermostat does not function correctly, the device is defective and must be replaced. If, however, the thermostat is passing power as it should, the cause for system malfunction lies elsewhere, and further system troubleshooting must be performed. For the purposes of our discussion, the heat pump thermostat is shown as a "black box" with its connection terminals identified, along with the internal connections made in each mode of heat pump system operation.

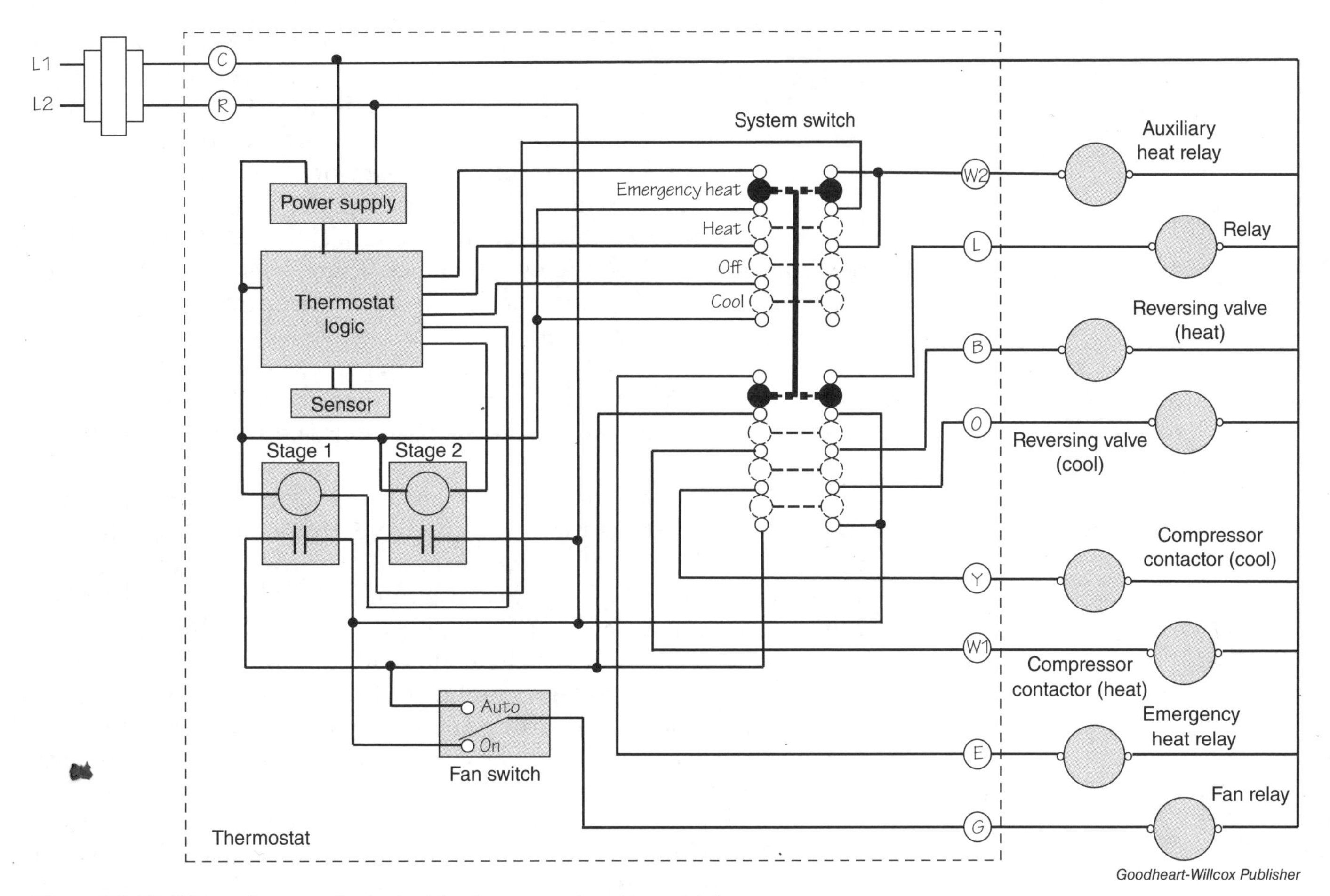

Figure 10-18. Wiring diagram of a typical heat pump system thermostat.

10.3.4 Lockout Relay

A ***lockout relay*** is part of a safety circuit found on both air-conditioning and heat pump systems. See **Figure 10-19.** This relay prevents the system's compressor from operating if the head pressure reaches an unsafe level. Unlike some pressure controls that automatically reset once the pressure has dropped to a safe level, the lockout relay must be reset manually.

A lockout relay circuit consists of a holding coil, a set of normally open contacts, and a set of normally closed contacts. The coil on the lockout relay, LOC, has a lower voltage rating than the compressor contactor coil, CC. During normal compressor operation, the compressor contactor coil, CC, is energized and the high-pressure switch, HPS, is in the closed position.

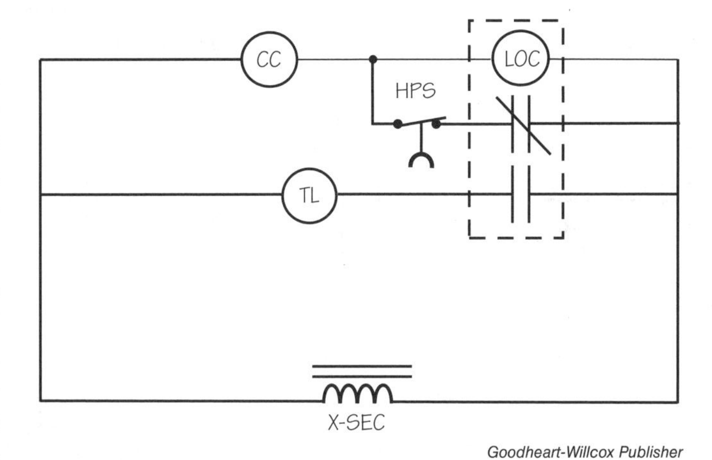

Figure 10-19. A lockout relay wiring diagram.

Current travels through the secondary winding of the transformer, the compressor contactor coil, the high-pressure switch, and the normally closed set of contacts on the lockout relay, **Figure 10-20**. The circuit is completed by the connection from the normally closed lockout relay contacts back to the transformer. The path of electric current is indicated by bold red lines in the schematic diagram.

The current flows through the normally closed contacts, avoiding the coil on the lockout relay altogether. There is no current through the trouble light, TL, during normal compressor operation because the lockout relay contacts in series with the trouble light are open. No current flows through the coil on the lockout relay because the LOC coil offers greater resistance to current flow than the low-resistance alternate path created by the high-pressure switch and the relay's normally closed contacts.

If the system pressure reaches an unsafe level, the high-pressure switch opens. There is then no available path for the current to take to bypass the LOC coil, so the current must flow through it. This is shown in **Figure 10-21**. In this situation, the CC coil and the LOC coil are now connected in series with each other. The 24-volt control voltage then splits between the contactor coil and the lockout relay coil. This makes the voltage supplied to the contactor coil too low to generate a magnetic field strong enough to keep the compressor contactor contacts closed. Thus, the contactor contacts open, turning the compressor off.

The lockout relay coil, LOC, has a lower voltage rating than the CC coil, and so it energizes. This causes the normally open contacts on the LOC relay to close, turning on the trouble light. At the same time, the normally closed contacts on the lockout relay opens. The trouble light remains on and the system stays in lockout until power to the control circuit is de-energized and then re-energized. Even if the pressure switch resets automatically while in lockout, the system remains in lockout because the normally closed contacts on the lockout relay are open.

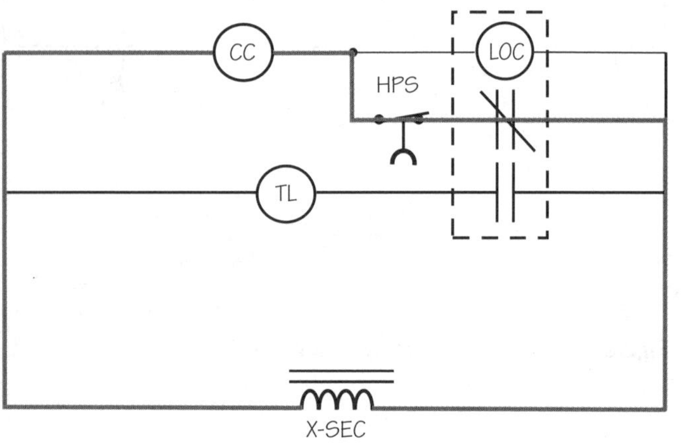

Goodheart-Willcox Publisher

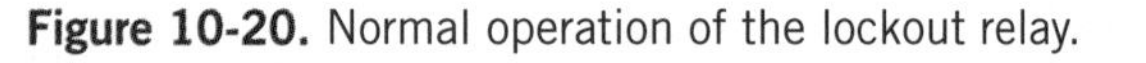
Figure 10-20. Normal operation of the lockout relay.

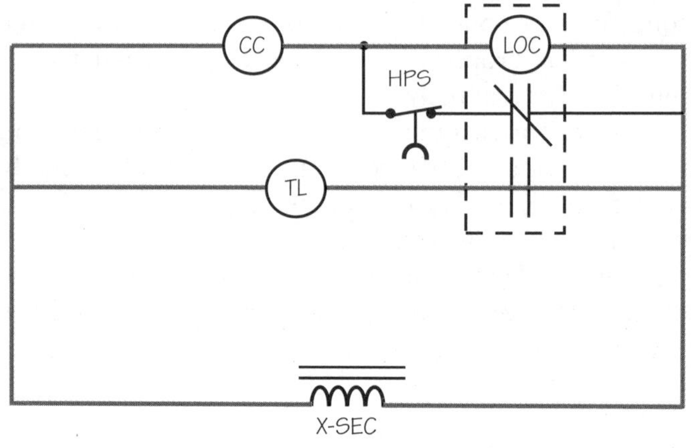

Figure 10-21. A lockout relay wiring diagram in the lockout position.

10.3.5 Electric Strip Heaters

Heat pumps have electric strip heaters that are used in three different modes of operation: as a supplementary heat source; a method to temper air during defrost mode; and emergency heat mode, where they function in place of the vapor-compression system. As a supplementary heat source, the strip heaters are energized when the second-stage heating contacts of the thermostat close. In this mode, the heaters work in conjunction with the vapor-compression heat-pump system.

Electric strip heaters are often controlled by a series of ***time-delay relays*** to manage the significant amount of current drawn from the heaters. The time-delay relay coils are controlled by a heat-pump thermostat. Time-delay relays operate similar to standard relays except there is a delay that occurs between the time the relay's coil is energized and the time the controlled contacts open or close. Most new-generation heat pump systems have incorporated the time delay feature into a solid-state circuit board that controls the operation of the heat pump.

A typical power circuit for electric strip heaters is shown in **Figure 10-22**. For an individual heater to be energized, two other safety devices—a limit switch and a fusible link—must be closed. Each heater has its own limit switch and fusible link. In this diagram, two of the heaters are controlled by the TDR1 contacts, while one heater is controlled by the TDR2 contacts.

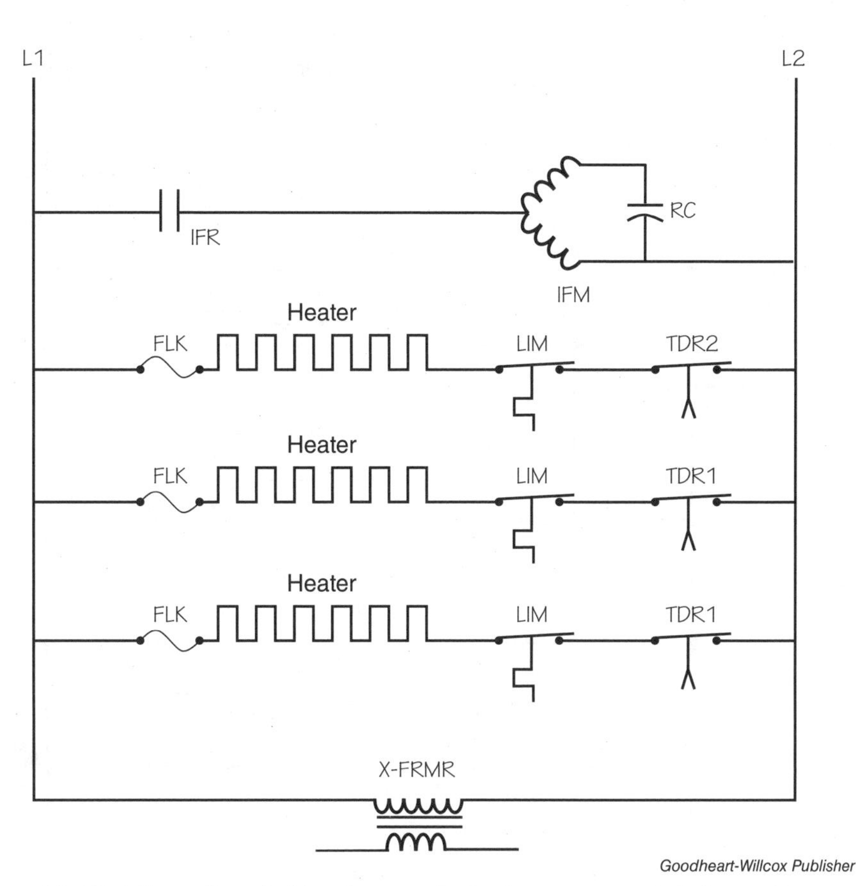

Goodheart-Willcox Publisher

Figure 10-22. A line-voltage ladder diagram of a heat pump system's indoor unit.

10.4 Heat Pump Modes of Operation

Heat pump systems have several modes of operation, including the following:

- Cooling
- First-stage heating
- Second-stage heating with a warm outdoor ambient temperature
- Second-stage heating with a cold outdoor ambient temperature
- Emergency heating with a warm outdoor ambient temperature
- Emergency heating with a cold outdoor ambient temperature
- Defrost with a warm outdoor ambient temperature
- Defrost with a cold outdoor ambient temperature

10.4.1 Cooling

In cooling mode, the compressor, outdoor fan motor, indoor fan motor, and the reversing valve are all energized. In this mode, the system is operating to cool the occupied space, so the compressor must operate. The indoor coil functions as the system evaporator, absorbing heat from the occupied space, while the outdoor coil works as the condenser, rejecting system heat. In cooling mode, the thermostat's power terminal, R, must make an electrical connection with the W1, Y1, and G terminals. These internal thermostat connections are shown on the left side of **Figure 10-23**.

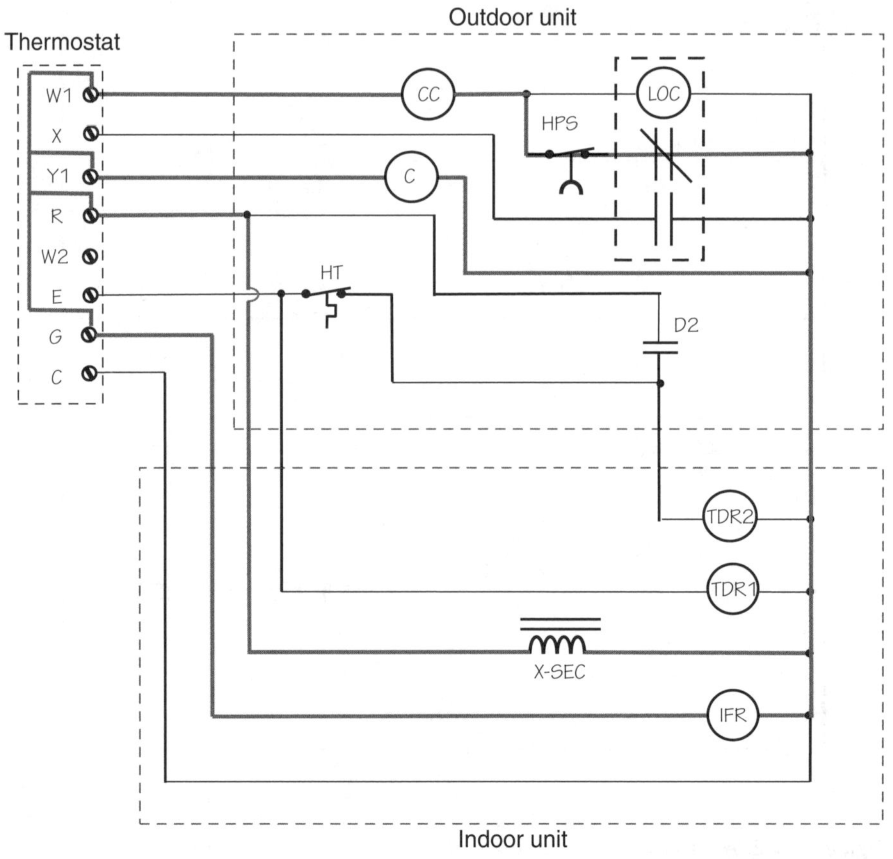

Goodheart-Willcox Publisher

Figure 10-23. Low-voltage control wiring for a heat pump system operating in cooling mode.

From terminal R, the current flows through three separate parallel circuits, as denoted by the bolded red lines in the same diagram:

- The circuit through the compressor contactor coil, CC
- The circuit through the cooling coil, C (reversing valve)
- The circuit through the indoor fan relay coil, IFR

In the cooling mode, thermostat terminals G, W1, and Y1 are energized. For indoor unit operation, the G terminal controls the indoor fan motor, IFM, by closing the IFR contacts on the indoor fan relay, **Figure 10-24A**. For outdoor unit operation, the Y1 terminal controls the cooling relay coil, C, closing the C1 contacts and energizing the reversing valve, RV. In addition, the W1 terminal energizes the CC coil, energizing the compressor and outdoor fan motor, **Figure 10-24B**.

10.4.2 First-Stage Heating

First-stage heating is the heat pump system mode of operation where only the reverse-cycle, vapor-compression refrigeration system is operating. No supplementary electric strip heaters are energized during stage-one heating. So, while operating in first-stage heating, the compressor, outdoor fan motor, and indoor fan motor are all energized. The reversing valve is in its normal, de-energized position, which configures the system to operate in heating mode. In first-stage heating, the thermostat's power terminal, R, must make an electrical connection with the W1 and G terminals. These internal thermostat connections are shown on the left side of **Figure 10-25**.

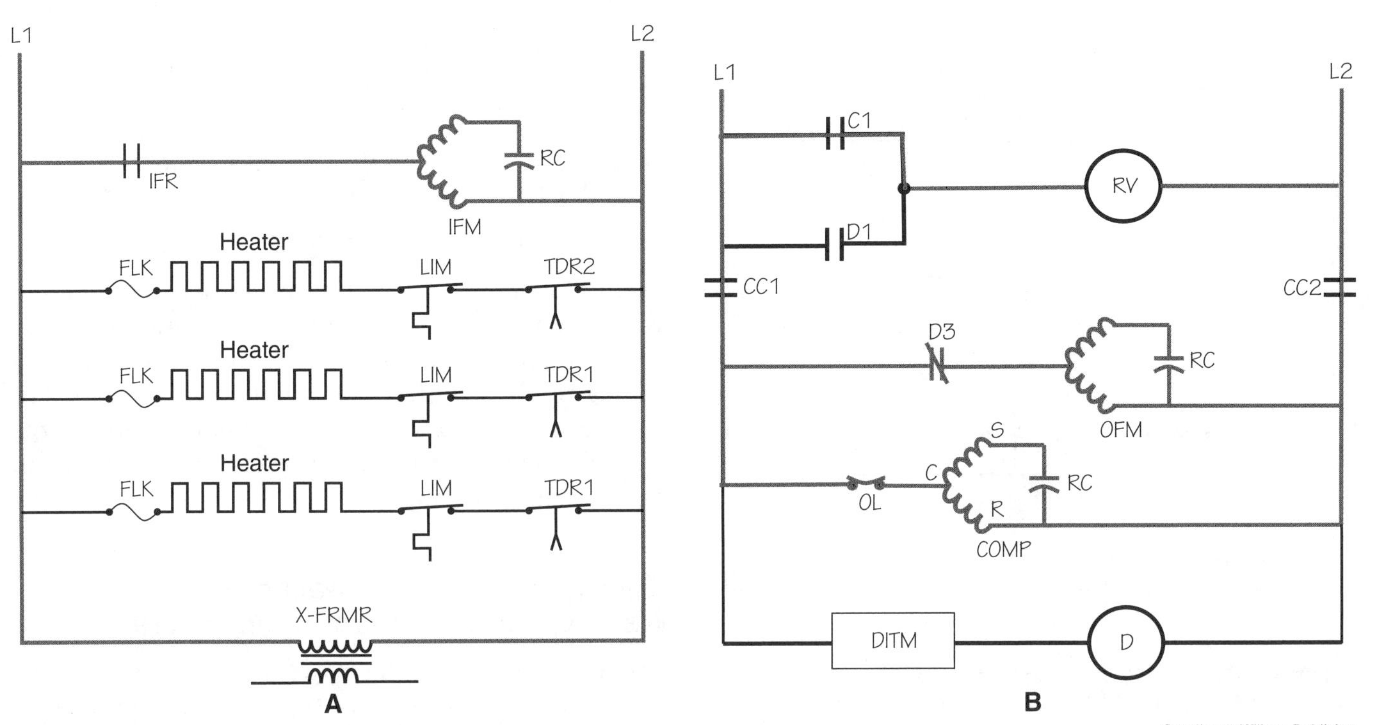

Goodheart-Willcox Publisher

Figure 10-24. A—Line-voltage ladder diagram of the heat pump system's indoor unit in cooling mode. B—Line-voltage ladder diagram of the heat pump system's outdoor unit in cooling mode.

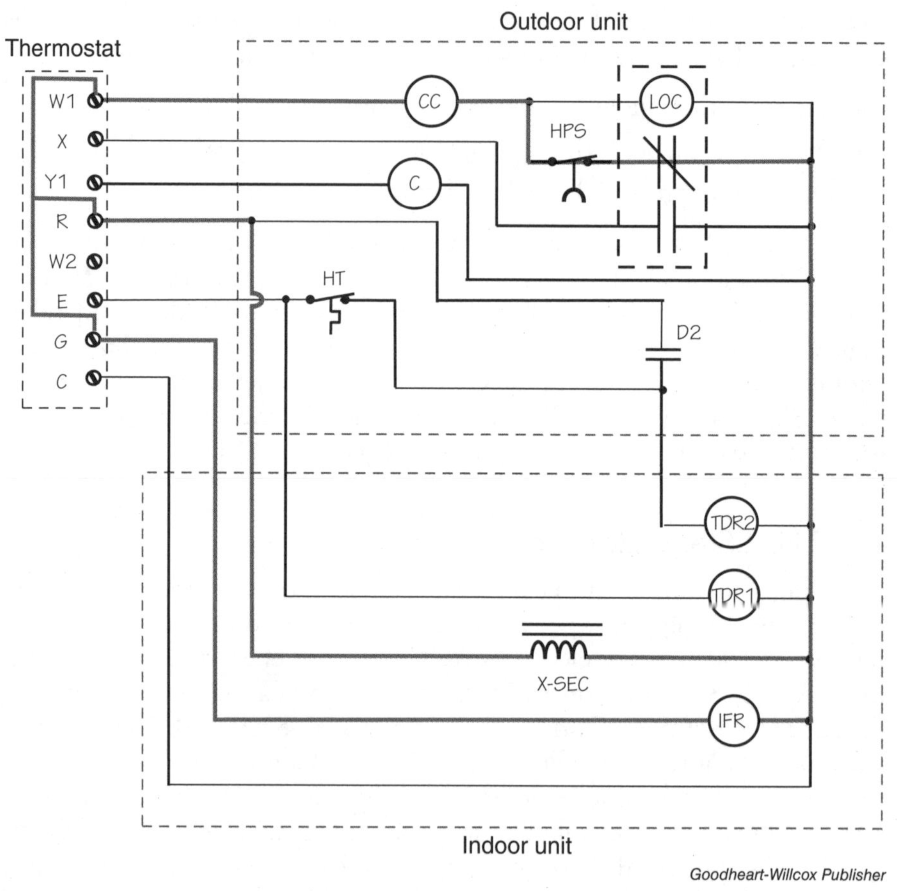

Goodheart-Willcox Publisher

Figure 10-25. Low-voltage control wiring for a heat pump system operating in first-stage heating.

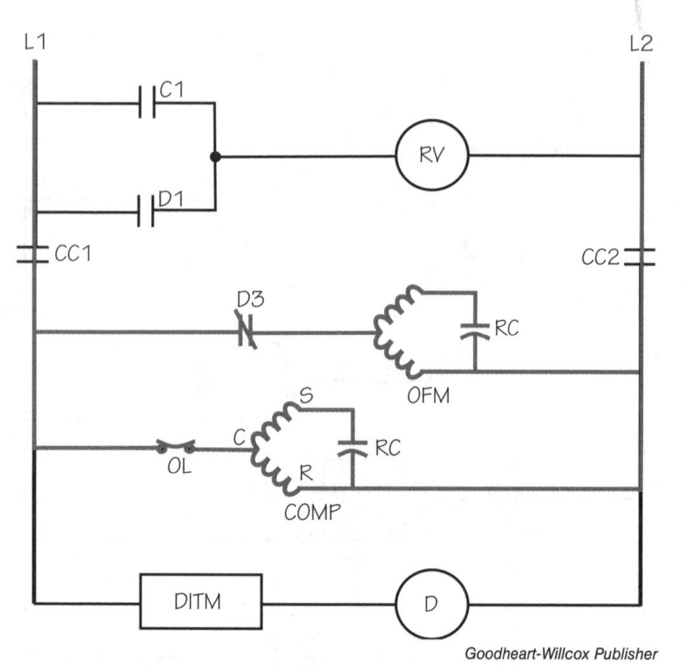

Goodheart-Willcox Publisher

Figure 10-26. Line-voltage ladder diagram of the heat pump system's outdoor unit in first-stage heating.

From terminal R, the current flows through two separate parallel circuits, as denoted by the bolded red lines in the same diagram:

- The circuit through the compressor contactor coil, CC
- The circuit through the indoor fan relay coil, IFR

In first-stage heating, thermostat terminals G and W1 are energized. For indoor unit operation, the G terminal controls the indoor fan motor operation by closing the IFR contacts on the indoor fan relay. For outdoor unit operation, the W1 terminal energizes the CC coil, energizing the compressor and outdoor fan motor, **Figure 10-26.**

10.4.3 Second-Stage Heating with Warm Outside Ambient Temperature

The second-stage heating mode is used if the first-stage heating mode cannot meet the heating requirements of the space. During second-stage heating, some or all of the supplementary electric strip heaters are energized, along with the reverse-cycle, vapor-compression refrigeration system. Second-stage heating is typically initiated on colder days when there is a malfunction in the heat-pump system or when the system has been off for a period of time.

While operating in second-stage heating, the compressor, outdoor fan motor, and indoor fan motor are energized. The reversing valve is not energized. In second-stage heating with a warm outside ambient temperature, the thermostat's power terminal, R, must make an electrical connection with the W1, W2, and G terminals. These internal thermostat connections are shown on the left side of **Figure 10-27.**

From terminal R, the current flows through three separate parallel circuits, denoted by the bolded red lines:

- The circuit through the compressor contactor coil, CC
- The circuit that feeds the TDR1/TDR2 time-delay relay coils
- The circuit through the indoor fan relay coil, IFR

In second-stage heating, thermostat terminals G, W1, and W2 are energized. For indoor unit operation, the G terminal controls the indoor fan motor by closing the IFR contacts on the indoor fan relay. Because the holdback thermostat is open due to the warm outside ambient conditions, only the TDR1 contacts on the heater bank are closed.

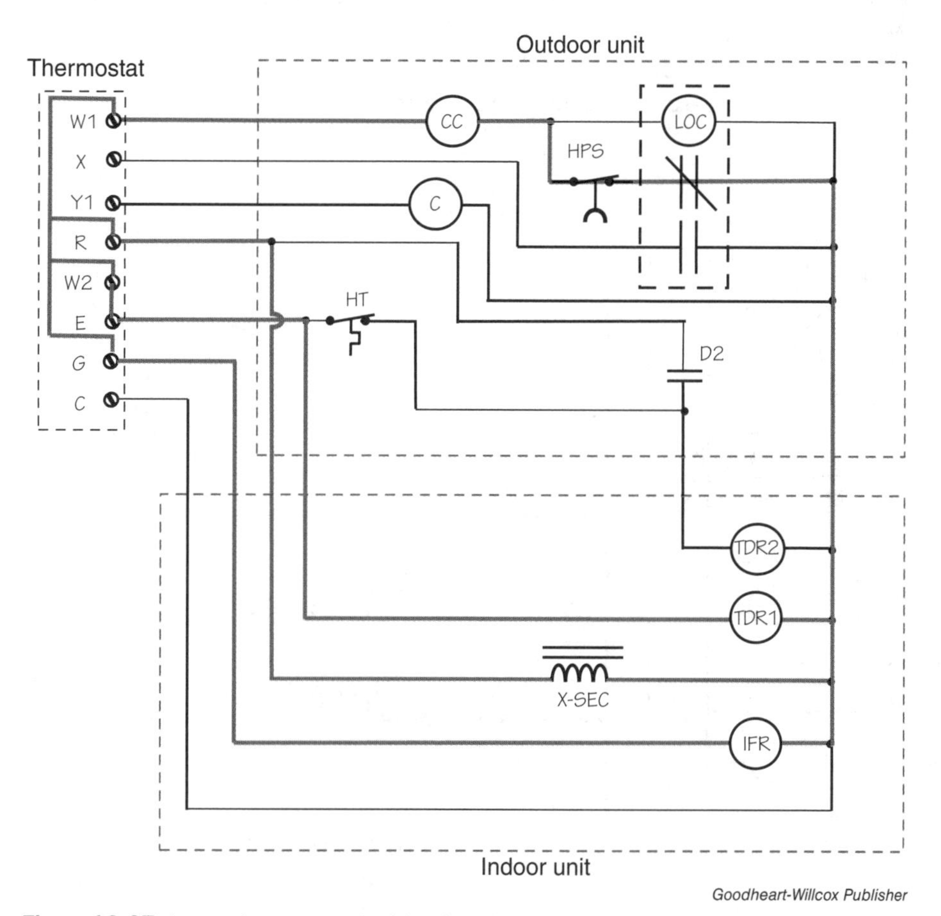

Goodheart-Willcox Publisher

Figure 10-27. Low-voltage control wiring for a heat pump system operating in second-stage heating with a warm outside ambient temperature.

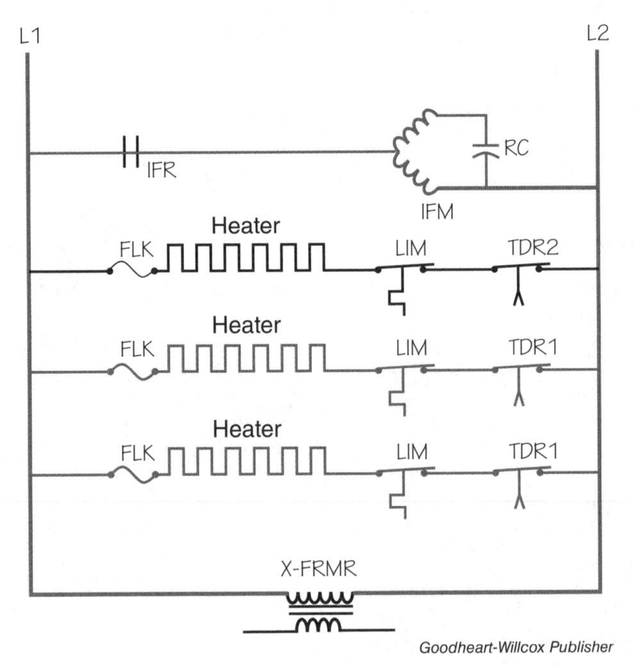

Figure 10-28. Line-voltage ladder diagram of the heat pump system's indoor unit in second-stage (or emergency) heating mode with a warm outside ambient temperature.

See **Figure 10-28**. For outdoor unit operation, the W1 terminal energizes the CC coil, which energizes the compressor and outdoor fan motor.

10.4.4 Second-Stage Heating with Cold Outside Ambient Temperature

Additional assistance for heating is required if the outside temperature is too cold for the reverse-cycle refrigeration system to satisfy the heating requirements of the structure by itself. To provide for this additional heating capacity, the holdback thermostat closes, allowing the additional heater, controlled by the TDR2 coil and contacts, to be energized.

In second-stage heating with a cold outside ambient temperature, the thermostat's power terminal, R, still makes an electrical connection with the W1, W2, and G terminals, similar to first-stage heating mode. The same three parallel control circuits are also created. The main difference between second-stage heating with a warm ambient and a cold ambient is the position of the holdback thermostat. Because the holdback thermostat is now closed due to the cold outside ambient conditions, both the TDR1 and the TDR2 coils are energized, **Figure 10-29A**. By energizing both TDR coils, all TDR contacts in the indoor unit are closed, and all supplementary electric strip heaters are energized, **Figure 10-29B**.

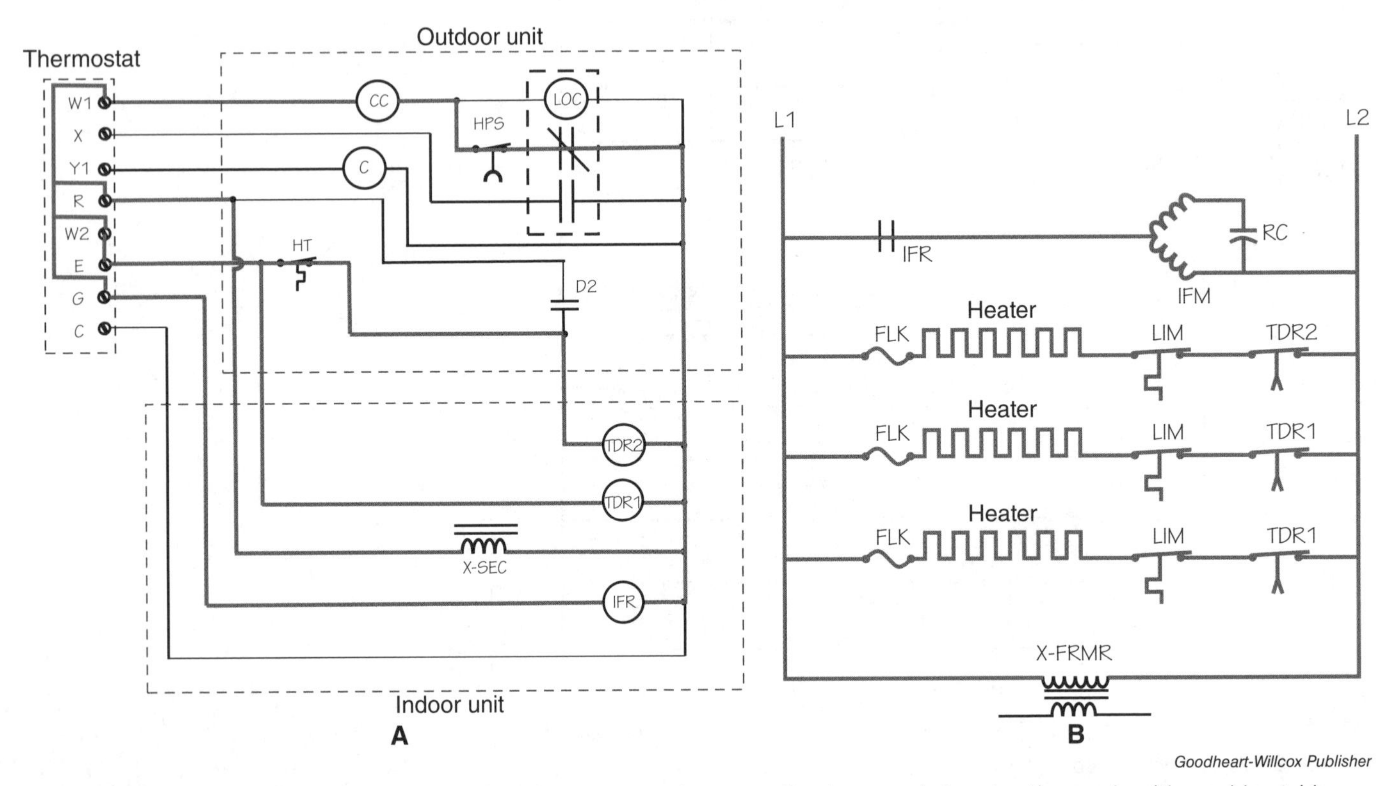

Figure 10-29. A—Low-voltage control wiring for a heat pump system operating in second-stage heating mode with a cold outside ambient temperature. B—Line-voltage ladder diagram of the heat pump system's indoor unit in second-stage (or emergency) heating mode with a cold outside ambient temperature.

10.4.5 Defrost Mode

During the heating operation of a heat pump, ice can form on the surface of the outdoor coil. This is because, during heating mode, the outdoor coil functions as the evaporator, which operates at a temperature lower than the outside ambient temperature. Defrost mode is activated to remove this ice. During defrost mode, the outdoor coil operates as the condenser coil. To concentrate the heat in the outdoor unit and speed the coil-defrosting process, the outdoor fan motor cycles off. Some of the electric strip heaters in the indoor unit are energized to raise the temperature of the air being introduced to the space.

Although manufacturers have different methods of initiating, terminating, and otherwise controlling defrost mode, the following sample system utilizes a line-voltage defrost relay. The defrost relay is located in the outdoor unit. See **Figure 10-30**. The coil of this relay is energized when the system is switched to defrost mode. The methods used to initiate and terminate defrost vary and most commonly consist of a combination of time, temperature, and pressure. These methods are identified on the wiring diagrams as a generic DITM, which stands for defrost initiation and termination mechanism.

In this example, the defrost relay coil controls three sets of contacts: D1, D2 and D3. When the defrost cycle is initiated and the defrost relay coil is energized, the following occurs:

- The compressor continues to operate.
- Defrost contacts D1 close, energizing the cooling coil (reversing valve), and switching the system into cooling mode.
- Defrost contacts D2 close, energizing the TDR control circuits. The D2 contacts are in the outdoor unit's low voltage control circuit.
- Defrost contacts D3 open, de-energizing the outdoor fan motor.

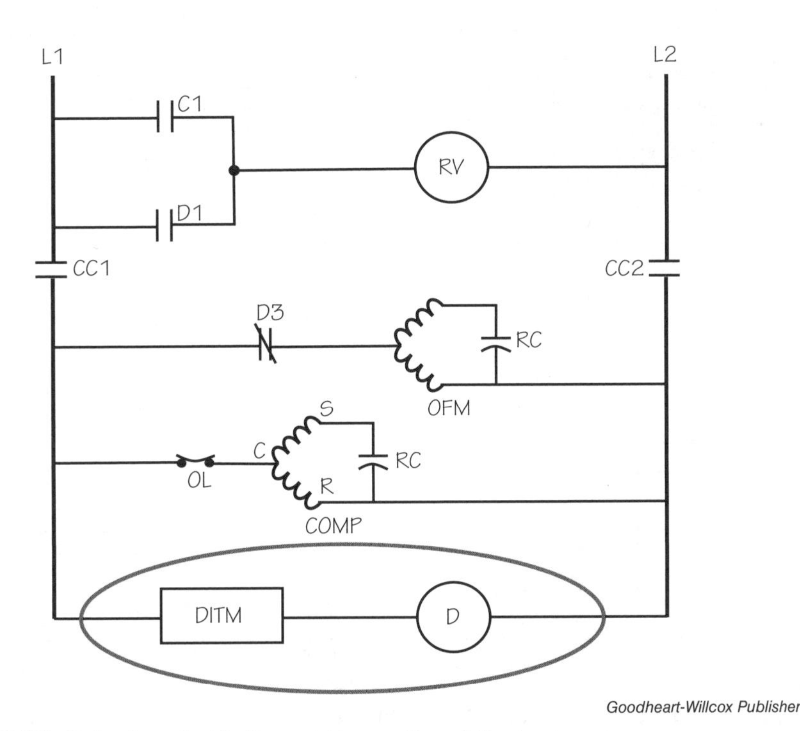

Goodheart-Willcox Publisher

Figure 10-30. Defrost control in the outdoor section of the heat pump.

At the indoor unit, the indoor fan motor continues to operate. One or more of the TDR sets of contacts close to energize some or all of the electric strip heaters. When a heat pump system operates in heating mode and calls for defrost, the system is, in essence, switching over to cooling mode.

Operating in cooling mode can cause confusion for the occupants in the space, especially if they feel cold air coming from the supply registers. To reduce the effects of a cool or cold draft, supplementary heaters are energized to temper the air being supplied to the space. The amount of heat added to the space is determined by the outside air temperature and the rate at which heat leaks out of the structure. When the outside ambient temperature is high, the rate of heat leakage from the structure is low and fewer electric strip heaters are energized.

Figure 10-31 shows a low-voltage wiring diagram for the heat pump system in defrost mode with a warm outside ambient temperature. When the outside ambient temperature is low, the rate of heat leakage from the structure to the outside is high and more supplementary heaters are energized. **Figure 10-32** shows a low-voltage wiring diagram for a heat pump system in defrost mode with a cold outside ambient temperature.

At the outdoor unit, when defrost is initiated, the defrost relay coil, D, is energized, **Figure 10-33A**. Once the defrost relay coil energizes, the normally open contacts on the defrost relay close, and the normally closed contacts open. This energizes the reversing valve coil by closing the normally open D1 contacts and de-energizes the outdoor fan motor by opening the D3 contacts, **Figure 10-33B**.

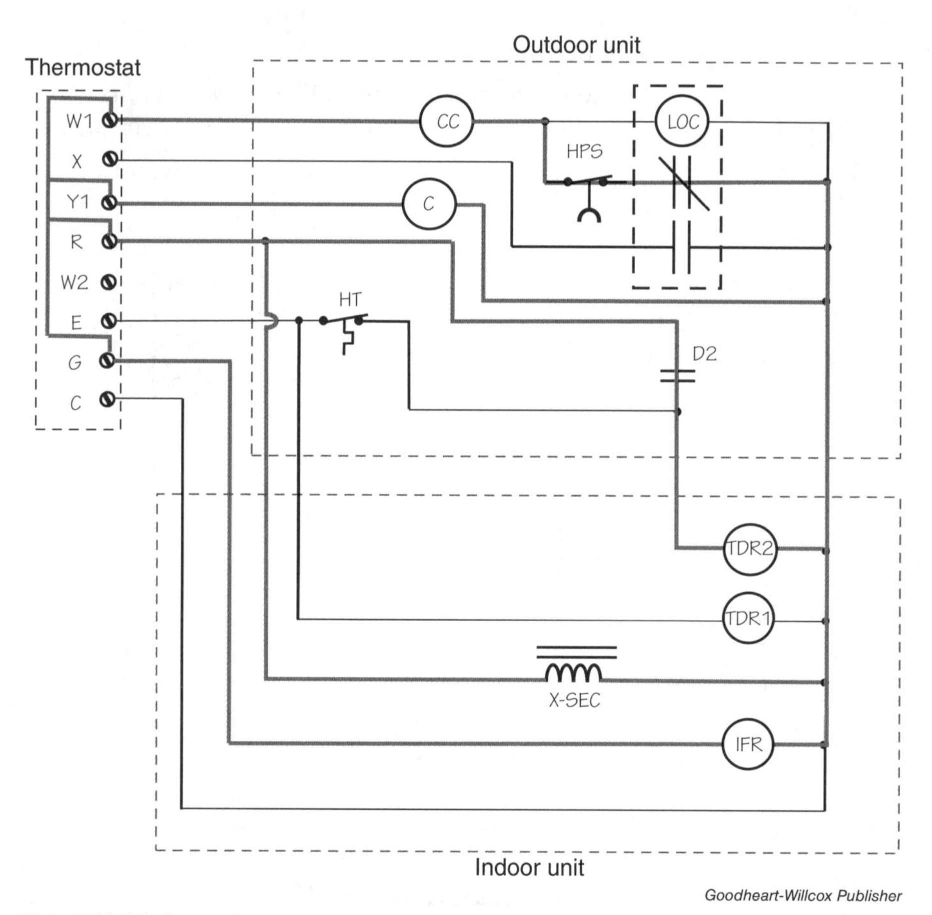

Goodheart-Willcox Publisher

Figure 10-31. Low-voltage control wiring for a heat pump system operating in defrost mode with a warm outside ambient temperature.

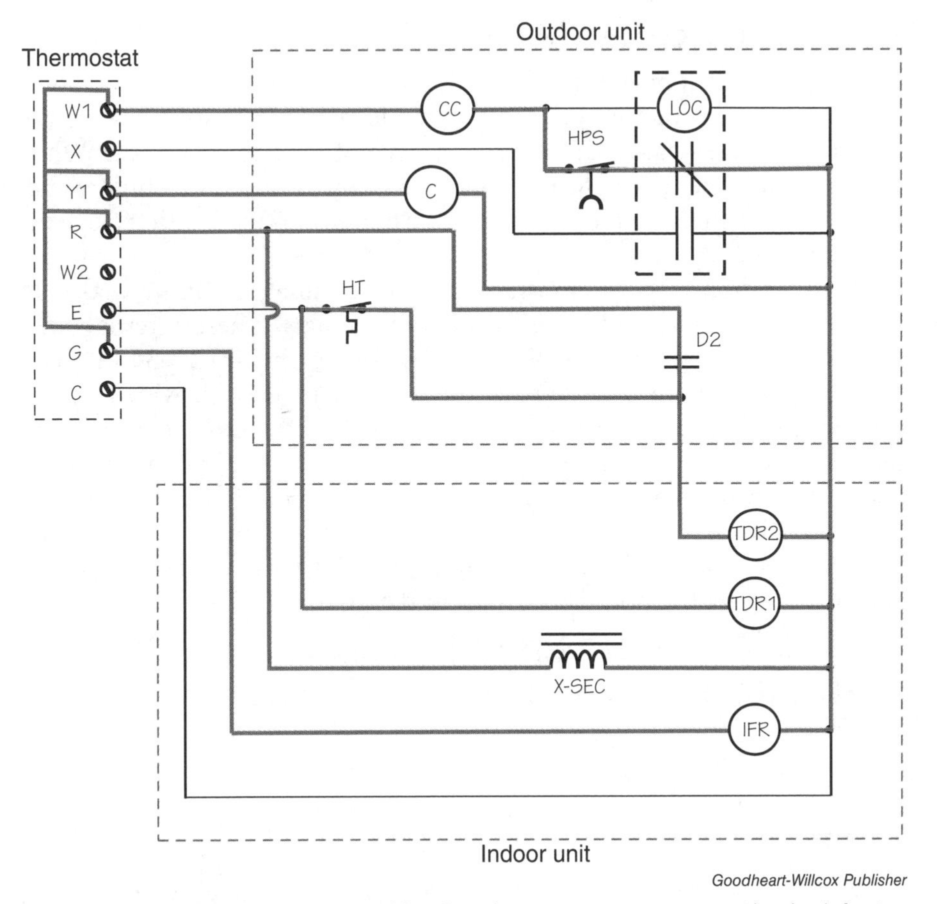

Figure 10-32. Low-voltage control wiring for a heat pump system operating in defrost mode with a cold outside ambient temperature.

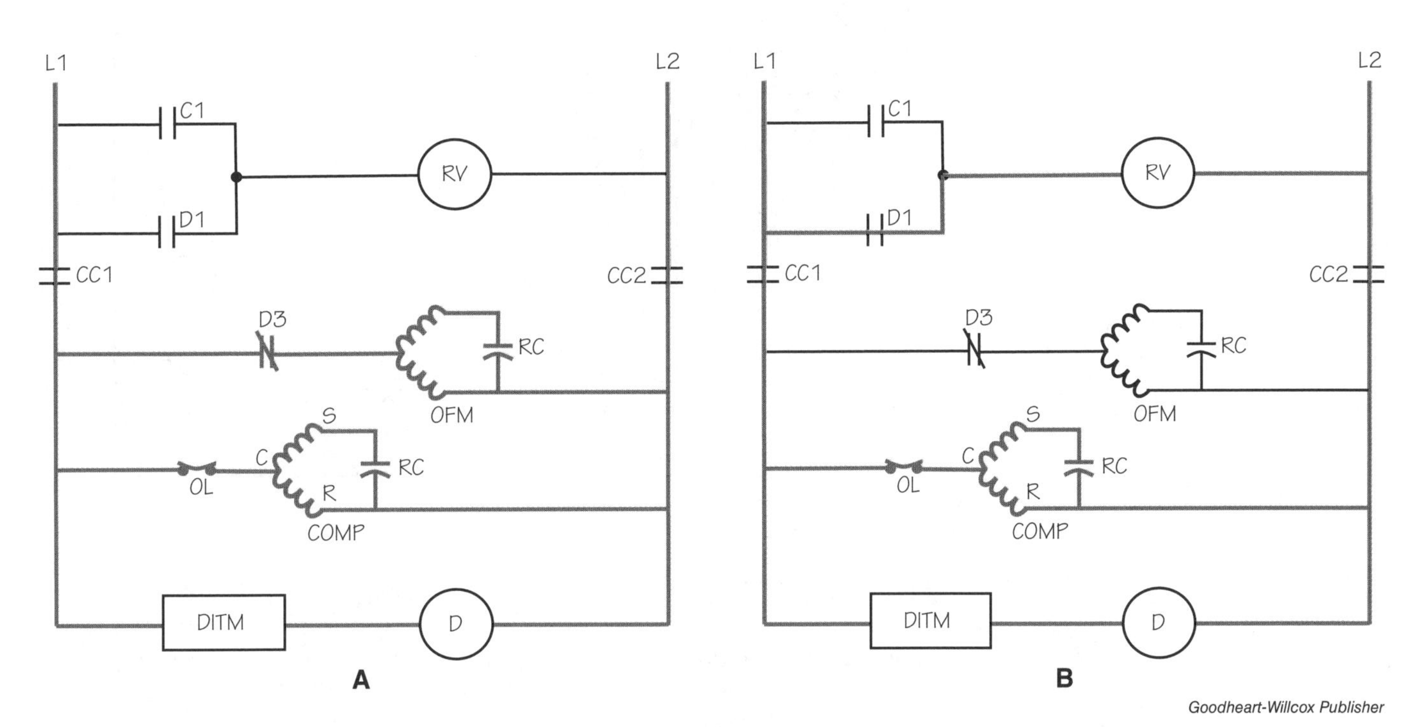

Figure 10-33. A—Initiation of the defrost mode. B—System is in defrost with the reversing valve energized and the outdoor fan motor de-energized.

10.4.6 Emergency-Heat Mode

Emergency-heat mode is used in the event of a failure in the vapor-compression heat-pump cycle. In this mode, all electric strip heaters are energized. These heaters are the sole source of heat for the space. While operating in emergency heat mode, the compressor and outdoor fan motor are not energized. Only the indoor fan motor and heating elements are energized.

In operation, the thermostat's power terminal, R, must make an electrical connection with the E and G terminals. These internal thermostat connections are shown on the left side of **Figure 10-34**. As with second-stage heating and defrost, the amount of heat provided by the supplementary electric strip heaters is determined by the outside ambient temperature. From this figure, we see the outside ambient temperature is warm. If it were colder outside, the HT contacts would be closed and the TDR2 coil would be energized, turning more heaters on.

From terminal R, the current flows through two separate parallel circuits denoted by the bolded red lines in the figure:

- The circuit that feeds the TDR1/TDR2 time-delay relay coils
- The circuit through the indoor fan relay coil, IFR

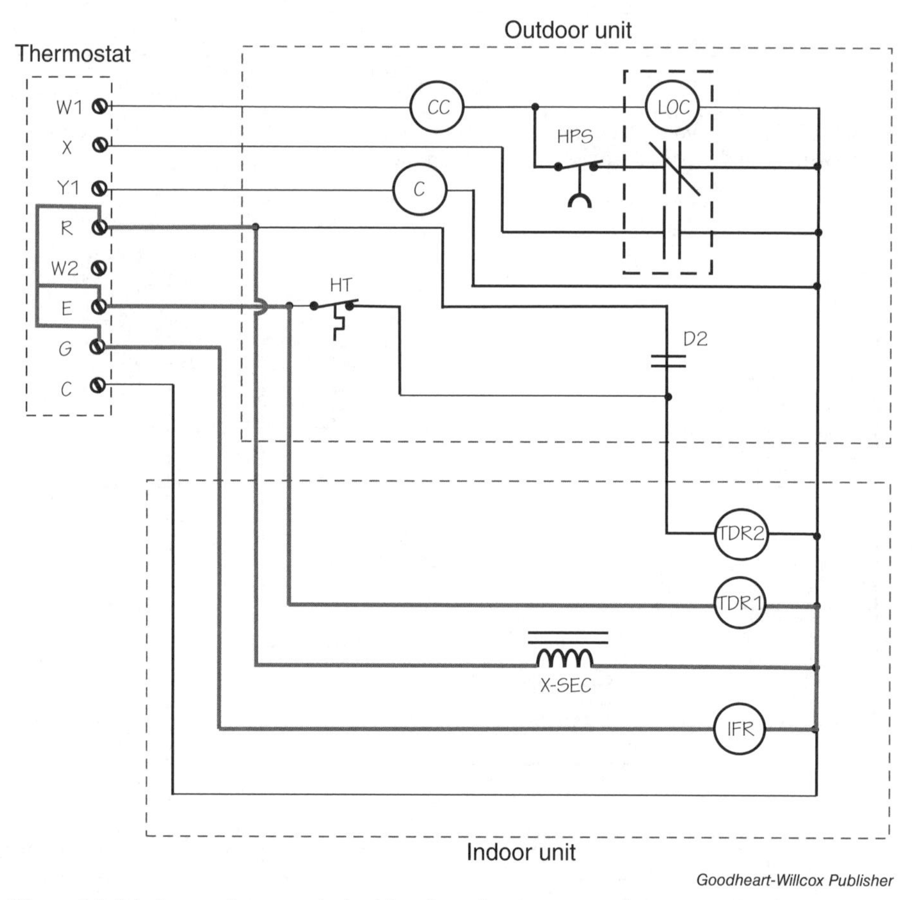

Goodheart-Willcox Publisher

Figure 10-34. Low-voltage control wiring for a heat pump system operating in emergency heat mode with a warm outside ambient temperature.

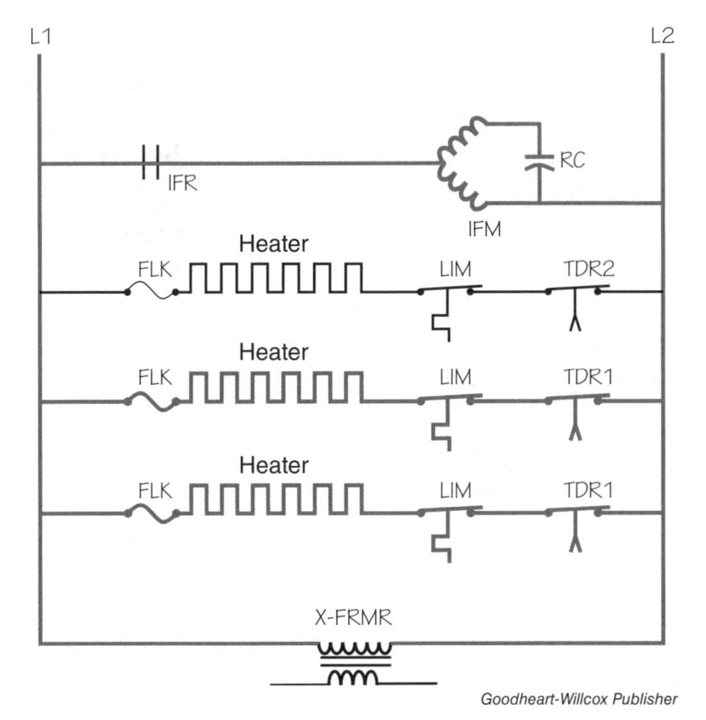

Figure 10-35. Line-voltage indoor unit ladder wiring diagram of a heat pump system operating in emergency heat mode with a warm outside ambient temperature.

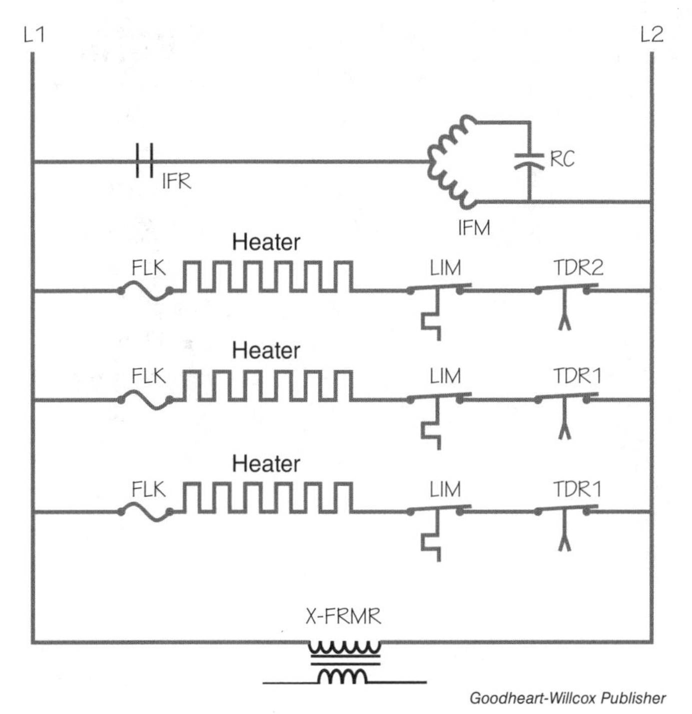

Figure 10-36. Line-voltage indoor unit ladder wiring diagram of a heat pump system operating in emergency heat mode with a cold outside ambient temperature.

The line-voltage ladder wiring diagram in **Figure 10-35** shows the indoor unit circuit when the outdoor ambient temperature is warm and the holdback thermostat is open. **Figure 10-36** shows what the circuit will look like when the contacts on the holdback thermostat close, energizing all the heater circuits. This maximizes the amount of heat introduced to the conditioned space. Although the electric strip heaters are energized in both emergency-heat mode and second-stage heating mode, operation of the strips in emergency mode is initiated manually by the system operator. Second-stage heating is initiated automatically by the thermostat.

Summary

- A furnace, a piece of HVACR equipment, can heat a space, and has the ability to provide comfort heating. A furnace uses a blower and duct system to remove and supply air.
- A furnace has multiple safety devices, including three switches, incorporated into its control systems. They must be wired correctly to ensure safe and efficient system operation.
- A hot water heating system uses boilers to heat the water, as well as a piping arrangement to carry the heated water to the occupied space.
- Four different wiring strategies are used for hot water heating systems, which depend on the boiler type and zoning configuration.
- A heat pump can add or remove heat from a conditioned space by using mechanical energy.
- A reversing valve changes the mode of operation from heating to cooling on a heat pump. The valve is typically energized when the system is in cooling or defrost mode.
- A holdback thermostat is an outdoor thermostat located within the outdoor unit of a heat pump system. It determines the number of heating elements energized.
- A lockout relay is part of a safety circuit found on both air-conditioning and heat pump systems.
- There are several modes of heat pump system operation, including cooling, first-stage heating, second-stage heating, emergency heating, and defrost.

Chapter Assessment

Name ______________________ Date ______________ Class ______________

Know and Understand

_______ 1. Which of the following is true regarding high-mass boilers?

A. They tend to heat up slowly and cool down quickly.

B. They tend to heat up and cool down slowly.

C. They tend to heat up quickly and cool down slowly.

D. They tend to heat up and cool down quickly.

_______ 2. Which of the following is typically true regarding the control of a low-mass boiler?

A. The temperature of the water in the boiler is controlled by the space thermostat.

B. The heat source is energized by an initial call for heat from the space thermostat.

C. The water in the boiler is maintained at the desired temperature all the time.

D. For the boiler to operate, the temperature of the water in the boiler must be higher than the set point on the limit switch.

_______ 3. The purpose of a lockout relay on a heat pump system is to prevent the _____.

A. compressor from operating if the system pressure drops too low

B. compressor from operating if the system pressure rises too high

C. electric strip heaters from operating if the space temperature rises too high

D. electric strip heaters from operating if the space temperature drops too low

_______ 4. The holdback thermostat on a heat pump system is most similar to a(n) _____.

A. compressor overload

B. high-pressure switch

C. outside ambient thermostat

D. low-pressure switch

_______ 5. Electric strip heaters on heat pump system are used for which of the following?

A. First-stage heat, second-stage heat, and defrost modes

B. Emergency-heat, second-stage heat, and defrost modes

C. First-stage heat, emergency-heat, and defrost modes

D. First-stage heat, second-stage heat, and emergency heat modes

_______ 6. When multiple switches are wired in series with each other, _____.

A. all switches must be off to de-energize the circuit

B. all switches must be closed to energize the circuit

C. closing any one switch energizes the circuit

D. opening any one switch energizes the circuit

_______ 7. When multiple switches are wired in parallel with each other, _____.

A. all switches must be off to energize the circuit

B. all switches must be closed to energize the circuit

C. closing any one switch energizes the circuit

D. opening any one switch de-energizes the circuit

_______ 8. Referring to the following figure, which of the following is a likely cause for the circulating pump to be operating when the heat source is not energized?

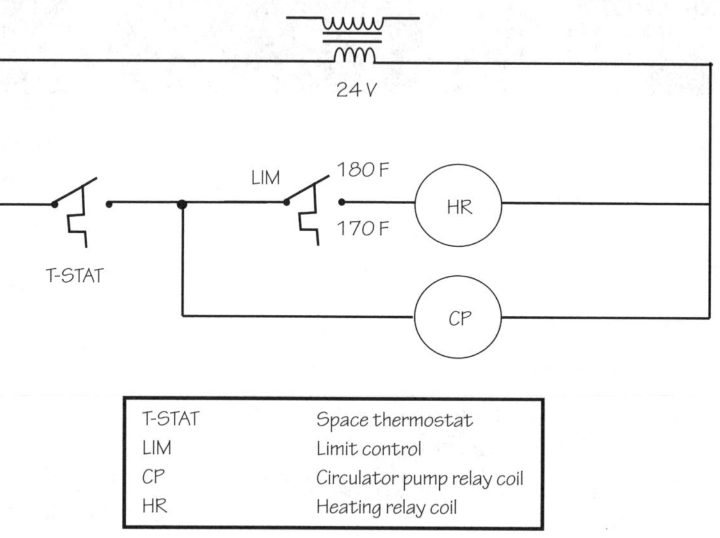

Goodheart-Willcox Publisher

A. The water temperature in the boiler is 150°F.
B. The water temperature in the boiler is 175°F and rising.
C. The water temperature in the boiler is 150°F and falling.
D. The water temperature in the boiler is 182°F.

Name ______________________________ Date ____________ Class ____________________

Apply and Analyze

_______ 1. Referring to the heat pump wiring diagram in the following figure, the C coil is energized by which thermostat terminal?

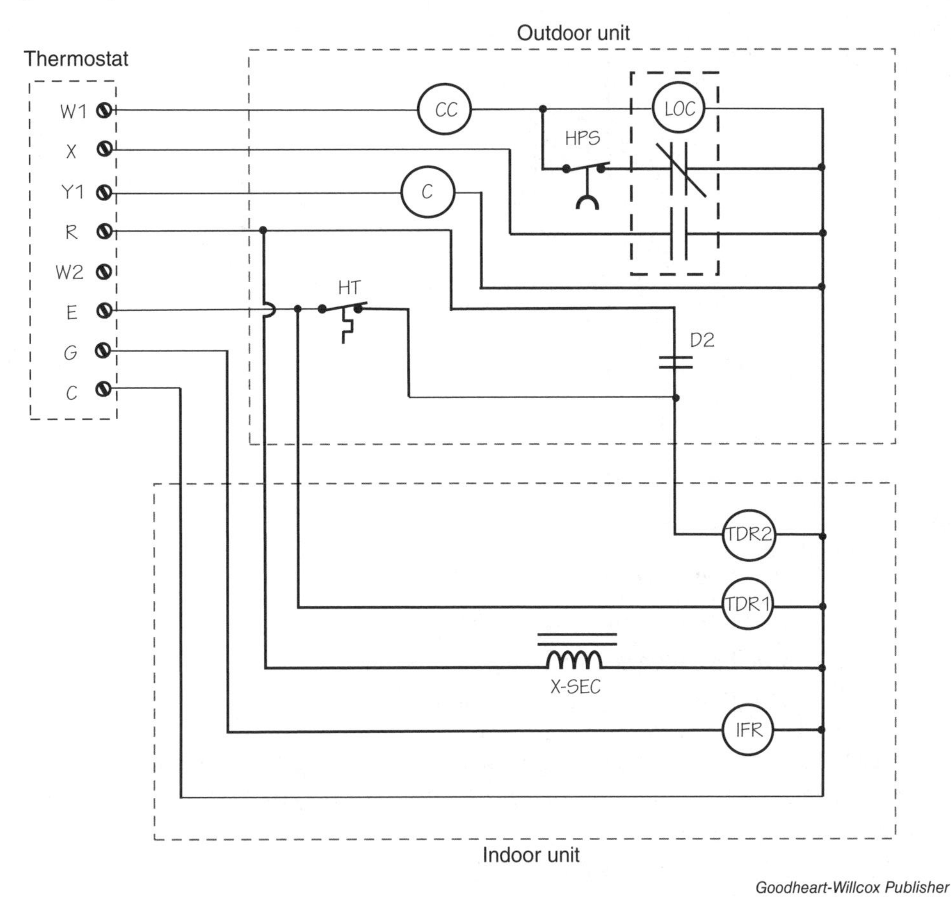

Goodheart-Willcox Publisher

A. W1
B. X
C. Y1
D. R

_______ 2. Referring to the heat pump wiring diagram in the following figure, if the compressor is operating but the outdoor fan motor is not, what is the most likely reason for this?

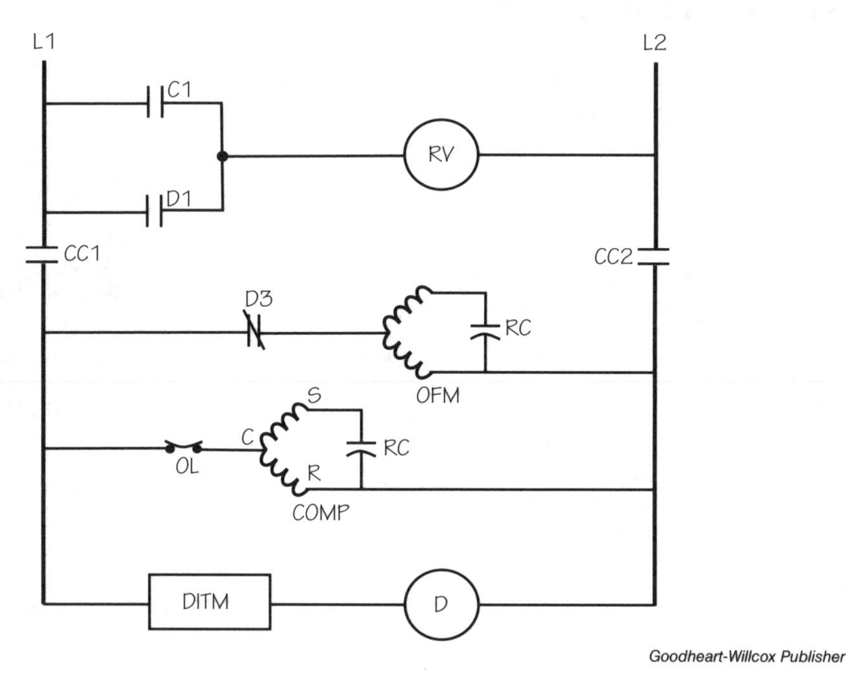

Goodheart-Willcox Publisher

A. The CC1 and CC2 contacts are open.
B. The D3 contacts are open.
C. The reversing valve coil is defective.
D. The compressor overload is open.

3. Regarding an outdoor unit, what are the main differences between a heat pump system operating in the cooling mode and a heat pump system operating in defrost mode?

____

____

____

____

____

____

____

____

Name ______________________ Date ____________ Class ______________

4. Referring to the following figure, what would be a likely cause for the outdoor unit of a heat pump system to be completely encased in a block of ice as the outdoor fan motor is operating?

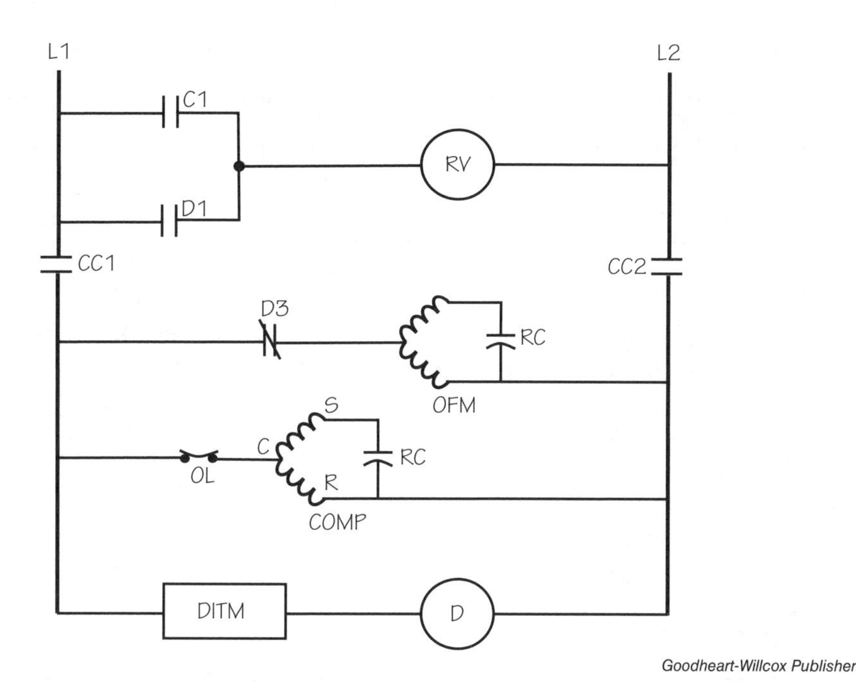

Goodheart-Willcox Publisher

____

____

____

____

____

____

_______ 5. Which of the following would be true if the outdoor fan motor remained running while the heat pump system entered the defrost cycle?

A. The outdoor coil would get too hot, and damage could occur.
B. The time required to defrost the coil would increase.
C. Liquid refrigerant could enter the compressor.
D. The defrost cycle would terminate too quickly.

Critical Thinking

1. How can a defective end switch on a zone valve prevent a system from heating a space effectively?

2. What would the most likely customer complaint be if a high-mass boiler was wired in a similar fashion to a low-mass boiler?

3. Why are heat pump thermostats equipped with an emergency heat indicator light?

4. Why are certain safety control devices, such as the lockout relay on a heat pump system, designed to be reset manually?

11 HVACR Schedules

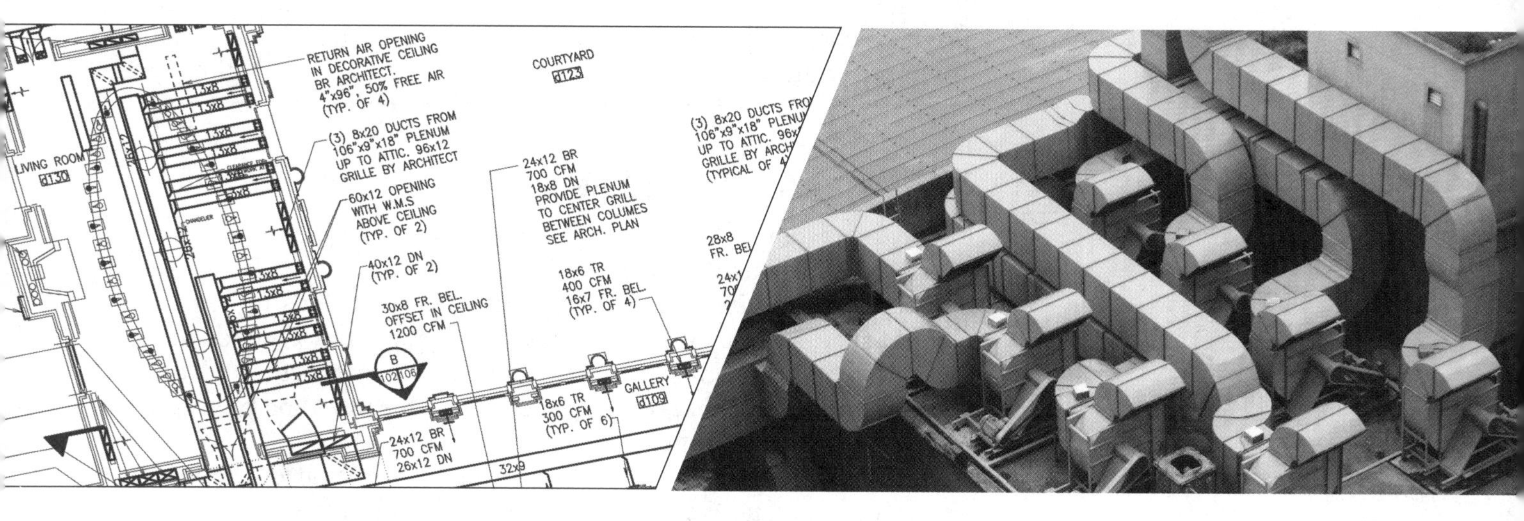

Chapter Outline

Learning Objectives

After studying this chapter, you will be able to:

- Define schedule as it applies to a set of project plans.
- Explain how schedules can be used to effectively and accurately help estimate project costs.
- Explain the purpose of tags on prints and how they relate to schedules.
- Extract information from common HVACR schedules.

Introduction

A complete set of job prints contains a vast amount of information in many different places and locations. It is easy to overlook some of the details on the plans. For tradespeople who must pull information from a set of plans, no piece of information can be missed. The omission of job-related details can result in improper project bidding and lost revenue for the contractor. The severity of making these omissions varies based on the cost of the item or component not detailed. For example, overlooking small details or low-cost items will have little or no effect on the project's cost. Low-cost omissions and similar oversights are often accounted for and even expected when the bid is being prepared. The omission of higher-cost items, however, such as equipment, is definitely not expected and can greatly affect the contractor's bottom line for the project.

To organize and simplify extracting information from a set of plans, schedules are included in project drawings. You must fully understand prints, their format, and how to interpret them to prevent mistakes when pulling installation- and equipment-related information from project drawings.

Technical Terms

brake horsepower (bhp)

horsepower (hp)

schedule

sound trap

static pressure

tag

variable frequency drive (VFD)

11.1 What Are Schedules?

A ***schedule***, **Figure 11-1**, lists required project components and provides information such as the item, size, and quantity required. Schedules are used to help organize and simplify a set of project plans to help extract key information. They list or group similar equipment, components, and other devices based on characteristics such as their function or location. There may be more than one schedule for the same type of equipment based on the size or scope of the project.

Schedules can also be used to provide information on how a component will operate. For example, a reset schedule shows data to a print reader about the relationship between the supply water temperature of a hot water heating system and the outdoor ambient temperature. See **Figure 11-2**. In this schedule, we can determine that, if the outdoor ambient temperature is 40°F, the boiler heats system water to 140°F. If the outdoor ambient temperature rises to 60°F, the system water is heated to 120°F.

11.2 Types of Schedules

Schedules contain different sets of information based on the project and work needed. An average schedule may list information such as the item, an identification mark, size, or number that cannot be written or expressed on a print due to space constraints. The building trades use many types of schedules, including electrical schedules, general construction schedules, and plumbing schedules.

HEATING COIL SCHEDULE BASED ON McQUAY													
TAG	MODEL No.	LOCATION	CFM	MBH TOTAL	FPM	FACE AREA SQ. FT.	FIN LEN. IN.	FIN HGT. IN.	AIR PD IN. H2O	WATER PD FT. H2O	GPM	ROWS FIN/IN
HC/H1-1	5BS0801G	PLAYHOUSE	200	6.5	400	0.5	12	6	.07	.10	0.7	1/8
HC/H1-2	5WQ0801G	PLAYHOUSE	1600	52.3	512	3.13	30	15	0.1	1.0	5.2	1/8
HC/H1-3	5BS0801G	PLAYHOUSE	200	6.5	400	0.5	12	6	.07	.10	0.7	1/8
HC/H1-4	5WQ0801G	PLAYHOUSE	1750	56.8	525	3.33	32	15	.11	1.2	5.7	1/8
HC/H1-5	5WQ0801G	PLAYHOUSE	600	19.4	450	1.33	16	12	.08	0.2	2.0	1/8
HC/H1-6	5WQ0801G	PLAYHOUSE	4000	127.3	571	7.0	42	24	.12	3.3	13.1	1/8
HC/H2-1	5WQ0801G	PLAYHOUSE	2700	89.1	514	5.25	36	21	.10	1.8	8.9	1/8

* BASED ON 55°EAT, 180°EWT, & 160°LWT 3-WAY VALVE (5/8" TUBE DIAMETER) SEE DETAIL FOR PIPING & VALVES

FAN SCHEDULE BASED ON COOK									
TAG	LOCATION	CFM	MODEL	HP	BHP	SP	RPM	VOLTAGE	REMARKS
F/H-1	PLAYHOUSE	7,200	CPV-300	2.0	1.35	0.75	567	460/3/60	VFD COMPATIBLE
F/H-2	PLAYHOUSE	1200	CPV-135	0.5	0.25	0.75	1300	460/3/60	
F/H-3	PLAYHOUSE	2,400	CPV-180	.75	.47	0.75	960	460/3/60	
F/H-4	PLAYHOUSE	2,500	CPV-180	.75	.47	0.75	960	460/3/60	VFD COMPATIBLE

* ACCESSORIES:
WEATHER COVER, BELT GUARD, STEEL DRAIN, VIBERATION ISOLATORS, FACTORY INSTALL DISCONNECT, MOTOR TO BE VFD ADAPTABLE, & CONTRACTOR TO CONFIRM FAN ROTATION

SOUND TRAP SCHEDULE BASED ON VIBRA ACOUSTICS						
TAG	DUCT SIZE	CFM	TRAP SIZE	LENGTH	AIR PRESS. DROP	MODEL
ST/H-2	30x12	3,500	30x12	6'-0"	0.2	RNM-MV-F4 (PACKLESS- NO MEDIA)

AREA REP: METRO AIR PRODUCTS (732) 906-9220
REFERENCE # MET-5846

HUMIDIFIER SCHEDULE BASED ON DRI STEAM						
TAG	MODEL No.	CAPACITY	KW	LBS/Hr	VOLTAGE	AMPS
H-7	SDU-VM25	75 lbs	25	75	480/3	30.1
H-8	SDU-VM25	75 lbs	25	75	480/3	30.1

*REMARKS: PROVIDE OPTIONAL FAN UNIT ON TOP OF MODULE, CONDENSATE PUMP, VAPOR-LOGIC TIME PROPORTIONING MODULATING CONTROL.

NOTE: HUMIDIFIERS SHALL BE MOUNTED HIGH WITH WALL BRACKETS, AS PER MANUFACTURERS CONDENSATE TRAP REQUIREMENTS.

CONDENSATE PUMP SCHEDULE BASED ON LITTLE GIANTS							
TAG	GPH	PUMP HEAD FT.	PSI	VOLTAGE	AMPS	TANK CAPACITY GAL.	MODEL
CP/H-1	25	15	8.6	115	1.5	0.3	VCC-20 (LOW PROFILE)
CP/H-2	25	15	8.6	115	1.5	0.3	VCC-20 (LOW PROFILE)
CP/H-3	25	15	8.6	115	1.5	0.3	VCC-20 (LOW PROFILE)
CP/H-4	25	15	8.6	115	1.5	0.3	VCC-20 (LOW PROFILE)
CP/H-5	25	15	8.6	115	1.5	0.3	VCC-20 (LOW PROFILE)
CP/H-6	25	15	8.6	115	1.5	0.3	VCC-20 (LOW PROFILE)

*REMARKS: STAINLESS STEEL SHAFT, ABS TANK, THERMAL OVERLOAD PROTECTION 6'-0" POWER CORD, & CHECK VALVE BUILT-IN

Lizardos Engineering Assoc., P.C.

Figure 11-1. Schedules accompany an HVACR print to provide information that cannot fit on the print itself.

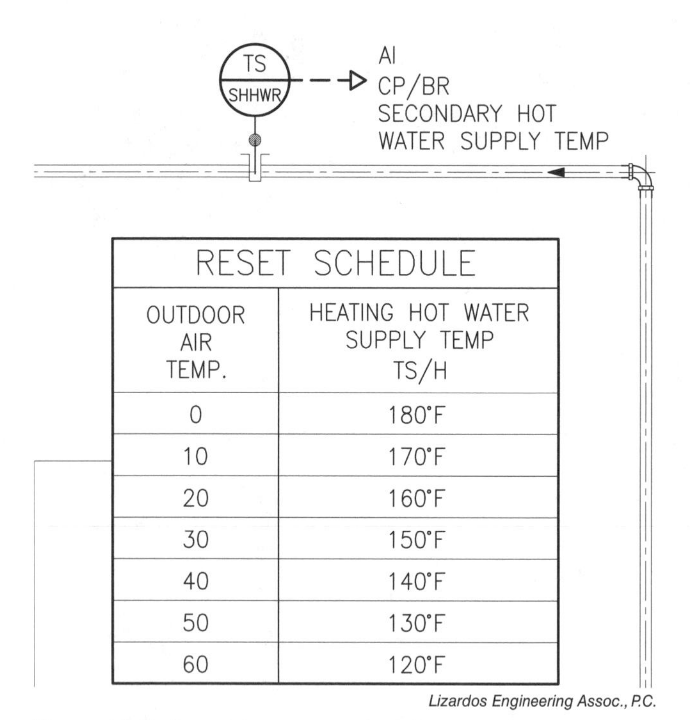

Figure 11-2. An outdoor reset schedule.

Electrical schedules can vary greatly from one project to another based on what needs to be done on that project, who prepared the schedules, and the size of the job. Fixture schedules and panel schedules are two common schedule types and include general information, such as type, manufacturer, and item quantities, as well as more specified information based on the component, **Figure 11-3.**

Construction schedules list information about general items required in a construction project, **Figure 11-4.** These may include door schedules, window schedules, or sky-light schedules. Information in these schedules may include type, dimensions, location, item quantity, and design details.

Plumbing schedules consist of water distribution systems and piping, which are also needed in heating and cooling systems. Some types of plumbing schedules are water closet schedules, lavatory schedules, or piping schedules, **Figure 11-5.**

Electrical Schedules	
Type	**Information**
Fixture schedule	Type, manufacturer, voltage, wattage, lamp type, lamp quantity
Panel schedule	Type, manufacturer, model number, circuit breaker frame size, number of circuits, number of phases, voltage rating, amperage rating

Goodheart-Willcox Publisher

Figure 11-3. Types of electrical schedules.

General Construction Schedules	
Type	**Information**
Door schedule	Type, height, width, interior/exterior, solid/hollow, left/right
Window schedule	Frame material, glass type, window style, number of panes, emissivity coatings, fixed or movable sash, U-factor, solar heat gain coefficient
Skylight schedule	Type, fixed/vented, glass type, frame type, manual/automatic

Goodheart-Willcox Publisher

Figure 11-4. Types of general construction schedules.

Plumbing Schedules	
Type	**Information**
Water closet (toilet) schedule	Make, model, color, flush type
Lavatory (sink) schedule	Make, model, faucet configuration, number of valve/stem holes, hole dimensions and spacing, lavatory type

Goodheart-Willcox Publisher

Figure 11-5. Types of plumbing schedules.

11.3 Using HVACR Schedules

HVACR schedules provide information on various mechanical equipment pieces in a project. The information that can be extracted from a schedule will vary based on the items referenced on each schedule and the purpose for each item.

For instance, the same model of equipment can be configured differently, or used for different applications on the same project. In other words, although the model number for multiple units on the job might be the same, the capacity, required flow rate, pressure drop, horsepower, and other variables can often vary.

The following sections introduce some of the common schedules used for HVACR projects. This will familiarize you with the types of information provided in these schedules.

11.3.1 Condensate Pump Schedule

The condensate pump schedule in **Figure 11-6** references six identical pumps. Upon first inspection, it appears the same information is repeated six times, which may be unnecessary. However, the information is repeated to ensure there are six pumps accounted for on this portion of the project. The schedule can be used as a checklist to identify the pumps already installed, as shown in **Figure 11-7**. This helps determine which pumps have not yet been installed and how long a typical installation might take.

CONDENSATE PUMP SCHEDULE

BASED ON LITTLE GIANTS

TAG	GPH	PUMP HEAD FT.	PSI	VOLTAGE	AMPS	TANK CAPACITY GAL.	MODEL
CP/H-1	25	15	8.6	115	1.5	0.3	VCC-20 (LOW PROFILE)
CP/H-2	25	15	8.6	115	1.5	0.3	VCC-20 (LOW PROFILE)
CP/H-3	25	15	8.6	115	1.5	0.3	VCC-20 (LOW PROFILE)
CP/H-4	25	15	8.6	115	1.5	0.3	VCC-20 (LOW PROFILE)
CP/H-5	25	15	8.6	115	1.5	0.3	VCC-20 (LOW PROFILE)
CP/H-6	25	15	8.6	115	1.5	0.3	VCC-20 (LOW PROFILE)

*REMARKS: STAINLESS STEEL SHAFT, ABS TANK, THERMAL OVERLOAD PROTECTION
6'-0" POWER CORD, & CHECK VALVE BUILT-IN

Lizardos Engineering Assoc., P.C.

Figure 11-6. A condensate pump schedule.

CONDENSATE PUMP SCHEDULE

BASED ON LITTLE GIANTS

TAG	GPH	PUMP HEAD FT.	PSI	VOLTAGE	AMPS	TANK CAPACITY GAL.	MODEL
✓ CP/H-1	25	15	8.6	115	1.5	0.3	VCC-20 (LOW PROFILE) ES 4/30/18
✓ CP/H-2	25	15	8.6	115	1.5	0.3	VCC-20 (LOW PROFILE) 05-02-2018 JR
CP/H-3	25	15	8.6	115	1.5	0.3	VCC-20 (LOW PROFILE)
CP/H-4	25	15	8.6	115	1.5	0.3	VCC-20 (LOW PROFILE)
CP/H-5	25	15	8.6	115	1.5	0.3	VCC-20 (LOW PROFILE)
CP/H-6	25	15	8.6	115	1.5	0.3	VCC-20 (LOW PROFILE)

*REMARKS: STAINLESS STEEL SHAFT, ABS TANK, THERMAL OVERLOAD PROTECTION
6'-0" POWER CORD, & CHECK VALVE BUILT-IN

Lizardos Engineering Assoc., P.C.

Figure 11-7. Using a schedule as a checklist. The condensate pump CP/H-1 was installed on 4/30/18 by someone with the initials ES. CP/H-2 was installed on 5/2/18 by JR.

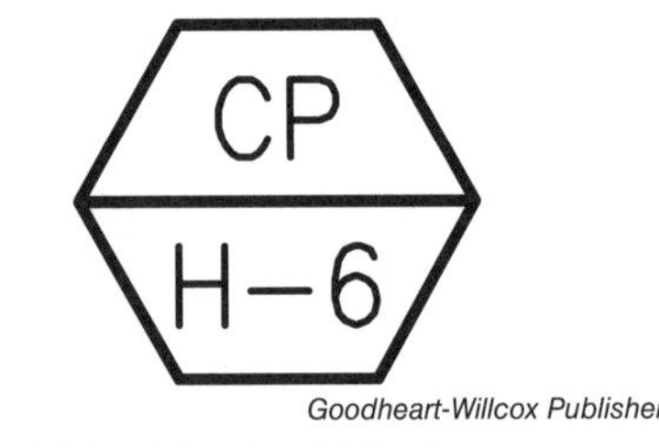

Goodheart-Willcox Publisher

Figure 11-8. A tag for CP/H-6.

Other information on this schedule includes gallons per hour (gph), pump head (feet), psi, voltage, amperage, and tank capacity. When extracting information such as the voltage, we can determine that six 115-V circuits must be provided, each sized to safely handle a current draw of 1.5 A.

When required, schedules can include remarks. These remarks are additional comments that provide more details about the items in the schedules. The additional remarks provide information about the tank and shaft construction, as well as other features, such as the overload, power cord, and check valve that condensate pumps are equipped with.

On the print, each condensate pump is given a tag. A *tag*, **Figure 11-8**, indicates where a specific device or component is to be located. The identification codes on a tag refer to the identifiers on a specific line of a schedule. **Figure 11-9** displays part of the print showing the location of the specific condensate pump, namely CP/H-6, which can be found on the sixth line of the condensate pump schedule.

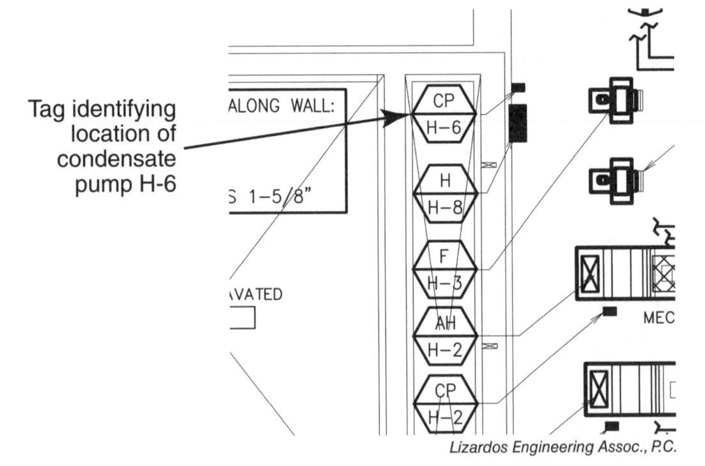

Lizardos Engineering Assoc., P.C.

Figure 11-9. The location of tag CP/H-6.

11.3.2 Humidifier Schedule

A humidifier schedule, like a condensate pump schedule, provides certain electrical data, as well as other information. See **Figure 11-10**. Electrical data are especially useful to the electricians on the job. These help determine how many electrical circuits must be provided, along with the voltage, phase, and current-carrying requirements.

Taking a quick glance at the humidifier schedule tells the electrician that two humidifier circuits need to be run, and each circuit must supply three-phase, 480-V power to the humidifier locations. In addition, each humidifier draws 30.1 amps of current, so the circuits must be installed appropriately to safely handle this load.

This schedule provides both remarks and notes. These remarks offer additional information about the equipment specifications, accessories, and features. This is also the case with remarks provided in the humidifier schedule. The notes provided typically refer to the installation, mounting, and operation of the equipment. The terms *notes* and *remarks* are often used interchangeably, but usage of the terms will vary depending on the intent of the drawing creator.

HUMIDIFIER SCHEDULE — BASED ON DRI STEAM

TAG	MODEL No.	CAPACITY	KW	LBS/Hr	VOLTAGE	AMPS
H-7	SDU-VM25	75 lbs	25	75	480/3	30.1
H-8	SDU-VM25	75 lbs	25	75	480/3	30.1

*REMARKS: PROVIDE OPTIONAL FAN UNIT ON TOP OF MODULE, CONDENSATE PUMP, VAPOR-LOGIC TIME PROPORTIONING MODULATING CONTROL.

NOTE: HUMIDIFIERS SHALL BE MOUNTED HIGH WITH WALL BRACKETS, AS PER MANUFACTURERS CONDENSATE TRAP REQUIREMENTS.

Lizardos Engineering Assoc., P.C.

Figure 11-10. A humidifier schedule.

SOUND TRAP SCHEDULE						BASED ON VIBRA ACOUSTICS
TAG	DUCT SIZE	CFM	TRAP SIZE	LENGTH	AIR PRESS. DROP	MODEL
ST/H-2	30x12	3,500	30x12	6'-0"	0.2	RNM-MV-F4 (PACKLESS- NO MEDIA)

AREA REP: METRO AIR PRODUCTS (732) 906-9220
REFERENCE # MET-5846

Lizardos Engineering Assoc., P.C.

Figure 11-11. A sound trap schedule.

11.3.3 Sound Trap Schedule

A ***sound trap*** is used on HVACR ductwork to reduce noise transmission through the air distribution system. This is accomplished by adding resistance to the airstream. Often when sound traps make up part of an HVACR project, a schedule is included. See **Figure 11-11**. This is because sound traps must be included in all duct system design calculations, given they affect the pressures at which the duct system operates.

Failure to account for the pressure drops associated with the sound traps can prevent the air distribution system from operating correctly. In addition, inclusion of a sound trap schedule calls attention to the device, while providing detailed design-related notes about it. Sound traps, or sound attenuators, can sometimes be difficult to identify on a project drawing and are often overlooked.

11.3.4 Heating Coil Schedules

A heating coil schedule identifies and provides information about the different heating coils used on a particular project. See **Figure 11-12**. This schedule provides only data that pertain to the seven heating coils used in the playhouse. Two different types of heating coils are used: the 5BS0801G coil, which is a lower-capacity coil; and the 5WQ0801G coil, which is a higher-capacity coil.

HEATING COIL SCHEDULE												BASED ON McQUAY
TAG	MODEL No.	LOCATION	CFM	MBH TOTAL	FPM	FACE AREA SQ. FT.	FIN LEN. IN.	FIN HGT. IN.	AIR PD IN. H2O	WATER PD FT. H2O	GPM	ROWS FIN/IN
HC/H1-1	5BS0801G	PLAYHOUSE	200	6.5	400	0.5	12	6	.07	.10	0.7	1/8
HC/H1-2	5WQ0801G	PLAYHOUSE	1600	52.3	512	3.13	30	15	0.1	1.0	5.2	1/8
HC/H1-3	5BS0801G	PLAYHOUSE	200	6.5	400	0.5	12	6	.07	.10	0.7	1/8
HC/H1-4	5WQ0801G	PLAYHOUSE	1750	56.8	525	3.33	32	15	.11	1.2	5.7	1/8
HC/H1-5	5WQ0801G	PLAYHOUSE	600	19.4	450	1.33	16	12	.08	0.2	2.0	1/8
HC/H1-6	5WQ0801G	PLAYHOUSE	4000	127.3	571	7.0	42	24	.12	3.3	13.1	1/8
HC/H2-1	5WQ0801G	PLAYHOUSE	2700	89.1	514	5.25	36	21	.10	1.8	8.9	1/8

* BASED ON 55°EAT, 180°EWT, & 160°LWT 3-WAY VALVE (5/8" TUBE DIAMETER) SEE DETAIL FOR PIPING & VALVES

Lizardos Engineering Assoc., P.C.

Figure 11-12. A heating coil schedule.

This schedule information can be used for installation purposes, as well as for setup and commissioning purposes. For example, consider heating coils HC/H1-2 and HC/H1-6. Although these heating coils have the same model number, their parameters are different.

Despite two heating coils being installed the same way, the operational differences between the two are significant. Additional information can be determined:

- The airflow through the HC/H1-6 heating coil is two and a half times that through the HC/H1-2 coil.
- The heat output of coil HC/H1-6, expressed in thousands of Btu/hour (mbh), is 127,300 Btu/hour, as compared to a heat output of only 52,300 Btu/hour from coil HC/H1-2.
- The air and water pressure differences or differentials across and through the coils are higher for the HC/H1-6 coil than for the HC/H1-2 coil.
- The speed of the air, expressed in feet per minute (fpm), is significantly greater through the HC/H1-6 coil than through the HC/H1-2 coil.
- The water flow rate, expressed in gallons per minute (gpm), for the HC/H1-6 coil is 13.1, compared to the 5.2 gpm flow rate through the HC/H1-2 coil.

Information contained in schedules plays a major role in ensuring that the system and its components are installed, adjusted, and commissioned properly. This then ensures that actual system performance matches the initial intent of the designer.

11.3.5 Fan Schedules

A fan schedule also provides valuable information. Similar fan models are designed to operate with different airflow (cfm) rates, static pressures, and motor speeds. These are common quantities associated with fan schedules and other HVACR-related schedules. It is the responsibility of the print reader to be able to understand and interpret these quantities.

Airflow is expressed in units of cubic feet per minute (cfm) and refers to the volume of air that passes a fixed point in a duct or other air-moving system per unit time. ***Static pressure*** refers to the resistance to airflow caused by the air moving through a duct or other air-moving channel. Static pressure is measured in inches of water column, where 27.7 inches = 1 pound per square inch (psi). Two other terms commonly found on fan schedules are horsepower (hp) and brake horsepower (bhp). ***Horsepower (hp)*** is used to describe the size of a motor, while ***brake horsepower (bhp)*** refers to the amount of energy used to achieve the desired output.

Consider, for example, F/F-2 and F/F-4 in the schedule shown in **Figure 11-13**. The data can help compare these two fans and make key determinations. Both fans have the same model number and the same size motor (1 hp). F/F-2 is intended to move more air than F/F-4 (3600 cfm compared to 3200 cfm). To accomplish this, fan F/F-2 must turn faster (1023 rpm) than fan F/F-4 (875 rpm). Because fan F/F-2 has to do more work, the brake horsepower of fan F/F-2 is higher (bhp = 0.89) than that of fan F/F-4 (bhp = 0.57). As a result of fan F/F-2 moving more air, the static pressure for fan F/F-2 is higher than that of fan F/F-4.

FAN SCHEDULE									BASED ON COOK
TAG	LOCATION	CFM	MODEL	HP	BHP	SP	RPM	VOLTAGE	REMARKS
F/F-1	GALLERY WING	2,150	CPV-150	1	0.63	1.0	1500	460/3/60	SEE NOTE BELOW
F/F-2	GALLERY WING	3,600	CPV-195	1	.89	0.75	1023	460/3/60	SEE NOTE BELOW
F/F-3	GALLERY WING	1,550	CPV-150	1/2	.28	0.75	1150	460/3/60	SEE NOTE BELOW
F/F-4	GALLERY WING	3,200	CPV-195	1	.57	0.62	875	460/3/60	SEE NOTE BELOW
F/F-5	GALLERY WING	3,200	CPV-195	1	.57	0.62	875	460/3/60	SEE NOTE BELOW
EF/F-1	GALLERY WING	1,650	CPV-165	1/2	.39	0.85	1080	460/3/60	SEE NOTE BELOW
EF/F-2	GALLERY WING	600	CPV-100	1/2	.15	1.0	1750	460/3/60	SEE NOTE BELOW

* ACCESSORIES:
WEATHER COVER, BELT GUARD, STEEL DRAIN, VIBRATION ISOLATORS, FACTORY INSTALL DISCONNECT,
MOTOR TO BE VFD ADAPTABLE, & CONTRACTOR TO CONFIRM FAN ROTATION

Lizardos Engineering Assoc., P.C.

Figure 11-13. A fan schedule.

Next take a look at the schedule's note regarding accessories. The fans must be adaptable to variable frequency drives (VFDs). The contractor is responsible for making certain the direction of rotation of the three-phase motors is correct. ***Variable frequency drives (VFDs)*** are control devices that vary the frequency of the power supplied to the motor to modulate the motor's speed.

11.3.6 Air Device Schedules

An air device schedule, **Figure 11-14,** provides detailed information on the grilles used to transition from the duct system to the occupied space.

AIR DEVICE SCHEDULE (BEDROOM WING)							
TAG #	ROOM#	MODEL #	LENGTH	WIDTH	SUPPLY OR RETURN	CFM	BORDER
1	E111/E116	CT-581	24"	4"	SUPPLY	175	TYPE 6
2	E-115	CT-581	36"	6"	SUPPLY	300	TYPE 4
3	E-115	CT-580	36"	6"	RETURN	1000	TYPE 4
4	E-114	CT-581	36"	6"	SUPPLY	300	TYPE 4
5	E-119S	CT-581	36"	6"	SUPPLY	250	TYPE 4
6	E-119S	CT-580	36"	6"	RETURN	600	TYPE 4
7	E-119CC	CT-581	12"	6"	SUPPLY	75	TYPE 12
8	E-110	CT-580	54"	4"	RETURN	340	TYPE 4
9	E-110	CT-581	54"	4"	SUPPLY	400	TYPE 4
10	E-109B	CT-580	12"	6"	EXHAUST	100	TYPE 12
11	E-110B	CT-580	12"	6"	EXHAUST	100	TYPE 12
12	E-119C	CT-580	12"	6"	RETURN	50	TYPE 12
13	E-110B	300-RL	12"	6"	SUPPLY	75	TYPE 12
14	E-109B	300-RL	12"	6"	SUPPLY	75	TYPE 12
15	E-109	CT-580	36"	6"	RETURN	300	TYPE 12
16	E-118F	CT-581	36"	6"	SUPPLY	350	TYPE 12
17	E-118A	CT-580	36"	6"	RETURN	400	TYPE 4
18	E-109	CT-581	36"	6"	SUPPLY	350	TYPE 12
19	E-108B	CT-580	12"	6"	RETURN	100	TYPE 12
20	E-108B	300-RL	12"	6"	SUPPLY	75	TYPE 12
21	E-108	CT-580	54"	4"	RETURN	400	TYPE 4
22	E-108	CT-581	54"	4"	SUPPLY	450	TYPE 4
23	E-107	CT-580	36"	6"	RETURN	500	TYPE 12
24	E-107	CT-581	36"	6"	SUPPLY	550	TYPE 12

Lizardos Engineering Assoc., P.C.

Figure 11-14. An air device schedule.

These grilles are, for the most part, decorative in nature. Many different configurations are available. Some grilles are stamped metal, whereas others are manufactured with adjustable dampers. Because there are technical differences between grilles, louvers, and registers, the general term often used for this type of schedule is *air device*. The information contained in an air device schedule includes the part number, dimensions, cfm rating, and the device's intended application. Air device schedules contain the most tags because there are many more grilles, louvers, and registers than there are pieces of equipment.

11.3.7 Air Handling Unit Schedules

Air handlers and condensing units are the most expensive components of a heating and cooling system. These pieces of equipment must be properly specified, properly selected, and not overlooked when pricing the project. Air handlers and condensing units must be properly matched to ensure proper and efficient system operation. The information contained on the schedules for air handlers and condensing units must, therefore, be carefully reviewed.

An example of a typical air handler schedule is shown in **Figure 11-15**. An air handling unit (AHU) schedule has much of the same information as other heating and cooling components, with the exception of the unit's weight. Because air handlers are often suspended overhead, the unit's weight is a concern and must be addressed when determining how the unit will ultimately be supported. Looking at the schedule, certain information can be extracted from this detailed list:

- AHU-1 weighs 255 pounds, has a total cooling capacity of 48,000 Btu/hour and a heating capacity of 113,000 Btu/hour.
- AHU-2 weighs 300 pounds, has a total cooling capacity of 60,000 Btu/hour and is equipped with duct-mounted heating coils.
- AHU-3 weighs 150 pounds, has a total cooling capacity of 24,000 Btu/hour, and has a heating capacity of 18,000 Btu/hour.
- HV weighs 300 pounds and is a heating-only unit with a heating capacity of 225,000 Btu/hour.

Other information that can be extracted from a schedule such as this includes total static pressure, motor speed, brake horsepower, water flow rates for both heating and cooling, and the unit's dimensions.

AIR HANDLING UNIT SCHEDULE												BASED ON McQUAY
AHU	CFM	MODEL	TSP	RPM	BHP	WEIGHT	COOLING CAPACITY TOTAL	GPM	HEATING CAPACITY TOTAL	GPM	DIMENSIONS	REMARKS
1	1400	SHB161B	1.25	1240	.58	255	48 MBH	9.7	113 MBH	7.8	48x44x17	
2	2025	SHB301B	1.5	1200	1.02	300	60 MBH	12	–	–	48x64x19	DUCT MOUNTED HEATING COILS
3	600	SHB081B	1.0	1120	.215	150	24 MBH	4.8	18 MBH	1.3	48x24x17	
HV	3000	SHB301B	1.5	1175	1.42	300	–	–	225 MBH	15	48x64x19	

Lizardos Engineering Assoc., P.C.

Figure 11-15. An air handling unit schedule.

Summary

- A schedule is a systematically created list that provides detailed information on specific system components and/or pieces of equipment.
- Schedules contain information useful for the pricing of a project, as well as for the proper installation and setup of system components.
- Schedules can be found on project plans that cover all of the skilled trades.
- Schedules can be used as checklists to keep track of the equipment that has been installed.
- Tags are used to identify a specific piece of equipment or system component on both schedules and project drawings.
- Schedules provide valuable information and notes for the contractors to use.
- Schedules should always be read carefully because they often contain additional items that the contractor is responsible for.

Chapter Assessment

Name ____________________ Date ____________ Class ______________

Know and Understand

_______ 1. On a typical fan schedule, all of the following pieces of information are likely to be included except the motor's _____.

A. voltage
B. dimensions
C. horsepower rating
D. speed

_______ 2. According to the AHU schedule shown in the following figure, the air handler that has the blower with the lowest motor speed is also the air handler that operates with the _____.

AIR HANDLING UNIT SCHEDULE												BASED ON McQUAY
AHU	CFM	MODEL	TSP	RPM	BHP	WEIGHT	COOLING CAPACITY TOTAL	GPM	HEATING CAPACITY TOTAL	GPM	DIMENSIONS	REMARKS
1	1400	SHB161B	1.25	1240	.58	255	48 MBH	9.7	113 MBH	7.8	48x44x17	
2	2025	SHB301B	1.5	1200	1.02	300	60 MBH	12	–	–	48x64x19	DUCT MOUNTED HEATING COILS
3	600	SHB081B	1.0	1120	.215	150	24 MBH	4.8	18 MBH	1.3	48x24x17	
HV	3000	SHB301B	1.5	1175	1.42	300	–	–	225 MBH	15	48x64x19	

Lizardos Engineering Assoc., P.C.

A. highest total static pressure
B. lowest brake horsepower
C. lowest heating capacity
D. highest cooling mode water flow

Refer to the heating coil schedule below to answer questions 3–5.

HEATING COIL SCHEDULE													BASED ON McQUAY
TAG	MODEL No.	LOCATION	CFM	MBH TOTAL	FPM	FACE AREA SQ. FT.	FIN LEN. IN.	FIN HGT. IN.	AIR PD IN. H2O	WATER PD FT. H2O	GPM	ROWS FIN/IN
HC/H1–1	5BS0801G	PLAYHOUSE	200	6.5	400	0.5	12	6	.07	.10	0.7	1/8
HC/H1–2	5WQ0801G	PLAYHOUSE	1600	52.3	512	3.13	30	15	0.1	1.0	5.2	1/8
HC/H1–3	5BS0801G	PLAYHOUSE	200	6.5	400	0.5	12	6	.07	.10	0.7	1/8
HC/H1–4	5WQ0801G	PLAYHOUSE	1750	56.8	525	3.33	32	15	.11	1.2	5.7	1/8
HC/H1–5	5WQ0801G	PLAYHOUSE	600	19.4	450	1.33	16	12	.08	0.2	2.0	1/8
HC/H1–6	5WQ0801G	PLAYHOUSE	4000	127.3	571	7.0	42	24	.12	3.3	13.1	1/8
HC/H2–1	5WQ0801G	PLAYHOUSE	2700	89.1	514	5.25	36	21	.10	1.8	8.9	1/8

* BASED ON 55°EAT, 180°EWT, & 160°LWT 3–WAY VALVE (5/8" TUBE DIAMETER) SEE DETAIL FOR PIPING & VALVES

Lizardos Engineering Assoc., P.C.

_______ 3. Which of the following represents the correct quantity and type of hot water heating coils needed for the playhouse?

A. Five 5BS0801G and two 5WQ0108G

B. Two 5BS0801G and five 5WQ0801G

C. Two 5BS0108G and five 5WQ0108G

D. Five 5BS0108G and two 5WQ0801G

_______ 4. What is the fin height on a 5WQ0801G coil designed to have between 500 and 700 cubic feet of air passing through it?

A. 6″

B. 9″

C. 12″

D. 21″

_______ 5. The total hot water requirement for the playhouse is closest to _____.

A. 3.63 gpm

B. 36.3 gpm

C. 363 gpm

D. 3630 gpm

Name ______________________________ Date ____________ Class ____________________

_______ 6. The three types of air devices referenced in the schedule shown in the following figure are _____.

AIR DEVICE SCHEDULE (BEDROOM WING)							
TAG #	ROOM#	MODEL #	LENGTH	WIDTH	SUPPLY OR RETURN	CFM	BORDER
1	E111/E116	CT-581	24"	4"	SUPPLY	175	TYPE 6
2	E-115	CT-581	36"	6"	SUPPLY	300	TYPE 4
3	E-115	CT-580	36"	6"	RETURN	1000	TYPE 4
4	E-114	CT-581	36"	6"	SUPPLY	300	TYPE 4
5	E-119S	CT-581	36"	6"	SUPPLY	250	TYPE 4
6	E-119S	CT-580	36"	6"	RETURN	600	TYPE 4
7	E-119CC	CT-581	12"	6"	SUPPLY	75	TYPE 12
8	E-110	CT-580	54"	4"	RETURN	340	TYPE 4
9	E-110	CT-581	54"	4"	SUPPLY	400	TYPE 4
10	E-109B	CT-580	12"	6"	EXHAUST	100	TYPE 12
11	E-110B	CT-580	12"	6"	EXHAUST	100	TYPE 12
12	E-119C	CT-580	12"	6"	RETURN	50	TYPE 12
13	E-110B	300-RL	12"	6"	SUPPLY	75	TYPE 12
14	E-109B	300-RL	12"	6"	SUPPLY	75	TYPE 12
15	E-109	CT-580	36"	6"	RETURN	300	TYPE 12
16	E-118F	CT-581	36"	6"	SUPPLY	350	TYPE 12
17	E-118A	CT-580	36"	6"	RETURN	400	TYPE 4
18	E-109	CT-581	36"	6"	SUPPLY	350	TYPE 12
19	E-108B	CT-580	12"	6"	RETURN	100	TYPE 12
20	E-108B	300-RL	12"	6"	SUPPLY	75	TYPE 12
21	E-108	CT-580	54"	4"	RETURN	400	TYPE 4
22	E-108	CT-581	54"	4"	SUPPLY	450	TYPE 4
23	E-107	CT-580	36"	6"	RETURN	500	TYPE 12
24	E-107	CT-581	36"	6"	SUPPLY	550	TYPE 12

Lizardos Engineering Assoc., P.C.

A. ventilation, exhaust, and supply
B. ventilation, exhaust, and return
C. return, ventilation, and supply
D. return, supply, and exhaust

_______ 7. The condensate pumps shown below are equipped with all of the following except _____.

CONDENSATE PUMP SCHEDULE — BASED ON LITTLE GIANTS							
TAG	GPH	PUMP HEAD FT.	PSI	VOLTAGE	AMPS	TANK CAPACITY GAL.	MODEL
CP/H-1	25	15	8.6	115	1.5	0.3	VCC-20 (LOW PROFILE)
CP/H-2	25	15	8.6	115	1.5	0.3	VCC-20 (LOW PROFILE)
CP/H-3	25	15	8.6	115	1.5	0.3	VCC-20 (LOW PROFILE)
CP/H-4	25	15	8.6	115	1.5	0.3	VCC-20 (LOW PROFILE)
CP/H-5	25	15	8.6	115	1.5	0.3	VCC-20 (LOW PROFILE)
CP/H-6	25	15	8.6	115	1.5	0.3	VCC-20 (LOW PROFILE)

*REMARKS: STAINLESS STEEL SHAFT, ABS TANK, THERMAL OVERLOAD PROTECTION 6'-0" POWER CORD, & CHECK VALVE BUILT-IN

Lizardos Engineering Assoc., P.C.

A. a 6′ power cord
B. a built-in check valve
C. an auto-leveling screw
D. thermal overload protection

_______ 8. A tag is used to _____.

A. identify a specific component or piece of equipment on a schedule and/or print

B. determine the model number of a specific component or piece of equipment

C. identify the number of pieces of equipment to be used on the job

D. determine whether or not a specific piece of equipment has been installed

Apply and Analyze

______ 1. If a particular piece of equipment is not available as described in the schedule, the contractor should immediately _____.

A. order the closest replacement to ensure they have enough equipment to complete the project without delay

B. notify the general contractor and architect about the situation and be prepared to offer possible alternative solutions

C. notify the project owner about the situation and be prepared to offer possible alternative solutions

D. identify an alternate manufacturer of equivalent or similar equipment and complete the job as specified

_______ 2. The total heating capacity for the playhouse, as indicated in the following figure, is closest to _____.

HEATING COIL SCHEDULE

BASED ON McQUAY

TAG	MODEL No.	LOCATION	CFM	MBH TOTAL	FPM	FACE AREA SQ. FT.	FIN LEN. IN.	FIN HGT. IN.	AIR PD IN. H2O	WATER PD FT. H2O	GPM	ROWS FIN/IN
HC/H1-1	5BS0801G	PLAYHOUSE	200	6.5	400	0.5	12	6	.07	.10	0.7	1/8
HC/H1-2	5WQ0801G	PLAYHOUSE	1600	52.3	512	3.13	30	15	0.1	1.0	5.2	1/8
HC/H1-3	5BS0801G	PLAYHOUSE	200	6.5	400	0.5	12	6	.07	.10	0.7	1/8
HC/H1-4	5WQ0801G	PLAYHOUSE	1750	56.8	525	3.33	32	15	.11	1.2	5.7	1/8
HC/H1-5	5WQ0801G	PLAYHOUSE	600	19.4	450	1.33	16	12	.08	0.2	2.0	1/8
HC/H1-6	5WQ0801G	PLAYHOUSE	4000	127.3	571	7.0	42	24	.12	3.3	13.1	1/8
HC/H2-1	5WQ0801G	PLAYHOUSE	2700	89.1	514	5.25	36	21	.10	1.8	8.9	1/8

* BASED ON 55°EAT, 180°EWT, & 160°LWT 3-WAY VALVE (5/8" TUBE DIAMETER) SEE DETAIL FOR PIPING & VALVES

Lizardos Engineering Assoc., P.C.

A. 3579 Btu/hour

B. 35,790 Btu/hour

C. 357,900 Btu/hour

D. 3,579,000 Btu/hour

Name ______________________ Date ____________ Class ______________

3. If an air-conditioning unit produces 1.5 pints of condensate per hour, and the condensate pump indicated in the following figure does not turn on until the pump's tank is completely filled, approximately how often does the condensate pump cycle on?

CONDENSATE PUMP SCHEDULE

BASED ON LITTLE GIANTS

TAG	GPH	PUMP HEAD FT.	PSI	VOLTAGE	AMPS	TANK CAPACITY GAL.	MODEL
CP/H-1	25	15	8.6	115	1.5	0.3	VCC-20 (LOW PROFILE)
CP/H-2	25	15	8.6	115	1.5	0.3	VCC-20 (LOW PROFILE)
CP/H-3	25	15	8.6	115	1.5	0.3	VCC-20 (LOW PROFILE)
CP/H-4	25	15	8.6	115	1.5	0.3	VCC-20 (LOW PROFILE)
CP/H-5	25	15	8.6	115	1.5	0.3	VCC-20 (LOW PROFILE)
CP/H-6	25	15	8.6	115	1.5	0.3	VCC-20 (LOW PROFILE)

*REMARKS: STAINLESS STEEL SHAFT, ABS TANK, THERMAL OVERLOAD PROTECTION
6'-0" POWER CORD, & CHECK VALVE BUILT-IN

Lizardos Engineering Assoc., P.C.

____

____

____

____

Critical Thinking

1. Why is the issue of proper fan rotation addressed in the fan schedule provided in the following figure?

FAN SCHEDULE

BASED ON COOK

TAG	LOCATION	CFM	MODEL	HP	BHP	SP	RPM	VOLTAGE	REMARKS
F/H-1	PLAYHOUSE	7,200	CPV-300	2.0	1.35	0.75	567	460/3/60	VFD COMPATIBLE
F/H-2	PLAYHOUSE	1200	CPV-135	0.5	0.25	0.75	1300	460/3/60	
F/H-3	PLAYHOUSE	2,400	CPV-180	.75	.47	0.75	960	460/3/60	
F/H-4	PLAYHOUSE	2,500	CPV-180	.75	.47	0.75	960	460/3/60	VFD COMPATIBLE

* ACCESSORIES:
WEATHER COVER, BELT GUARD, STEEL DRAIN, VIBERATION ISOLATORS, FACTORY INSTALL DISCONNECT, MOTOR TO BE VFD ADAPTABLE, & CONTRACTOR TO CONFIRM FAN ROTATION

Lizardos Engineering Assoc., P.C.

____

____

____

____

2. How do reset strategies, such as the one depicted in the following figure, help conserve energy?

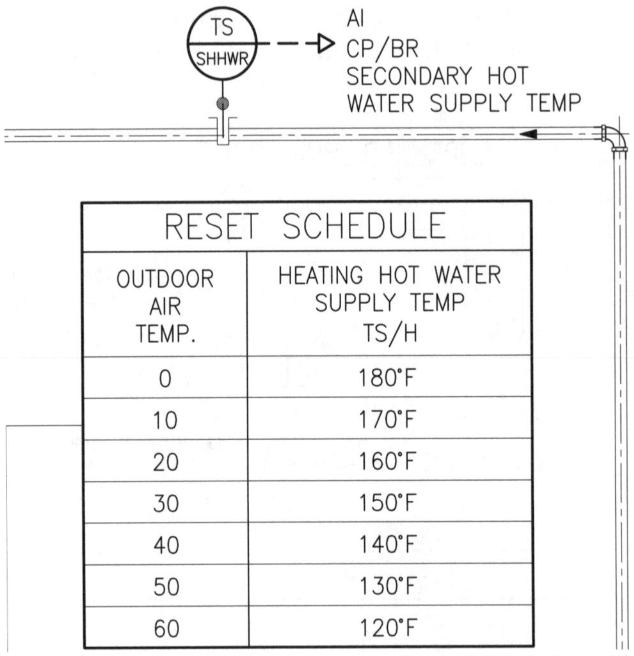

TS
SHHWR
AI
CP/BR
SECONDARY HOT
WATER SUPPLY TEMP

RESET SCHEDULE	
OUTDOOR AIR TEMP.	HEATING HOT WATER SUPPLY TEMP TS/H
0	180°F
10	170°F
20	160°F
30	150°F
40	140°F
50	130°F
60	120°F

Lizardos Engineering Assoc., P.C.

Large Prints Activity

Name ______________________ Date ____________ Class ______________

Practice Using Large Prints

Refer to Print-21 in the Large Prints supplement to answer the following question.

1. Referring to the humidifier schedule provided, explain what must be taken into account when determining what types of wall brackets and associated hardware should be used to mount the humidifiers.

____

____

____

____

Refer to Print-18 in the Large Prints supplement to answer the following question.

2. Referring to the Service Wing schedules provided, what air handler installation considerations not otherwise listed can add considerable cost to the project?

____

____

____

____

Refer to Print-09 in the Large Prints supplement to answer the following question.

3. Referring to the Bedroom Wing schedule provided, what additional research will need to be done by the mechanical contractor prior to submitting the bid for the project?

____

____

____

____

12 Estimating HVACR Jobs

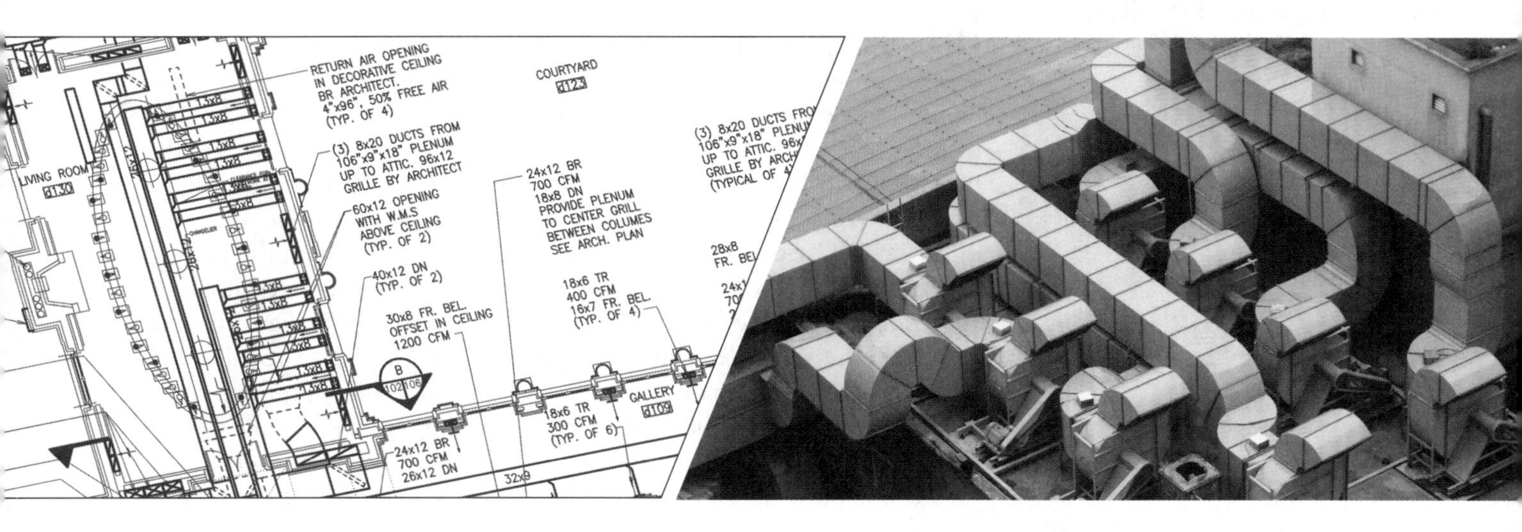

Chapter Outline

Learning Objectives

After studying this chapter, you will be able to:

- Differentiate between an estimate, quote, bid, and proposal.
- Explain contractor or trade bid overlapping.
- Discuss the differences between the project cost and project price.
- Calculate an overrun and an underrun on a project.
- Determine the labor costs and the amount of labor required to complete a project.
- Assess the list of required equipment for a project.
- Prepare a list of required materials for a project.
- Determine additional overhead costs included in a project estimate.
- Establish a final bid price for a project.

Technical Terms

bid

bid overlapping

estimate

estimator

overhead

overrun

profit

project cost

proposal

quote

total labor cost

underrun

Introduction

Contracting companies are in business to make money. Basic economics dictates that, in order for a company to make a ***profit***, it must be able to provide a good or service to its customer at a price point that is higher than the cost of producing or providing that good or service. If the cost of producing or providing a good or service is greater than the price point at which the good or service is offered to the customer, the company providing the good or service loses money and fails to realize a profit. Establishing the correct price for a job is an underlying factor that can make or break a company. In the world of contracting, a quoted price that is too high will typically result in the contractor not being awarded the project. A quoted price that is too low will result in a monetary loss for the company in the event the contractor is awarded the contract for the job.

A project ***estimator*** is responsible for weighing these factors for a contracting company. It is an integral position in any successful contracting business. This chapter discusses the general processes used to create accurate and profitable estimates and provides insight into the elements that must be factored in and accounted for. Often specific mathematical formulas are required for project estimates, which will be introduced. Software packages are available to assist in the process of estimating a project.

12.1 Estimates, Quotes, Bids, and Proposals

When it comes to performing work for hire, the terms estimate, quote, bid, and proposal are commonly used, **Figure 12-1**. These terms have different meanings and should be used properly. Because each term is distinct, always ensure a client is aware of what is being presented and the implications of each.

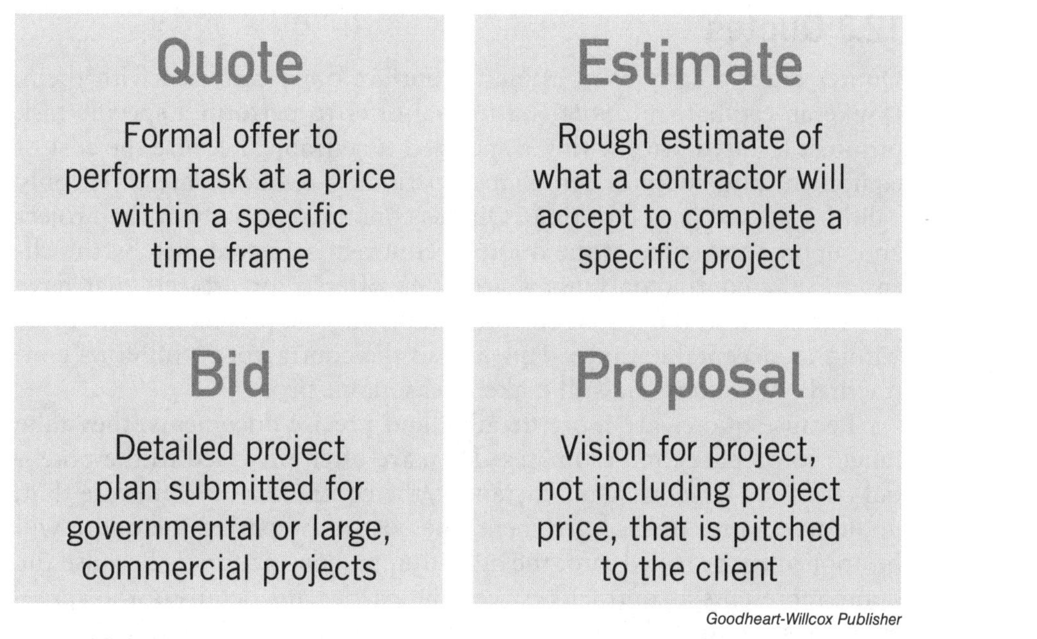

Goodheart-Willcox Publisher

Figure 12-1. Quotes, estimates, bids, and proposals are all specific terms normally discussed in work-for-hire agreements.

12.2 Estimates

In construction and other industries, an ***estimate*** is a rough approximation of the amount a contractor is willing to accept to perform a given project. This estimate is presented to the customer of a particular project. The key word in this definition is *approximation*. Although estimates should provide a rough idea or reasonable guess as to the final cost of the project, it should be understood that the actual cost may differ from the dollar figure originally presented.

BETWEEN THE LINES

Actual Project Price

In many cases, the actual price of a project can vary, on average, from 10% to 15% of the presented price. Always make sure the client is aware of the difference between an estimate and the actual project price. If it becomes known the actual price is more than 15% greater than the original estimate, the client should be notified as soon as possible.

Consider a situation where a wall needs to be relocated in a private home. An estimate for a project such as this might include the base price of the wall relocation, but extra compensation is added should unforeseen problems arise. For example, a contractor may realize during the project that there was black mold in a wall cavity that must first be remediated. The contractor could not have possibly known what was behind the wall until it was opened. This would be an additional expense for the client. In reality, the vast majority of jobs are rarely completed at the price point indicated in the estimate.

12.3 Quotes

Quotes differ greatly from estimates but are often confused with them. Unlike an estimate, a ***quote*** is a formal offer to perform a specific task or job at a stated price within a specified time frame. Because the cost of equipment and supplies can change periodically, quotes are often only valid for a fixed period of time. Quotes contain details about the project and, upon acceptance of the quote, a contract is entered into by the client and the contractor. Quotes are often offered for projects that have few variables, so the actual cost of completing the project is likely to be within an acceptable range. This allows the contractor to fulfill its contractual obligations and still make a reasonable profit.

Because quotes are more detailed and precise documents, they take longer to prepare than estimates. They are often provided to the potential client for a fee. In many instances, the contractor may indicate that, upon acceptance of the agreement, the cost of preparing the quote will be applied to, or rolled into, the agreed-upon contract price. Because the quote represents a contract between the parties, the details of the agreement must be made clear, such as if the contract price includes all taxes and other pertinent fees and additional charges.

12.4 Bids

Bids are commonly submitted on governmental or larger commercial projects. They are more detailed than quotes and are prepared based on a specific project plan that is to be completed exactly as indicated. The architects, engineers, and designers have already worked out all of the details. It is the sole job of the contractor to execute the project.

Bids typically include project time lines, which are set to determine when each major project task will be completed. Project tasks may include items such as equipment delivery/rigging, building penetrations, piping rough-in, duct testing, and equipment startup or commissioning. Bids can also include detailed price breakdowns, incorporating labor rates, as well as equipment and material costs. Because bids are most often submitted on larger projects, awarding the project to a contractor can take a significant amount of time. Thus, the figures presented in bids, as opposed to those in quotes, do not have an expiration date.

Large projects often have a strict process for submitting a bid on a job. For example, contractors typically will need to express, in writing, their desire to bid on a project before seeing a set of job plans from which they will price the project. There is also a bid submission deadline, which must be met to have the bid accepted for consideration.

In some instances, a fee must be paid to participate in the bidding process. This fee is referred to as a *bid bond*. A bid bond helps ensure that the bidding contracting firm follows through on the project, at the price presented and within the proposed time line, if it is awarded the contract. The bid bond amount varies based on the estimated value of the project and can be as high as 10% to 20% of the project bid. Given the nature of bids, it is absolutely necessary to visit the jobsite before a bid is submitted. This gives the contractor the opportunity to access the space and also identify any logistical problems that might affect the workflow.

12.4.1 Bid Overlapping

Bid overlapping is a problem that arises when multiple trades are bidding on their portion of a construction project. It occurs when a contractor includes work in the project bid that is actually the responsibility of another contractor. Consider, for example, a task such as running control wiring for an air-conditioning system, or installing a condensate drain line from an air handler to its termination point in a waste line. These two tasks are usually performed by the HVACR contractor, but they can be performed by either an electrical or plumbing contractor. Bid overlapping results in a final project bid that is higher than it would have been had the additional items not been included. Depending on the amount of work improperly included in the estimate, the effect of overlapping on the final proposal quote can be significant.

BETWEEN THE LINES

Reducing Bid Overlap

To reduce the possibility of bid overlapping, the individual responsible for extracting information from the set of plans should be completely aware of the scope of work to be performed by each of the trades. If there is any doubt regarding which trade is responsible for each aspect of the project, it is far better to inquire before preparing and submitting the final bid.

12.5 Proposals

A ***proposal*** is a document that presents a contractor's vision for a project or task, **Figure 12-2**. It is intended to sell the client on the contractor's interpretation and solution for that project or task, rather than be a document that outlines pricing information.

For example, consider the situation where a homeowner wants to install an HVACR system in his home. The homeowner is not certain what type of system he wants (furnace with a cooling coil, heat pump, ducted, ductless, etc.), how many systems are needed, the efficiency of the equipment, the location of the equipment, or the required capacity of the equipment. The homeowner might then call upon multiple contractors to provide their expertise on how to best meet the HVACR needs of the structure. These contractors will also work within the client's budget for the project.

Proposals also offer potential contractors a way to showcase their work by providing written recommendations from past clients, manufacturer's literature, photos of the proposed concepts, or detailed renderings of how the contractor envisions the completed project. Whereas project bids are more formal and follow specific guidelines, proposals allow a contractor to be creative.

Pormezz/Shutterstock.com

Figure 12-2. A contractor discusses his project vision in the proposal submitted to a potential client.

Proposals often combine many items, such as estimates, quotes, and bids. They also include a marketing aspect, which helps the client visualize the project, learn more about the contracting firm, and understand its ability to complete the project.

12.6 Overruns and Underruns

Overruns and underruns describe how much a project is under or over budget, **Figure 12-3**. When a project is ***underrun***, the actual cost to complete a project is less than 90% of the project's original estimated cost. A project that is ***overrun***, however, exceeds the estimated cost by more than 10%.

The calculation to determine whether the budget was an overrun or underrun is expressed as a percentage:

$$\text{budget percentage} = \frac{\text{actual cost} - \text{estimated cost}}{\text{estimated cost}} \times 100\%$$

Consider a project that had a projected cost of $500,000. If the actual cost of completing that project was $475,000, the project came in $25,000 ($500,000 – $475,000) under budget. We know that the project was under budget from the formula. So we can calculate:

$$\begin{aligned}\frac{\$475{,}000 - \$500{,}000}{\$500{,}000} \times 100\% &= \frac{-\$25{,}000}{\$500{,}000} \times 100\% \\ &= -0.05 \times 100\% \\ &= -5\%\end{aligned}$$

This project came in 5% under budget, or at 95% of budget. However, if the project's actual cost was $400,000, the budget percentage can be calculated as:

$$\begin{aligned}\frac{\$400{,}000 - \$500{,}000}{\$500{,}000)} \times 100\% &= \frac{-\$100{,}000}{\$500{,}000} \times 100\% \\ &= -0.2 \times 100\% \\ &= -20\%\end{aligned}$$

In this case, the project had a 20% underrun.

Term	Actual Cost of Project
Under budget	Between 90% and 100% of the estimated cost
Over budget	Between 100% and 110% of the estimated cost
Underrun	Less than 90% of the estimated cost
Overrun	More than 110% of the estimated cost

Goodheart-Willcox Publisher

Figure 12-3. The table depicts the differences between a project that is under budget, over budget, underrun, or overrun.

Although project underruns often equate to higher profits, consistently overpricing future projects can result in contracts being awarded to other contractors, whose bids were lower and more indicative of the actual value of the project. Consistent project overruns can be costly to the contractor, because overruns can reduce or eliminate the profit a company relies on to stay in business. If a company consistently experiences overruns, the project estimating process must be carefully evaluated to determine what pricing adjustments must be made, given the prices being quoted are too low. When a contractor consistently underprices projects, the company's profits are at risk.

The most common contractor-related cause for cost overruns is an inaccurate estimation of the time required to complete the project. Not all overruns and underruns result from deficiencies in the contractor's ability to accurately price projects. Problems often arise from inconsistencies or inaccuracies in the plans of a project. In many instances, the project design or layout is not realistic, or does not accurately represent the intended final product. In any event, a contractor should continually compare the project plans to the work as it was ultimately completed.

12.7 Project Cost and Project Price

Project cost is the sum of all of the costs associated with executing a particular job or project. This term is different from the total *project price*, also called the *project quote*, which refers to the price at which the provider of the project, or HVACR contractor, is willing to complete it. This includes the cost of the job plus profit. See **Figure 12-4**. The project cost comprises many factors including the total labor cost, total equipment cost, total material cost, project filing fees, and additional overhead costs.

For example, consider an HVACR project that includes the following costs:

Total labor cost = $56,000
Total material cost = $22,000
Total equipment cost = $38,000
Project filing fees = $1500
Overhead costs = $27,000

Figure 12-4. The project price comprises the project cost plus profit.

To determine the total cost of this project, each expense is added together. The total cost of completing the project is $144,500. If the contractor presented this figure as the bid price for this project, the contractor would break even or, more likely, lose money on the project. Quite often, a project encounters unforeseen expenses and costs.

The contractor must make a profit to stay in business. So, to make that profit, the cost of completing the project must be marked up. For instance, if the contractor marks up the project cost 15%, the total bid price for this particular project will be $166,175. This is completed by calculating $144,500 + $21,675. The $21,675 figure represents 15% of the total project cost ($144,500 × 0.15). Thus, the profit made by the contractor on this job will be 13%. This is calculated by dividing the markup on the project by the total job quote ($21,675 ÷ $166,175), and then multiplying this result by 100%.

If the contractor intends to make a profit of 15% on the job, this means 15% of the quoted price will be profit, or in other words, 85% of the quoted price represents the cost of the project. So, the cost of the project with a 15% profit can be determined by dividing the cost of the project by 85% or 0.85. This will result in a project quote of $170,000 ($144,500 ÷ 0.85).

The result can be checked by multiplying the project quote of $170,000 by 15% to get $25,500, which is the difference between the quoted job price and the cost to complete the project. The following sections provide more detailed information on each of the project costs identified and how they are calculated.

12.7.1 Total Labor Cost

The ***total labor cost*** for a job is the total cost of all aspects of the project that require labor. This includes more than the manual labor required to perform the project. A number of other tasks must be completed that do not directly contribute to the project-specific tasks. These items include:

- Travel time to the jobsite.
- Unloading materials, tools, and equipment from trucks.
- Transporting materials, tools, and equipment to installation locations.
- Unpacking and preparing equipment for installation.

Travel Time to the Jobsite

The travel time to the jobsite may not be a concern, or it may be a major contributing factor in the pricing of a job.

Assume a particular contractor requires that all employees report to the home office or shop at the start of each workday. Employees would then report to the office at the assigned start time, get their work assignments for the day, gather the materials and supplies they need to complete their tasks, and then obtain their work truck and proceed to the jobsite. Sometimes this can compute to 1–2 hours or more of lost time each day. When calculating the number of hours required to complete the project, this daily loss must be accounted for.

On some projects, a contractor might require that the workers report to the jobsite at the beginning of the workday. When such is the work arrangement, the tradespeople travel to the jobsite on their own time and are expected to be ready to start work upon clocking in at the jobsite. This option increases the number of hours the tradesmen are physically on the job, but often requires that workers take their service trucks home with them. This is seen as a valuable benefit to the worker, especially because the fuel costs and vehicle wear-and-tear are attributed to the company's truck and not the worker's personal vehicle. This work schedule model thus includes additional vehicle-related overhead expenses, such as additional fuel, repair, and maintenance expenses.

Loading and Unloading of Equipment, Tools, and Materials

Unless the equipment and required materials are delivered directly to the jobsite, these items must be loaded on trucks, which is labor supplied by the contractor. Depending on the amount of equipment and materials, this can be a significant cost and must be taken into account. Sometimes, this process requires multiple people and the use of forklifts, or similar lifting machines, to properly and safely load the equipment and materials onto the truck.

Depending on the quantity and weight of the equipment being installed, unloading it may also be a cost to consider when estimating the project. Heavy machinery such as forklifts may be required to offload equipment from the truck, which can make the process expensive and time-consuming.

BETWEEN THE LINES

Factoring Loading and Unloading into Project Bids

To determine if the unloading of equipment and materials should be factored into the project bid, it is good practice to inquire about the accommodations available at the job location. In some instances, depending on the project location, deliveries may need to be accepted at a loading dock. This might entail lengthy wait times should multiple trucks be attempting delivery to the site simultaneously. It is also a good idea to visit the location to physically observe conditions at the site.

Equipment, Tool, and Material Transport to Installation Locations

There are some questions that need to be answered when equipment and supplies are unloaded at a jobsite:

- What is the distance between the location where the equipment was unloaded and the location(s) where the equipment will be installed?
- Is an elevator required to bring the equipment to the installation location?
- Are there any physical barriers that can affect the movement of the equipment?
- Is a rigging company needed to bring equipment to the roof or other elevated platform?

The only way to determine the potential effects of these items on the project bid is to, once again, inquire about these factors and, better yet, pay a visit to the site. Relying solely on a verbal response to an inquiry will provide no financial relief if significant expenses are incurred as a result of these items. Most general contractors will include, in their request for proposals, a strong suggestion that subcontractors physically visit the jobsite before preparing and submitting bids.

Unpacking and Preparing Equipment for Installation

There are situations where costs for uncrating, unpacking, and preparing the equipment for installation are considered. For instance, consider a project where a large number of crates must be opened. These crates, once the equipment is removed and inspected, must be discarded, which is often the responsibility of the contractor. The waste disposal process can have large costs associated with it, given that dumpsters or other receptacles for the refuse may need to be rented.

When uncrating and unpacking equipment, it may also be necessary to alter the configuration of equipment prior to installation. For example, a multipoise, or multidirectional, air handler may need to have the evaporator coil repositioned to facilitate the desired airflow direction. This typically will not take much time if only a few units need to be reconfigured, but it can become quite time-consuming if there are a lot of them.

12.7.2 Total Equipment Cost

Project estimating requires preparing and compiling complete and accurate equipment lists. Air handlers, condensing units, and similar items are among the most expensive line items in an estimate. The elimination of a single piece of equipment from an estimate has the potential to be devastating from a financial standpoint. So, always check each piece of equipment shown on the project drawings. Then compare the actual renderings to the schedules and any other lists that have been created.

It is the responsibility of the estimator to ensure that every piece of equipment is accounted for. Errors found after bid submission have the potential to be very costly. Even if the estimator is careful, mistakes can be made. Compare the two schedules shown in **Figure 12-5.** The condensate pump schedule, is relatively straightforward because it can be clearly seen that this schedule calls for six identical VCC-20 Little Giant, Low Profile condensate pumps.

The heating coil schedule lists seven heating coils that are not all identical. Although it may look like there are seven identical McQuay heating coils, there are five McQuay 5WQ0801G coils and two McQuay 5BS0801G coils. If seven of the same coils are ordered, there are two potential problems. First, the two types may differ in cost, which creates an inaccurate equipment cost calculation. There is also the possibility that the installers will mount and pipe-in the incorrect coils. This can lead to additional work to uninstall the incorrect coils and reinstall the correct ones.

Because the coils were installed, they cannot be returned and resold as new, so additional contractor expenses will be incurred. If, however, the coil order is checked prior to installation, the incorrect coils will still

HEATING COIL SCHEDULE

BASED ON McQUAY

TAG	MODEL No.	LOCATION	CFM	MBH TOTAL	FPM	FACE AREA SQ. FT.	FIN LEN. IN.	FIN HGT. IN.	AIR PD IN. H20	WATER PD FT. H20	GPM	ROWS FIN/IN
HC/H1-1	5BS0801G	PLAYHOUSE	200	6.5	400	0.5	12	6	.07	.10	0.7	1/8
HC/H1-2	5WQ0801G	PLAYHOUSE	1600	52.3	512	3.13	30	15	0.1	1.0	5.2	1/8
HC/H1-3	5BS0801G	PLAYHOUSE	200	6.5	400	0.5	12	6	.07	.10	0.7	1/8
HC/H1-4	5WQ0801G	PLAYHOUSE	1750	56.8	525	3.33	32	15	.11	1.2	5.7	1/8
HC/H1-5	5WQ0801G	PLAYHOUSE	600	19.4	450	1.33	16	12	.08	0.2	2.0	1/8
HC/H1-6	5WQ0801G	PLAYHOUSE	4000	127.3	571	7.0	42	24	.12	3.3	13.1	1/8
HC/H2-1	5WQ0801G	PLAYHOUSE	2700	89.1	514	5.25	36	21	.10	1.8	8.9	1/8

* BASED ON 55°EAT, 180°EWT, & 160°LWT 3-WAY VALVE (5/8" TUBE DIAMETER) SEE DETAIL FOR PIPING & VALVES

A

CONDENSATE PUMP SCHEDULE

BASED ON LITTLE GIANTS

TAG	GPH	PUMP HEAD FT.	PSI	VOLTAGE	AMPS	TANK CAPACITY GAL.	MODEL
CP/H-1	25	15	8.6	115	1.5	0.3	VCC-20 (LOW PROFILE)
CP/H-2	25	15	8.6	115	1.5	0.3	VCC-20 (LOW PROFILE)
CP/H-3	25	15	8.6	115	1.5	0.3	VCC-20 (LOW PROFILE)
CP/H-4	25	15	8.6	115	1.5	0.3	VCC-20 (LOW PROFILE)
CP/H-5	25	15	8.6	115	1.5	0.3	VCC-20 (LOW PROFILE)
CP/H-6	25	15	8.6	115	1.5	0.3	VCC-20 (LOW PROFILE)

*REMARKS: STAINLESS STEEL SHAFT, ABS TANK, THERMAL OVERLOAD PROTECTION
6'-0" POWER CORD, & CHECK VALVE BUILT-IN

B

Lizardos Engineering Assoc., P.C.

Figure 12-5. A—Condensate pump schedule. B—Heating coil schedule.

have to be removed from the jobsite and returned. New coils will have to be ordered. Delivery charges are also additional costs regarding equipment. This can delay the project and lose valuable man-hours on the job.

Always plan for any price changes that might occur between the time the bid is prepared and the time the project is awarded. A bidding contractor may request an equipment quote from the manufacturer or distributor, which locks in the prices of the equipment for a period of time. If the project has not been awarded on the date the quote expires, the bidding contractor can request that the price-lock period be extended. Because manufacturers and distributors are aware of the nature of bids, they are typically flexible with the pricing they offer. Failure to request a quote from the manufacturer or supplier eliminates the protection that quotes provide to the contractor.

12.7.3 Total Material Cost

Materials and equipment cannot be used interchangeably. Equipment refers to HVACR components, such as blowers, pumps, air handlers, condensing units, and other items installed on the job. Materials refer to the nuts, bolts, steel channel, beam clamps, threaded rods, brackets, and other items used to install equipment.

Because materials are small and not typically expensive, some inexperienced estimators may underestimate the number of materials needed or omit them from a bid calculation altogether. An individual piece of material, such as a threaded rod, might not be very expensive, but can be costly when 100 lengths are required for a project. Two methods for helping ensure that installation materials are accurately accounted for in the bid are to refer to the project's detail drawings and obtain the installation manuals for each piece of equipment.

Detail drawings provide specific information regarding installation and mounting procedures that must be followed on the job. Referring to these detail drawings will help ensure that all hardware and other mounting and supporting materials are accounted for. Consider the typical hanger detail in **Figure 12-6**. In this drawing, a C-clamp, steel retaining strap, threaded rod, and clevis hanger are required for each hanger. This information, along with knowing the overall length of each piping run and the required maximum spacing between hangers, can help the estimator determine the total number of hardware items required.

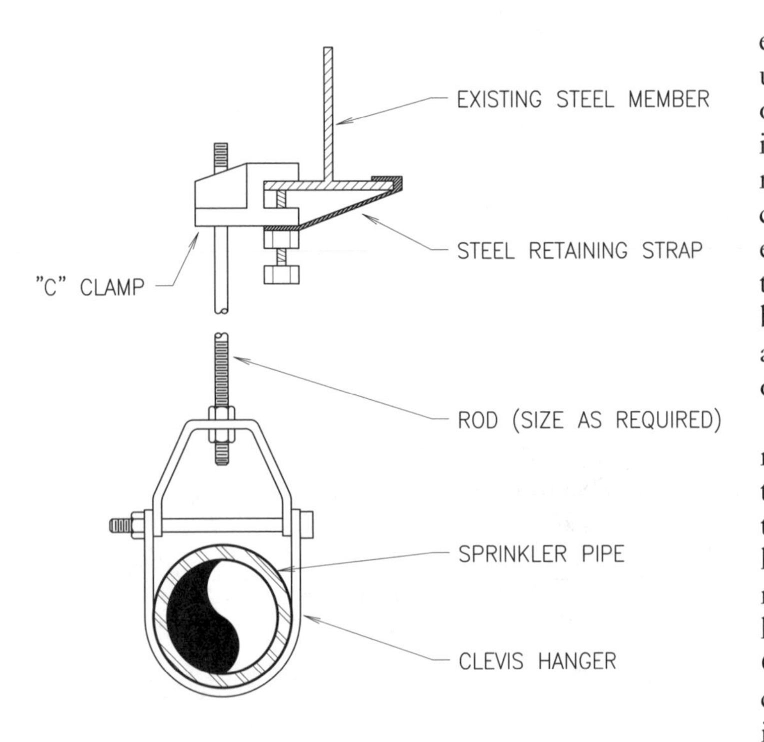

Figure 12-6. Typical hanger detail.

BETWEEN THE LINES

Calculating Material

It is strongly recommended that an additional 10% to 20% of these hardware items be factored into the material list for the project. These installation hardware items are not project-specific and can, if not used on this project, be put into stock and used on future jobs.

Piping details, such as the one shown in **Figure 12-7**, can also be used to create material lists and account for all required fittings, valves, and control devices. Because the actual installation can vary from the detail drawings offered, additional fittings should be provided in the event the piping runs need to be installed differently.

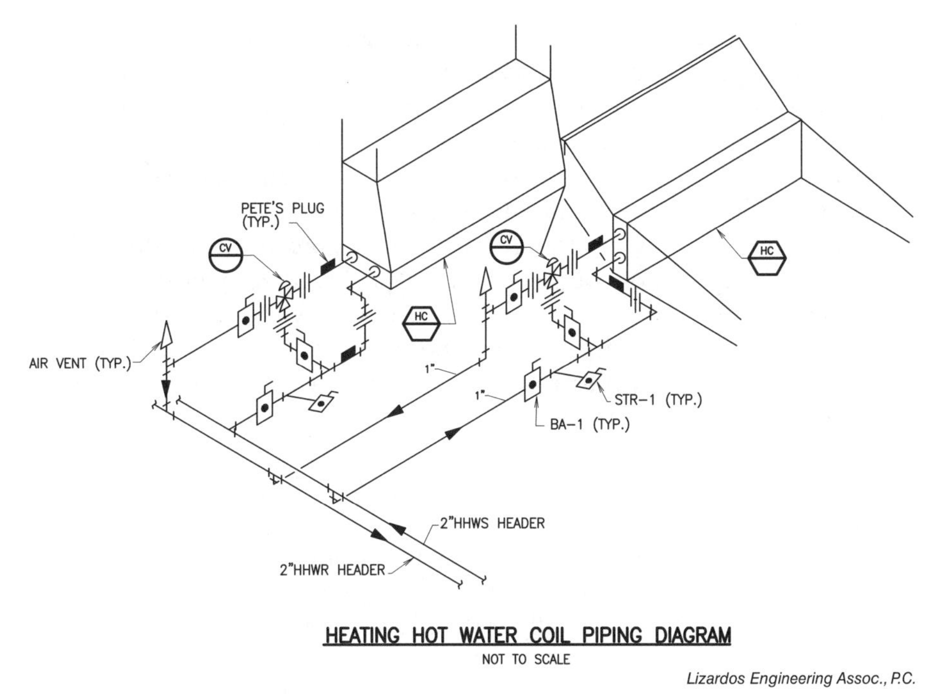

Lizardos Engineering Assoc., P.C.

Figure 12-7. Piping diagrams can be used to create accurate material lists.

Installation manuals can be used to verify a project has all the required installation materials. These manuals often have installation guidelines and details not included in the project drawings. If it is determined the manufacturer's installation guidelines contradict the installation guidelines in the project plans, always bring this discrepancy to the attention of the project coordinator, general contractor, or architect.

12.7.4 Project Filing Fees

There are fees regarding filing or permit fees that must be paid. If not specified in the scope of the project, determine who is responsible for pulling the job permits and paying any fees associated with the work performance.

12.7.5 Additional Project Costs

Often a contractor chooses not to complete all of the tasks in the contract. An HVACR contractor might not be the contractor who physically installs the ductwork, and instead brings in an additional subcontractor responsible for the installation of the job's air distribution system. This practice can be risky, because the HVACR contractor must then rely on the subcontractor to complete that part of the project according to plan and for the dollar figure quoted. To cover this, an additional cost is factored into the bid calculation to allow for any reasonable monetary overages encountered.

Other job-related costs, such as equipment rigging, are included in the overall cost. The rigging process and its associated costs vary greatly depending on the job location, access to the equipment landing location, and rigging methods used. In some cases, equipment rigging requires street closures, which often involves the levying of additional fees.

If street closures are required, the local authorities often establish a time frame for the closures to minimize disruptions to traffic flow. These time frames are often on weekends, late at night, or early in the morning and require the rigging company to perform their work outside of normal business hours. In addition, representatives from the installation company must be on the job to ensure each piece of equipment is properly located and oriented correctly.

Roof penetrations are included in an HVACR system installation project, but they are often overlooked in a project bid. Roof penetrations include installing roof curbs, pitch pockets, and chases through which ducts are run. Pitch pockets are roof penetrations through which electrical or plumbing lines are run. A roofing contractor is needed to make any penetrations to a roof, especially if the roof is bonded or insured.

12.8 Hourly Billing Rate for Labor

One challenge a contractor faces when it comes to billing is determining how much to charge for each hour of a worker's time. Employee-related expenses have to be taken into account. There are also significant overhead expenses that must be offset in order for the contractor to remain in business and make a profit. Overhead expenses typically refer to the non-labor expenses a company pays to operate the business. The most common employee-related expenses include items such as an employee hourly rate, payroll taxes, Social Security, and employee benefits.

The cost of having an employee greatly exceeds the hourly wages paid. In addition to an hourly wage, various taxes and insurances must be paid on the employee's behalf. These additional expenses can add approximately 25% to the amount the company must pay to have an employee.

12.8.1 Overhead Cost per Hour

The ***overhead*** cost per hour must be determined once all expenses have been tabulated. This cost is the minimum hourly charge that will allow a service company to meet its expenses and does not include profit or the hourly cost of the service personnel.

To calculate the overhead cost per hour, the annual overhead cost is divided by the number of annual billable hours:

$$\text{overhead cost/hour (\$/hr)} = \frac{\text{annual overhead cost}}{\text{annual billable hours}}$$

This is the hourly rate that must be charged to offset the annual, nonservice-personnel operating expenses for the company. Consider a scenario where the total annual overhead expenses, excluding service personnel, is $1,000,000. Assume the company has 10 members on the service team, and that each individual works 2000 hours per year. First the amount of work hours for one member is calculated.

$$40 \text{ hours/week} \times 50 \text{ weeks/year} = 2000 \text{ hours/year}$$

Then the result is multiplied by the number of members, which in this case is 10.

$$10 \times 2000 \text{ hours/year} = 20{,}000 \text{ hours/year}$$

Once the annual hours are calculated, the overhead cost per hour is determined:

$$\text{overhead cost/hour} = \frac{\text{annual overhead cost}}{\text{annual billable hours}}$$

$$\frac{\$1{,}000{,}000}{20{,}000 \text{ hours}} = \$50/\text{hr}$$

This calculation does not take into account any profit. This cost is the bare minimum hourly charge that will allow the service company to meet its expenses. The previous calculation assumes that each member of the service team will be working in a billable situation for each one of the 2000 hours per year. This is not realistic, as there will always be hours included in a workday that are not billed to the client, such as travel between jobsites, picking up materials at a supply house, fueling the service vehicle, or coordinating work with a service manager. As a result, the number of billable hours is often lower than the number of hours worked.

Depending on the company, the daily percentage of unbillable time can range from 10% to 15%. Thus, when the overhead cost per hour is calculated, only a percentage of the total number of annual hours should be classified as billable. This causes the overhead cost per hour to increase, allowing the company to meet its overhead expenses even when all daily hours worked are not billable.

Therefore, the formula for overhead cost per hour can be revised to include the billable hours percentage:

$$\text{overhead cost/hour} = \frac{\text{annual overhead cost}}{(\text{annual worked hours} \times \text{percentage billable work hours})}$$

12.8.2 Service Personnel Cost per Hour

The cost of a worker per hour is also factored into an hourly billing rate for labor. The actual hourly cost for each employee depends on the employee's hourly rate. Employees do not always have the same hourly rate. To simplify the calculation, most companies have billing rates dependent on the skill level of the employees. For instance, the levels can be a helper, a junior technician, and a senior technician.

Consider a company that wants to determine the hourly billing rate for a helper. This helper makes $30,000 per year, based on an hourly wage of $15 per hour and a 2000-hour work year. The actual cost to keep the employee on staff is approximately $37,500 when including wages paid to the employee and additional payments made on the employee's behalf. Remember, additional payments can amount to 25% of the employee's base salary. Thus, to calculate the hourly cost, we divide $37,500 by 2000 hours, which gives us $18.75. This is calculated by simply multiplying the employee's hourly rate by 25%, or 0.25, and adding this amount to the original hourly wage, as shown:

$$\begin{aligned} \text{employee cost per hour} &= \text{hourly wage} + (\text{hourly wage} \times 0.25) \\ &= \$15/\text{hr} + (\$15/\text{hr} \times 0.25) \\ &= \$15/\text{hr} + \$3.75/\text{hr} \\ &= \$18.75/\text{hr} \end{aligned}$$

This rate represents the hourly cost of a helper. It does not include any profit because this figure is the actual cost the employer must pay to, and on behalf of, the employee.

12.8.3 Service-Related Billing Rate per Hour

The service-related billing rate per hour is based on a combination of the overhead cost per hour and the service personnel cost per hour. By adding these two figures together, the company can determine the total hourly cost that must be collected to meet the operating and payroll expenses of the company. Thus, the equation to calculate the hourly cost is simple addition:

$$\text{total hourly cost for helper} = \text{overhead cost per hour} + \text{cost per hour}$$

Assume that the overhead cost per hour for a company is $62.50/hr, while the cost per hour for a technician is $37.50/hr. To find the total hourly cost for a technician, these two costs are added together.

$$\$62.50/\text{hr} + \$37.50/\text{hr} = \$100/\text{hr}$$

Thus, it costs the employer approximately $100/hr to keep a technician working. This does not take into account any profit the company must make to stay in business. If the company expects to make a profit of 20%, this means that 80% of the billable rate is the cost and is expressed as:

$$\text{total hourly cost} = 0.80 \times \text{billable rate for technician}$$

If the total hourly cost is $100 as previously indicated, the billable rate per hour for a technician is calculated as follows:

$$\begin{aligned} 0.80 \times \text{billable rate for technician} &= \$100 \\ \$100 \div 0.80 &= \$125 \end{aligned}$$

This result can be checked by multiplying $125 by 0.20 to obtain $25 profit per hour, which is equal to the billing hourly rate of $125 minus the hourly cost of $100.

A company usually charges the final hourly billing rates to its customers to meet both employee-related expenses and the company's operating expenses, while still making a profit.

12.9 Estimating Labor Costs for a Project

Knowing how many hours will be required to complete the project is another factor necessary to prepare an accurate bid. The more the estimator is able to break down each task into smaller tasks, within reason, the more accurate the time estimate will be. The details of each of these tasks include how long each of these subtasks will take, how many individuals are required to complete them, with the skill level of each individual included in the calculations.

Consider the process of hanging an air handler from an existing overhead steel beam grid. This task can be broken down into the following:

- Uncrating the air handler
- Reconfiguring the evaporator coil
- Measuring and determining the final location of the air handler
- Securing (4) beam C-clamps to the beams
- Cutting (4) lengths of threaded rod to length
- Cutting (2) lengths of steel channel to length
- Securing threaded rod to beam clamps
- Assembling and positioning unit jack/lift
- Lifting unit to its final location
- Installing steel channel, vibration pads, and other hardware

This is not intended to be a comprehensive list of the required tasks. It is intended as an example of how to break down a larger task into smaller ones. A chart can be used to identify who will perform each of the subtasks and how long each subtask will take, **Figure 12-8**. In the chart, the estimator assumes that one helper can uncrate an air handler in 30 minutes. It can also be shown it will take 30 minutes for a technician and a helper, working together, to measure and identify the location of the air handler.

TASK	Number of Technicians	Number of Hours	Number of Tech-Hours	Number of Helpers	Number of Hours	Number of Helper-Hours
Uncrate				1	0.50	
Reposition Coil	1	0.50				
Measuring	1	0.50		1	0.50	
Securing Clamps				1	0.50	
Cutting Rod				1	0.25	
Cutting Channel				1	0.25	
Securing Rod				1	0.50	
Jack Assembly				1	0.75	
Unit Lifting	1	0.50		1	0.50	
Unit Securing				2	0.50	

Goodheart-Willcox Publisher

Figure 12-8. A chart that outlines how many tradespeople and time is required for each project task.

Once all the subtasks have been evaluated, a worker's hours can be calculated. This is done by simply multiplying the number of workers required for each subtask by the number of hours required for each subtask. This information is entered into the chart as shown in **Figure 12-9.**

For example, this chart shows it will take one helper-hour (two helpers for 0.50 hours) to secure the unit once it has been raised into place. By adding all of the values in the number of tech-hours and number of helper-hours columns, the total number of tech-hours and helper-hours can be determined. In this example, 1.5 technician-hours and 4.75 helper-hours are required to hang an air handler.

Because we have already calculated the hourly cost for technicians and helpers, the total cost for hanging an air handler is:

$$\text{cost to hang air handler} = (\text{tech cost/hr} \times \#\ \text{tech-hours}) + (\text{helper cost/hr} \times \#\ \text{helper-hours})$$

$$(\$100/\text{hr} \times 1.5\ \text{hours}) + (\$81.25/\text{hr} \times 4.75\ \text{hours}) = \$150 + \$386 = \$536$$

This figure represents the cost to the contractor to hang one air handler. It does not include any profit that must be made. Assuming the contractor wishes to make a 20% profit on the project, this means that 80% (100% – 20%) of the price will be the cost to hang the air handler:

$$\text{cost to hang air handler} = 0.80 \times \text{price to hang air handler}$$

The billable figure for hanging the air handler will be:

$$0.80 \times \text{price to hang air handler} = \$536$$

$$\$536 \div 0.80 = \$670$$

TASK	Number of Technicians	Number of Hours	Number of Tech-Hours	Number of Helpers	Number of Hours	Number of Helper-Hours
Uncrate				1	0.50	0.50
Reposition Coil	1	0.50	0.50			
Measuring	1	0.50	0.50	1	0.50	0.50
Securing Clamps				1	0.50	0.50
Cutting Rod				1	0.25	0.25
Cutting Channel				1	0.25	0.25
Securing Rod				1	0.50	0.50
Jack Assembly				1	0.75	0.75
Unit Lifting	1	0.50	0.50	1	0.50	0.50
Unit Securing				2	0.50	1.00

Goodheart-Willcox Publisher

Figure 12-9. A tradesperson's hours on the job must be calculated and documented. This can be determined based on the information provided in the chart.

Multiplying $670 by 0.20 indicates that the profit will be $134, which is also the difference between the billing price of $670 and the cost of $536. If the project involves hanging 10 air handlers, the figure for a single unit can simply be multiplied by 10. This process can then be applied to all tasks that make up the project to accurately determine the amount of labor required and how much this labor will cost the client.

12.10 Establishing a Bid for a Project

The final bid for the project can be expressed as the sum of all of the individual aspects of the bid. The final bid should include the following items:

- Total labor cost
- Total equipment cost
- Total materials cost
- Project filing fees
- Additional overhead
- Profit

Only time determines whether a contractor's estimating practices are effective. If the contractor continually bids on projects, but is not awarded any, the bids are too high, and the figures need to be adjusted downward. However, if the contractor gets every job that is bid, and the company loses money on every job, it seems obvious that the calculated figures are too low. Estimating is a craft that, once mastered, will yield significant profits for the contractor.

Summary

- Estimates are a rough approximation of the amount a contractor is willing to accept to perform a given project.
- Quotes are formal offers to perform a task at a set price on a fixed time line.
- Bids are more detailed than quotes. They are prepared based on a very specific project plan that is to be completed exactly as presented.
- Bid overlapping occurs when a contractor includes items in a bid that the contractor is not responsible for.
- Proposals are documents that present a contractor's vision for a project to the client. They attempt to sell the client on the contractor's interpretation of a project.
- Underrun is used when the project cost is less than 90% of the estimate cost. Overrun is used when the project cost is more than 110% of the estimate cost.
- Project cost is the sum of all of the costs associated with executing a project. This cost includes the total labor, material, and equipment cost, as well as filing fees and other overhead.
- Project price is the price at which a contractor is willing to complete a project. This price includes profit in addition to all of the elements included in project cost.
- Total labor cost includes the time spent performing both project-specific tasks and other important tasks that cannot be directly associated with a project task.
- All pieces of equipment must be accounted for in a project estimate.
- Referring to manufacturer's installation literature and detail drawings help ensure that all installation materials are accounted for.
- The hourly billing rate is established to allow for the company's operating expenses to be met, while providing for a fair profit to be made.
- Breaking a task down into smaller subtasks helps accurately determine the time required to complete a task.

Chapter Assessment

Name ______________________ Date ____________ Class ______________

Know and Understand

_______ 1. Which of the following overhead expenses has the greatest effect on determining the difference in the hourly billing rate for two different employees at the same company?

A. The hourly wage of each employee

B. The number of miles driven by each employee each year

C. The number of service calls performed each week by each employee

D. The amount of training each employee requires

_______ 2. Which of the following is true regarding labor and materials/parts pricing?

A. Profit should not be included in either labor or materials/parts pricing.

B. Profit should be included in labor pricing, but not materials/parts pricing.

C. Profit should be included in materials/parts pricing but not labor pricing.

D. Profit should be included in both materials/parts and labor pricing.

_______ 3. A(n) _____ is a document that is most likely to have an expiration date.

A. estimate

B. quote

C. proposal

D. bid

_______ 4. A(n) _____ is a document that is most likely to include recommendation letters from previous clients.

A. estimate

B. quote

C. proposal

D. bid

_______ 5. Which of the following statements best describes a bid bond?

A. A fee paid to the general contractor to ensure the contract is awarded on time

B. A fee paid by the contractor to ensure the contractor fulfills its obligations if awarded the contract

C. A fee paid to the contractor to ensure the contractor takes on the project

D. A fee paid by the general contractor to the architect

_______ 6. How can an estimator help ensure that all installation materials are accounted for?

A. Refer to the manufacturer's installation literature.

B. Refer to the detail drawings.

C. Both A and B.

D. Neither A nor B.

_______ 7. How much additional installation material should be on hand to ensure adequate supplies?

A. 0%–10%

B. 10%–20%

C. 20%–30%

D. 30%–40%

Apply and Analyze

_______ 1. If the estimated cost of a project was $50,000 and the actual cost to complete the job was $58,000, which of the following statements is correct?

A. There was a cost underrun of $8,000 or 16%.

B. There was a cost overrun of $8,000 or 16%.

C. There was a cost underrun of $58,000 or 16%.

D. There was a cost overrun of $58,000 or 16%.

_______ 2. What is a company's overhead cost per hour if its annual overhead cost is $2,000,000 and the company has 25 service employees? Assume that 10% of the hours worked by the service employees are not billable.

A. $40.00

B. $44.44

C. $400.00

D. $444.44

_______ 3. If a particular task takes one technician and two helpers one hour to complete, how much will be billed if the technician's billable hourly rate is $120 and the helper's billable hourly rate is $90?

A. $120

B. $210

C. $300

D. $420

4. If the total cost of installing one exhaust fan is $500 and the company wishes to make a 20% profit on the project, how much will the project quote be if the job calls for the installation of 10 exhaust fans?

____

____

____

____

5. A contractor submits a project quote to a client. The cost of the job is $7500 and the contractor will make a profit of $2000 on the job. What is the quoted project price? What is the profit percentage that the contractor will make? By what percentage was the project marked up to make the profit as outlined?

____

____

____

____

Name ______________________ Date ____________ Class ____________________

Critical Thinking

1. Why is hourly overhead a cost that should be reevaluated and adjusted on a regular basis?

____

____

____

____

2. Why would it be impossible for a company to make a 100% profit on any installation job or project?

____

____

____

____

Large Prints Activity

Name ______________________ Date ____________ Class ______________

Practice Using Large Prints

Refer to the supply air duct system connected to AHU-1 on Print-25 in the Large Prints supplement and the information below to answer the following questions. Note: The large prints are created at 50% of their original size. When extracting information from the print using an architect's scale, use a scale that is one-half of the one indicated on the drawing.

- Cost per foot of 22 × 10 duct straight duct sections: $15
- Cost per foot of 12 × 6 duct straight duct sections: $10
- Cost per balancing damper: $80
- Cost per 22 × 10 90° elbow (Overall measurements 36″ × 36″): $100
- Cost per 4-way supply register: $30
- Cost of miscellaneous hanging supplies and other materials: $400
- Required hours: 10 hours at a cost of $75/hr

1. Determine the total cost of installing the supply ductwork for AHU-1.

____

____

2. What will be the price given to the customer if 20% of the total price is to be profit?

____

____

3. Using the information gathered in the previous question, determine the price that will be quoted for wrapping the supply ductwork for AHU. Assume the following:

- Duct wrap material costs $\$2/ft^2$
- There will be no waste
- The labor cost will be $25/hr for 8 hours
- 20% of the price quoted will be profit
- Calculations should reflect the overall length of the duct runs, so determining the amount of material required to wrap the elbow is not necessary.

____

____

Appendix

The following pages contain a number of tables, charts, and worksheets that are useful references when reading and extracting information from HVACR-related prints. The material in this section is listed below, along with the page number.

Duct Chart

1	2	3	4	5	6	7	8
H + W (inches)	Longest side (inches)	Gage	Square Feet of Material per 1′ of Duct	Hot Rolled Weight/Foot (pounds)	Galvanized Weight/Foot (pounds)	Stainless Weight/Foot (pounds)	Aluminum Weight/Foot (pounds)
12	ALL	26	2.00	1.50	1.81	1.51	0.45
13	ALL	26	2.17	1.63	1.96	1.64	0.49
14	ALL	26	2.33	1.75	2.11	1.76	0.52
15	ALL	26	2.50	1.88	2.27	1.89	0.56
16	ALL	26	2.67	2.00	2.42	2.02	0.60
17	ALL	26	2.83	2.13	2.57	2.14	0.63
18	ALL	26	3.00	2.25	2.72	2.27	0.67
19	0–12″	26	3.17	2.38	2.87	2.39	0.71
	13″–30″	24	3.17	3.17	3.66	3.20	0.90
20	0–12″	26	3.33	2.50	3.02	2.52	0.75
	13″–30″	24	3.33	3.33	3.85	3.36	0.95
21	0–12″	26	3.50	2.63	3.17	2.65	0.78
	13″–30″	24	3.50	3.50	4.05	3.53	0.99
22	0–12″	26	3.67	2.75	3.32	2.77	0.82
	13″–30″	24	3.67	3.67	4.24	3.70	1.04
23	0–12″	26	3.83	2.88	3.47	2.90	0.86
	13″–30″	24	3.83	3.83	4.43	3.86	1.09
24	0–12″	26	4.00	3.00	3.62	3.02	0.90
	13″–30″	24	4.00	4.00	4.62	4.03	1.14
25	13″–30″	24	4.17	4.17	4.82	4.20	1.18
26	13″–30″	24	4.33	4.33	5.01	4.37	1.23
27	13″–30″	24	4.50	4.50	5.20	4.54	1.28
28	13″–30″	24	4.67	4.67	5.39	4.70	1.33
29	13″–30″	24	4.83	4.83	5.59	4.87	1.37
30	13″–30″	24	5.00	5.00	5.78	5.04	1.42
31	13″–30″	24	5.17	5.17	5.97	5.21	1.47
32	13″–30″	24	5.33	5.33	6.17	5.38	1.51
33	13″–30″	24	5.50	5.50	6.36	5.54	1.56
34	13″–30″	24	5.67	5.67	6.55	5.71	1.61
35	13″–30″	24	5.83	5.83	6.74	5.88	1.66
36	13″–30″	24	6.00	6.00	6.94	6.05	1.70
	31″–60″	22	6.00	7.50	8.44	7.56	2.14
37	13″–30″	24	6.17	6.17	7.13	6.22	1.75
	31″–60″	22	6.17	7.71	8.68	7.77	2.20
38	13″–30″	24	6.33	6.33	7.32	6.38	1.80
	31″–60″	22	6.33	7.91	8.90	7.98	2.26
39	13″–30″	24	6.50	6.50	7.51	6.55	1.85
	31″–60″	22	6.50	8.13	9.14	8.19	2.32

Duct Chart *(continued from previous page)*

1	2	3	4	5	6	7	8
H + W (inches)	Longest side (inches)	Gage	Square Feet of Material per 1′ of Duct	Hot Rolled Weight/Foot (pounds)	Galvanized Weight/Foot (pounds)	Stainless Weight/Foot (pounds)	Aluminum Weight/Foot (pounds)
40	13″–30″	24	6.67	6.67	7.71	6.72	1.89
	31″–60″	22	6.67	8.34	9.38	8.40	2.38
41	13″–30″	24	6.83	6.83	7.90	6.89	1.94
	31″–60″	22	6.83	8.54	9.60	8.61	2.44
42	13″–30″	24	7.00	7.00	8.09	7.06	1.99
	31″–60″	22	7.00	8.75	9.84	8.82	2.50
43	13″–30″	24	7.17	7.17	8.28	7.22	2.04
	31″–60″	22	7.17	8.96	10.08	9.03	2.56
44	13″–30″	24	7.33	7.33	8.48	7.39	2.08
	31″–60″	22	7.33	9.16	10.31	9.24	2.62
45	13″–30″	24	7.50	7.50	8.67	7.56	2.13
	31″–60″	22	7.50	9.38	10.55	9.45	2.68
46	13″–30″	24	7.67	7.67	8.86	7.73	2.18
	31″–60″	22	7.67	9.59	10.78	9.66	2.74
47	13″–30″	24	7.83	7.83	9.06	7.90	2.22
	31″–60″	22	7.83	9.79	11.01	9.87	2.80
48	13″–30″	24	8.00	8.00	9.25	8.06	2.27
	31″–60″	22	8.00	10.00	11.25	10.08	2.86
49	13″–30″	24	8.17	8.17	9.44	8.23	2.32
	31″–60″	22	8.17	10.21	11.49	10.29	2.92
50	13″–30″	24	8.33	8.33	9.63	8.40	2.37
	31″–60″	22	8.33	10.41	11.71	10.50	2.97
51	13″–30″	24	8.50	8.50	9.83	8.57	2.41
	31″–60″	22	8.50	10.63	11.95	10.71	3.03
52	13″–30″	24	8.67	8.67	10.02	8.74	2.46
	31″–60″	22	8.67	10.84	12.19	10.92	3.10
53	13″–30″	24	8.83	8.83	10.21	8.90	2.51
	31″–60″	22	8.83	11.04	12.41	11.13	3.15
54	13″–30″	24	9.00	9.00	10.40	9.07	2.56
	31″–60″	22	9.00	11.25	12.65	11.34	3.21
55	13″–30″	24	9.17	9.17	10.60	9.24	2.60
	31″–60″	22	9.17	11.46	12.89	11.55	3.27
56	13″–30″	24	9.33	9.33	10.79	9.41	2.65
	31″–60″	22	9.33	11.66	13.12	11.76	3.33
57	13″–30″	24	9.50	9.50	10.98	9.58	2.70
	31″–60″	22	9.50	11.88	13.36	11.97	3.39
58	13″–30″	24	9.67	9.67	11.17	9.74	2.75
	31″–60″	22	9.67	12.09	13.60	12.18	3.45

Duct Chart *(continued from previous page)*

1	2	3	4	5	6	7	8
H + W (inches)	Longest side (inches)	Gage	Square Feet of Material per 1′ of Duct	Hot Rolled Weight/Foot (pounds)	Galvanized Weight/Foot (pounds)	Stainless Weight/Foot (pounds)	Aluminum Weight/Foot (pounds)
59	13″–30″	24	9.83	9.83	11.37	9.91	2.79
	31″–60″	22	9.83	12.29	13.82	12.39	3.51
60	13″–30″	24	10.00	10.00	11.56	10.08	2.84
	31″–60″	22	10.00	12.50	14.06	12.60	3.57
61	31″–60″	22	10.17	12.71	14.29	12.81	3.63
62	31″–60″	22	10.33	12.92	14.53	13.02	3.69
63	31″–60″	22	10.50	13.13	14.76	13.23	3.75
64	31″–60″	22	10.67	13.33	15.00	13.44	3.81
65	31″–60″	22	10.83	13.54	15.23	13.65	3.87
66	31″–60″	22	11.00	13.75	15.47	13.86	3.93
67	31″–60″	22	11.17	13.96	15.70	14.07	3.99
	61″–90″	20	11.17	16.76	18.50	16.89	5.03
68	31″–60″	22	11.33	14.17	15.93	14.28	4.05
	61″–90″	20	11.33	17.00	18.76	17.13	5.10
69	31″–60″	22	11.50	14.38	16.17	14.49	4.11
	61″–90″	20	11.50	17.25	19.04	17.39	5.18
70	31″–60″	22	11.67	14.58	16.40	14.70	4.17
	61″–90″	20	11.67	17.51	19.33	17.65	5.25
72	31″–60″	22	12.00	15.00	16.87	15.12	4.28
	61″–90″	20	12.00	18.00	19.87	18.14	5.40
74	31″–60″	22	12.33	15.42	17.34	15.54	4.40
	61″–90″	20	12.33	18.50	20.42	18.64	5.55
76	31″–60″	22	12.67	15.83	17.81	15.96	4.52
	61″–90″	20	12.67	19.01	20.98	19.16	5.70
78	31″–60″	22	13.00	16.25	18.28	16.38	4.64
	61″–90″	20	13.00	19.50	21.53	19.66	5.85
80	31″–60″	22	13.33	16.67	18.75	16.80	4.76
	61″–90″	20	13.33	20.00	22.07	20.15	6.00
82	31″–60″	22	13.67	17.08	19.22	17.22	4.88
	61″–90″	20	13.67	20.51	22.64	20.67	6.15
84	31″–60″	22	14.00	17.50	19.68	17.64	5.00
	61″–90″	20	14.00	21.00	23.18	21.17	6.30
86	31″–60″	22	14.33	17.92	20.15	18.06	5.12
	61″–90″	20	14.33	21.50	23.73	21.67	6.45
88	31″–60″	22	14.67	18.33	20.62	18.48	5.24
	61″–90″	20	14.67	22.01	24.29	22.18	6.60
90	31″–60″	22	15.00	18.75	21.09	18.90	5.36
	61″–90″	20	15.00	22.50	24.84	22.68	6.75

Duct Chart *(continued from previous page)*

1	2	3	4	5	6	7	8
H + W (inches)	Longest side (inches)	Gage	Square Feet of Material per 1′ of Duct	Hot Rolled Weight/Foot (pounds)	Galvanized Weight/Foot (pounds)	Stainless Weight/Foot (pounds)	Aluminum Weight/Foot (pounds)
92	31″–60″	22	15.33	19.17	21.56	19.32	5.47
	61″–90″	20	15.33	23.00	25.39	23.18	6.90
94	31″–60″	22	15.67	19.58	22.03	19.74	5.59
	61″–90″	20	15.67	23.51	25.95	23.69	7.05
96	31″–60″	22	16.00	20.00	22.50	20.16	5.71
	61″–90″	20	16.00	24.00	26.50	24.19	7.20
98	31″–60″	22	16.33	20.42	22.96	20.58	5.83
	61″–90″	20	16.33	24.50	27.04	24.69	7.35
100	31″–60″	22	16.67	20.83	23.43	21.00	5.95
	61″–90″	20	16.67	25.01	27.61	25.21	7.50
102	31″–60″	22	17.00	21.25	23.90	21.42	6.07
	61″–90″	20	17.00	25.50	28.15	25.70	7.65
104	31″–60″	22	17.33	21.67	24.37	21.84	6.19
	61″–90″	20	17.33	26.00	28.70	26.20	7.80
106	31″–60″	22	17.67	22.08	24.84	22.26	6.31
	61″–90″	20	17.67	26.51	29.26	26.72	7.95
108	31″–60″	22	18.00	22.50	25.31	22.68	6.43
	61″–90″	20	18.00	27.00	29.81	27.22	8.10
110	31″–60″	22	18.33	22.92	25.78	23.10	6.55
	61″–90″	20	18.33	27.50	30.35	27.71	8.25
112	31″–60″	22	18.67	23.33	26.25	23.52	6.66
	61″–90″	20	18.67	28.01	30.92	28.23	8.40
114	31″–60″	22	19.00	23.75	26.71	23.94	6.78
	61″–90″	20	19.00	28.50	31.46	28.73	8.55
116	31″–60″	22	19.33	24.17	27.18	24.36	6.90
	61″–90″	20	19.33	29.00	32.01	29.23	8.70
118	31″–60″	22	19.67	24.58	27.65	24.78	7.02
	61″–90″	20	19.67	29.51	32.57	29.74	8.85
120	31″–60″	22	20.00	25.00	28.12	25.20	7.14
	61″–90″	20	20.00	30.00	33.12	30.24	9.00
122	61″–90″	20	20.33	30.50	33.67	30.74	9.15
124	61″–90″	20	20.67	31.00	34.22	31.25	9.30
126	61″–90″	20	21.00	31.50	34.78	31.75	9.45
128	61″–90″	20	21.33	32.00	35.33	32.26	9.60

Fraction and Decimal Equivalents

Fraction	Inches	mm
1/64	0.01563	0.397
1/32	0.03125	0.794
3/64	0.04688	1.191
1/16	0.0625	1.588
5/64	0.07813	1.984
3/32	0.09375	2.381
7/64	0.10938	2.778
1/8	0.12500	3.175
9/64	0.14063	3.572
5/32	0.15625	3.969
11/64	0.17188	4.366
3/16	0.18750	4.763
13/64	0.20313	5.159
7/32	0.21875	5.556
15/64	0.23438	5.953
1/4	0.25000	6.350
17/64	0.26563	6.747
9/32	0.28125	7.144
19/64	0.29688	7.541
5/16	0.31250	7.938
21/64	0.32813	8.334
11/32	0.34375	8.731
23/64	0.35938	9.128
3/8	0.37500	9.525
25/64	0.39063	9.922
13/32	0.40625	10.319
27/64	0.42188	10.716
7/16	0.43750	11.113
29/64	0.45313	11.509
15/32	0.46875	11.906
31/64	0.48438	12.303
1/2	0.50000	12.700

Fraction	Inches	mm
33/64	0.51563	13.097
17/32	0.53125	13.494
35/64	0.54688	13.891
9/16	0.56250	14.288
37/64	0.57813	14.684
19/32	0.59375	15.081
39/64	0.60938	15.478
5/8	0.62500	15.875
41/64	0.64063	16.272
21/32	0.65625	16.669
43/64	0.67188	17.066
11/16	0.68750	17.463
45/64	0.70313	17.859
23/32	0.71875	18.256
47/64	0.73438	18.653
3/4	0.75000	19.050
49/64	0.76563	19.447
25/32	0.78125	19.844
51/64	0.79688	20.241
3/16	0.81250	20.638
53/64	0.82813	21.034
27/32	0.84375	21.431
55/64	0.85938	21.828
7/8	0.87500	22.225
57/64	0.89063	22.622
29/32	0.90625	23.019
59/64	0.92188	23.416
15/16	0.93750	23.813
61/64	0.95313	24.209
31/32	0.96875	24.606
63/64	0.98438	25.003
1	1.00000	25.400

Print Reading Abbreviations

A

AB Above
ABS Acrylonitrile Butadiene Styrene (pipe)
AC Alternating current
AD Access door, Automatic damper
AFC Automatic flow control
AFD Adjustable frequency drive
AFF Above finished floor
AFG Above finished grade
AH Air handler
AHU Air handling unit
AP Access panel
A/C Air-conditioning
ACC Air-cooled condensing unit
AL Acoustical lining
ARCH Architectural
AWG American Wire Gauge

B

BB Bottom of beam
BDD Back draft damper
BEL Below
BF Bottom is flat
BHP Brake horsepower
BOP Bottom of pipe
BTU British thermal unit
BTUH British thermal units per hour
BU Bottom up from floor

C

CAV Constant air volume
CEG Ceiling exhaust grille
CER Ceiling exhaust register
CHWR Chilled water return
CHWS Chilled water supply
CLG Ceiling, Cooling
CONC Concrete
CRG Ceiling return grille
CRR Ceiling return resister
CSD Ceiling supply diffuser
CNT Continuation
CP Control panel, Condensate pump
CC Cooling coil
CFM Cubic feet per minute
CT Cooling tower
CU Condensing unit

D

DC Direct current
DDC Direct digital control
DHW Domestic hot water
DIA Diameter
DIFF Diffuser
DN Down
DWG Drawing
DB Dry-bulb
DBT Dry-bulb temperature
DWH Domestic water heater
DX Direct expansion

E

EA Exhaust air
EAT Entering air temperature
EB Exhaust blower
EER Energy efficiency ratio
EF Exhaust fan
EG Exhaust grille
EL Elevation
ELEC Electrical
ET Expansion tank
EX Exhaust
EXIST Existing
(E) Existing
EXT Exterior

F

F Fan
FB Flat on bottom
FC Flexible connector
FCU Fan coil unit
FD Fire damper
FF Finished floor
FT Flat on Top
FC Flexible connection
FLR Floor
FOB Flat on bottom
FOT Flat on top
FPM Feet per minute

G

GPM Gallons per minute
GWR Ground loop water return
GWS Ground loop water supply

H

H	Humidifier
HC	Heating coil
HHWR	Heating hot water return
HHWS	Heating hot water supply
HP	Horsepower
HZ	Hertz (Frequency)

I

ID	Inside diameter
IWC	Inches of water column

K

kW	Kilowatt
kWh	Kilowatt hour

L

LAT	Leaving air temperature
LBS	Pounds
LD	Linear diffuser
LWT	Leaving water temperature

M

MA	Mixed air
MBH	Thousands of Btu per hour
MUA	Make-up air
MAX	Maximum
MIN	Minimum

N

(N)	New
N/A	Not applicable
NC	Normally closed
NO	Number, Normally open
NIC	Not in contract
NTS	Not to scale

O

OA	Outside air
OAI	Outside air intake
OBD	Opposed blade damper
OD	Outside diameter

P

PD	Pressure drop
Ph	Phase
PG	Pressure gauge
PRV	Pressure-reducing valve
PSI	Pounds per square inch
PSIG	Pounds per square inch gauge
PVC	Polyvinyl chloride pipe

Q

QTY	Quantity

R

R	Rise, Relocate
RA	Return air
RED	Reducer
RG	Return grill
RHC	Reheat coil
RHG	Refrigerant hot gas
RL	Refrigerant liquid
RPM	Rotations per minute
RS	Refrigerant suction
RTU	Rooftop unit

S

SA	Supply air
SD	Splitter damper
SEER	Seasonal energy efficiency ratio
SG	Supply grille
SP	Static pressure
SPEC	Specification
ST	Sound trap

T

TBD	To be determined
TEMP	Temperature, Temporary
TF	Top is flat
TG	Transfer grille
TH	Thermometer
THRU	Through
TL	Transfer louver
TU	Top up from floor
TV	Turning vane
TYP	Typical

V

V	Volt
VAV	Variable air volume
VD	Volume damper
VFD	Variable frequency drive

W

W/	With
W/O	Without
W	Watt
WB	Wet-bulb
WBT	Wet-bulb temperature
WC	Water column
WH	Water heater
WRT	With reference to

Ductwork Takeoff Worksheet

Project Name:		Project #
Location/Area:		Drawing #
Date:	Worksheet # ____________ of ____________	Prepared by:

Ref #	Size	Type	Length	Gage	Material	Connection Type	Wrapped (Y/N)	Lined (Y/N)	Lb/ft	Total Weight
							Total Weight:			

HVACR Piping Takeoff Worksheet

Project Name:		Project #
Location/Area:		Drawing #
Date:	Worksheet # ____________ of ____________	Prepared by:

			Pipe			Fitting		
Ref #	Size	ID/OD	Length	Type	Material	Type	Material	Quantity

HVACR Piping Installation Takeoff Worksheet

Project Name:		Project #
Location/Area:		Drawing #
Date:	Worksheet # ________ of ________	Prepared by:

Clamps				Mounting Channel			
Ref #	Type	Size	Quantity	Ref #	Type	Size	Length
Threaded Rod				Straps/Brackets			
Ref #	Size	Length	Quantity	Ref #	Type	Size	Quantity
Anchors				Nuts/Bolts/Washers			
Ref #	Type	Size	Quantity	Ref #	Type	Size	Quantity
Clamps				Soldering/Brazing Material			
Ref #	Type	Size	Quantity				
Insulation				Compound/Dope/Teflon			
Type	Size	Thickness	Quantity				
Other:				Other:			

HVACR Equipment Takeoff Worksheet

Project Name:		Project #
Location/Area:		Drawing #
Date:	Worksheet # ____________ of ____________	Prepared by:

Manufacturer	Model #	Quantity	Print Tag #	Notes

HVACR Air Device (Grille, Register, and Louver) Takeoff Worksheet

Project Name:		Project #
Location/Area:		Drawing #
Date:	Worksheet # ____________ of ____________	Prepared by:

Model #	**Dimensions**	**Type (S/R/E)**	**Quantity**	**CFM**	**Tag #**

Project Estimating Worksheet

Project Name:		Project #
Location/Area:		
Date:	Worksheet # ____________ of ____________	Prepared by:
Project Comments:		

Equipment

Manufacturer	Model	Quantity	Worksheet Ref #	Unit Cost	Total Cost	Profit %	Proposal Price
Equipment Total Proposal Price							$

Materials and Supplies

Item	Part (Description)	Quantity	Unit Cost	Total Cost	Profit %	Proposal Price
Materials Total Proposal Price						$

Labor

Type	Billable Hourly Rate	Number of Hours	Worksheet Ref #s	Comments	Proposal Price
Labor Total Proposal Price					$

Administrative Expenses		
Type	Description/Worksheet Ref #s	Proposal Price
Administrative Total Proposal Price		$

Other Project Related Costs/Fees		
Type	Description/Worksheet Ref #s	Proposal Price
Other Project Expenses Total Proposal Price		$

Subcontractor/Additional Work		
Type	Description/Worksheet Ref #s	Proposal Price
Subcontractor/Additional Work Total Proposal Price		$

Total Project Estimate		
Subsection	Comments/Reference	Proposal Price
Equipment		
Materials and Supplies		
Labor		
Administrative		
Other Project Costs/Fees		
Subcontractor/Additional Work		
Total Project Proposal Price		$

Electrical Symbols

Switches and Contacts

Single-pole single-throw switch

Heating thermostat

Cooling thermostat

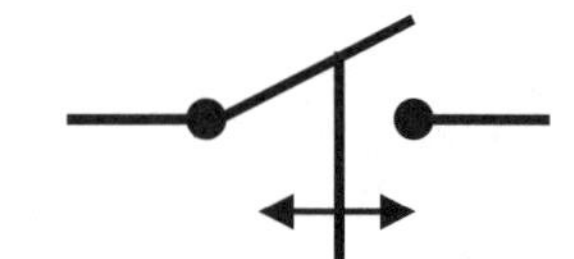
Sail switch

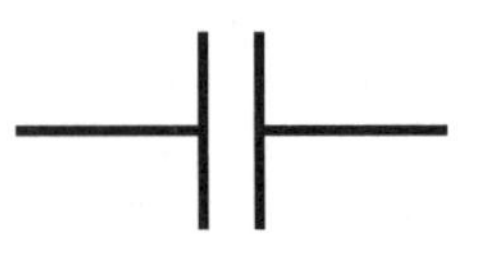
Normally open (NO) contacts

Normally closed (NC) contacts

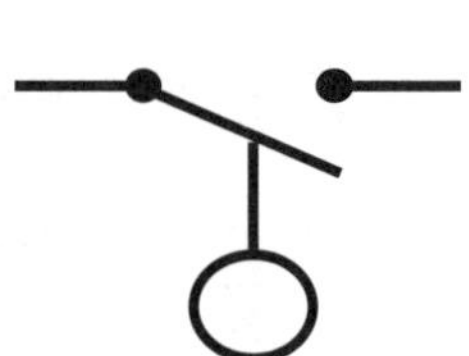
Float switch (close on level rise)

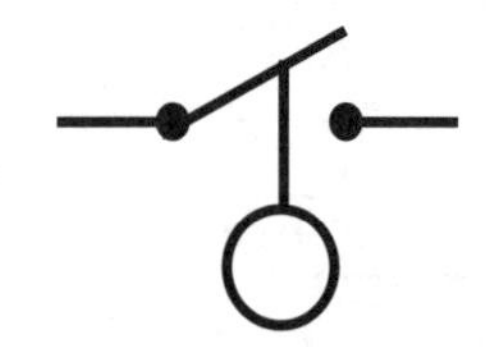
Float switch (open on level rise)

Time-controlled contacts

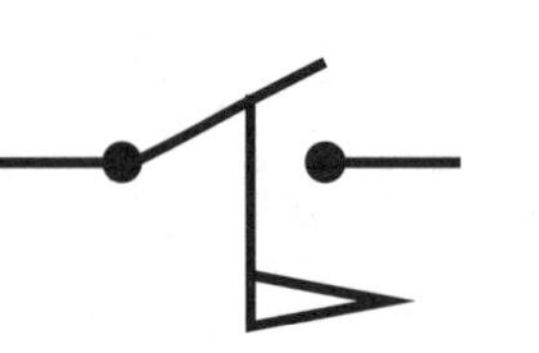
Flow switch

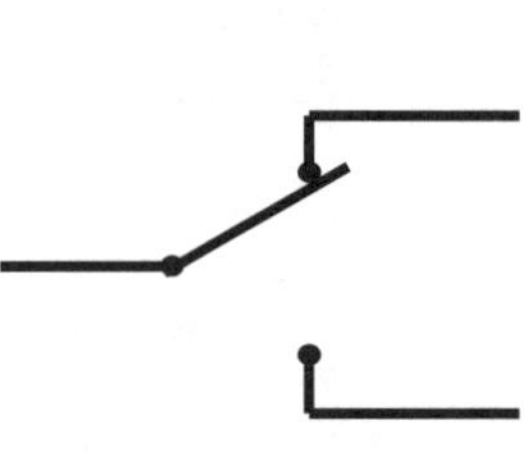
Single-pole double-throw (SPDT) switch

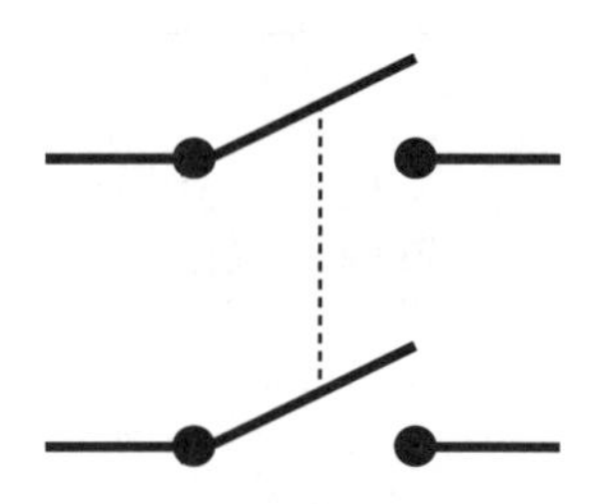
Double-pole single-throw (DPST) switch

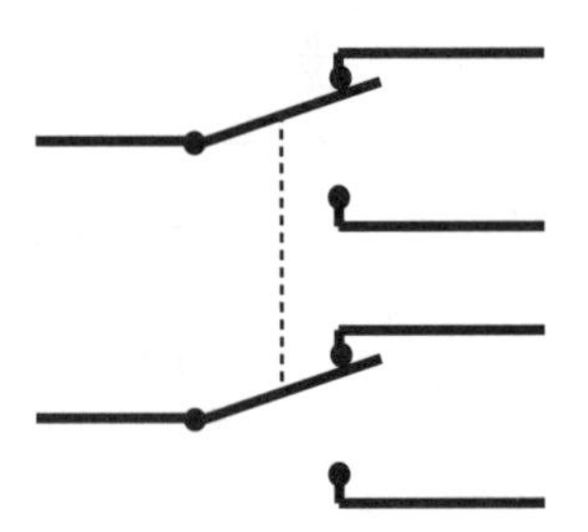
Double-pole double-throw (DPDT) switch

High-pressure switch

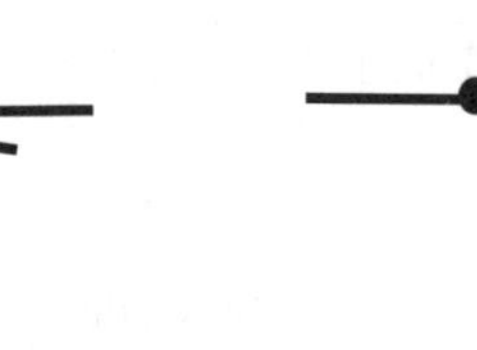
Low-pressure switch

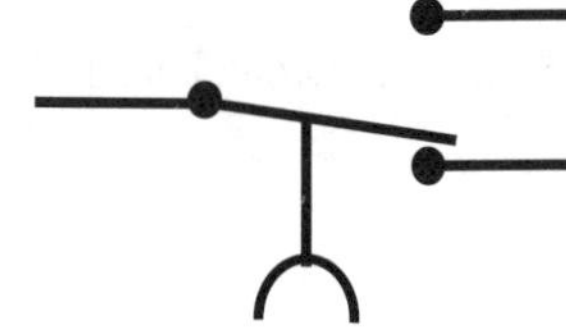
Single-pole double-throw (SPDT) pressure switch

Goodheart-Willcox Publisher

Relays, Contactors, and Starters

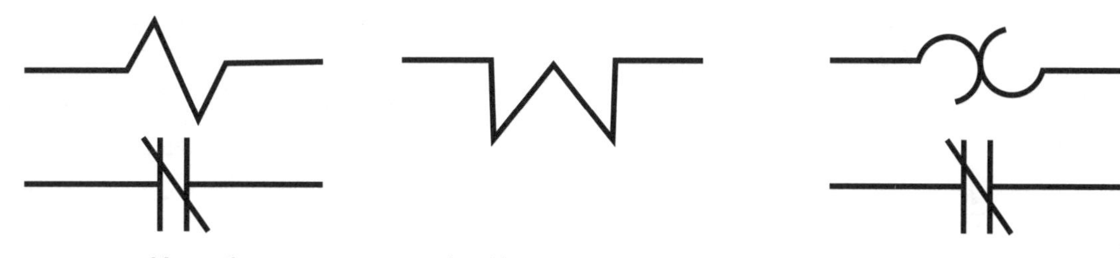

Magnetic motor starter overload heaters

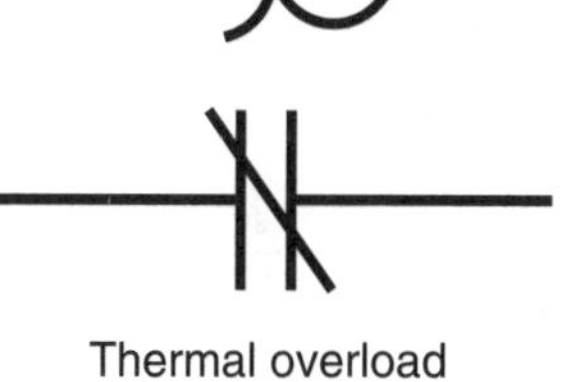

Thermal overload

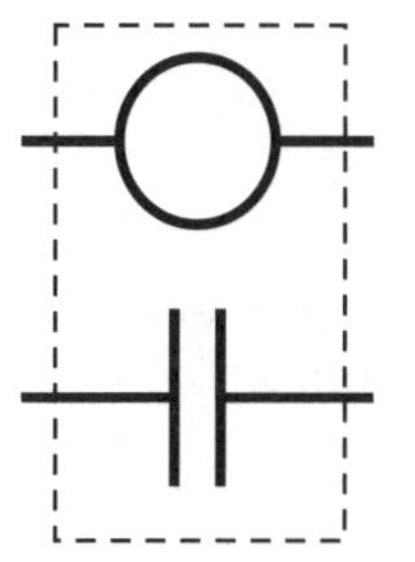

Relay with one NO set of contacts

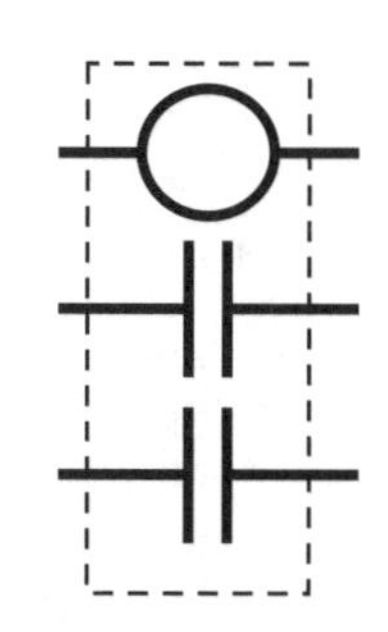

Relay with two NO sets of contacts

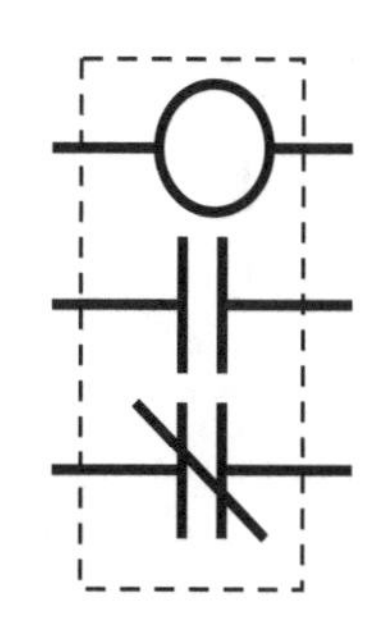

Relay with one NO set of contacts and one NC set of contacts

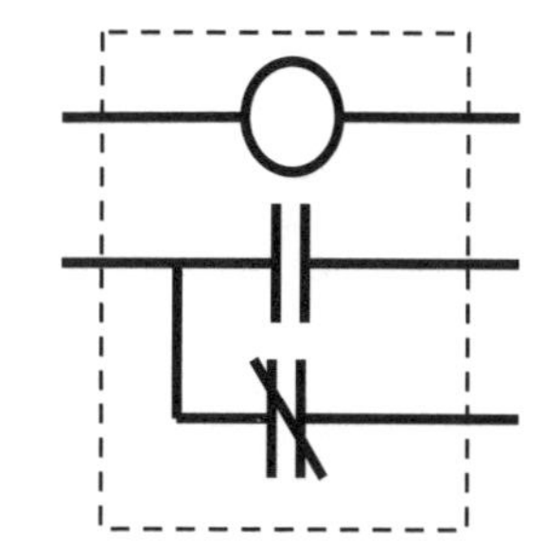

Single-pole double-throw relay

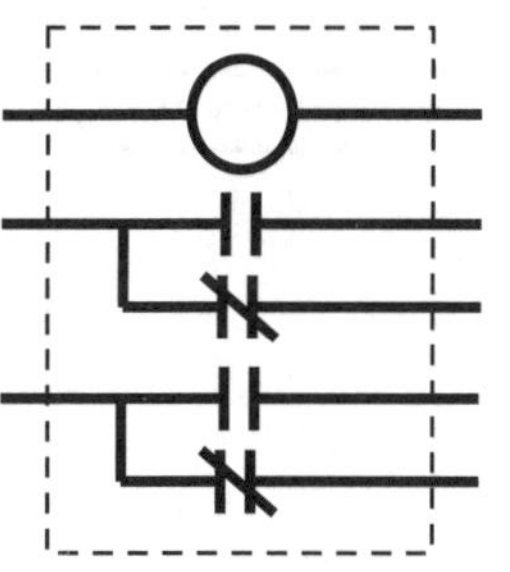

Double-pole double-throw relay

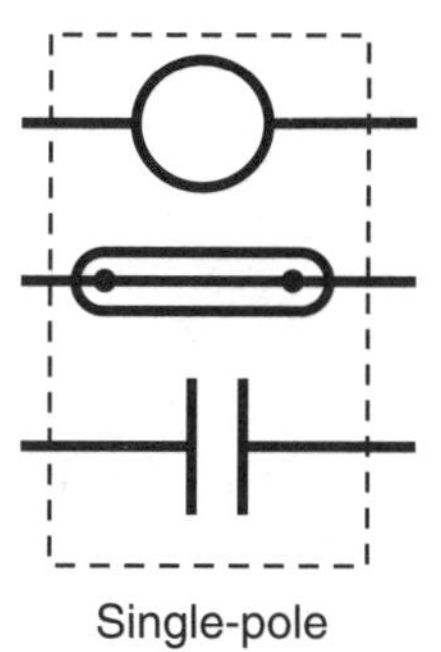

Single-pole contactor

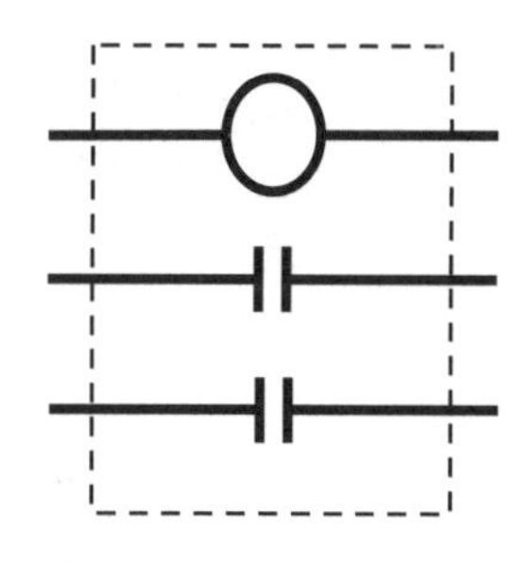

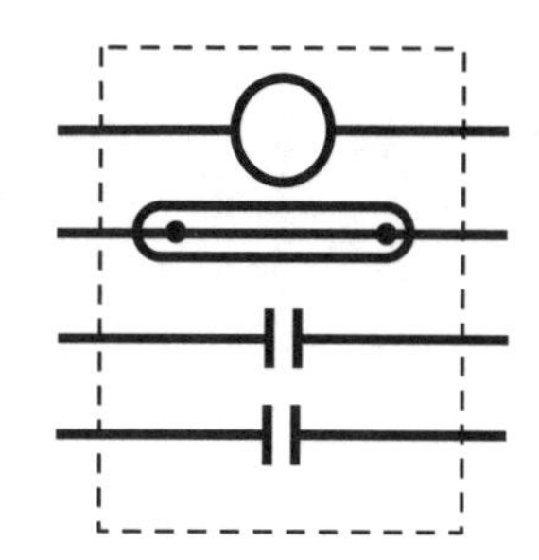

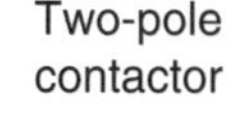

Two-pole contactor

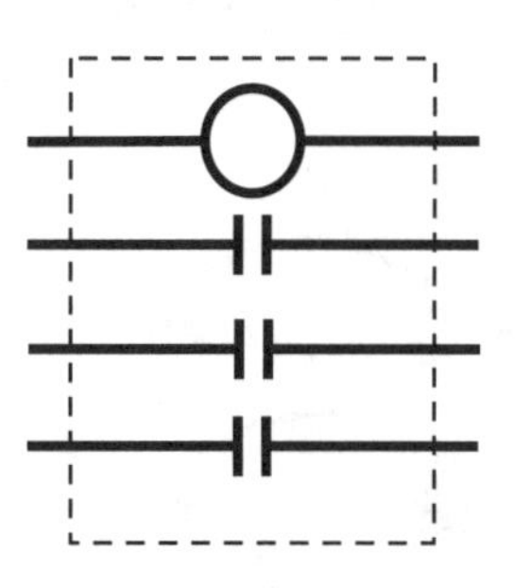

Three-pole contactor

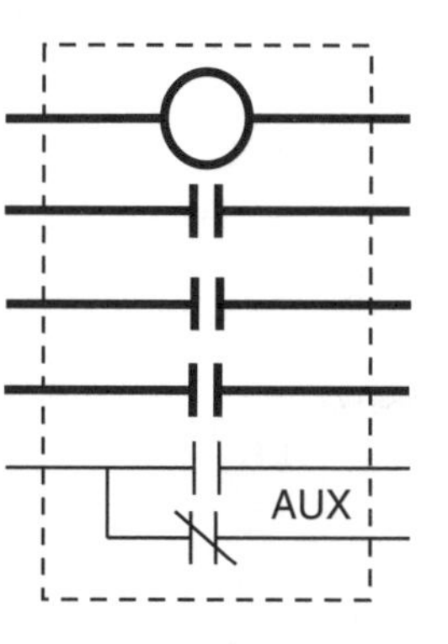

Three-pole contactor with auxiliary contacts

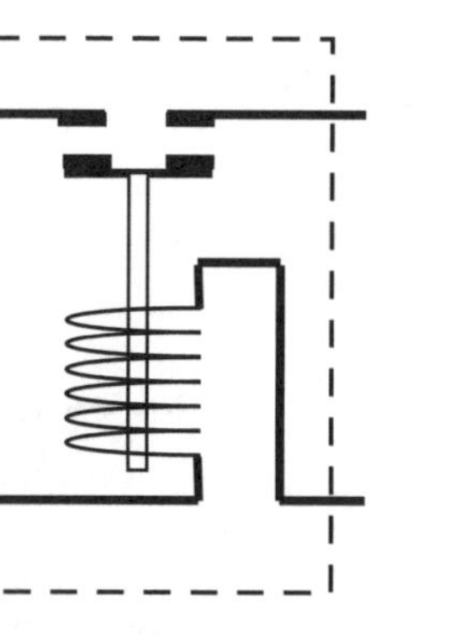

Current magnetic relay (CMR)

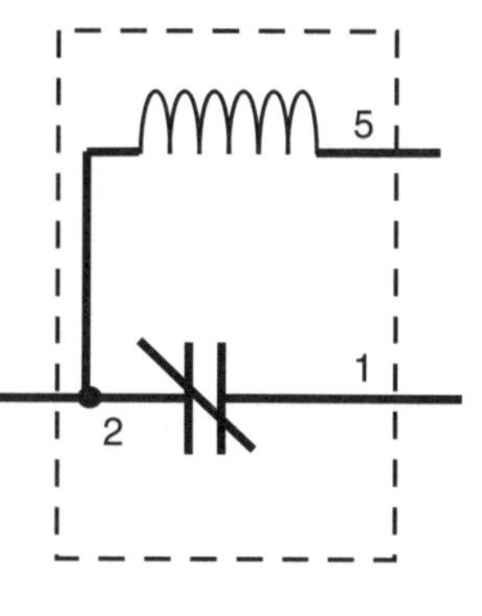

Potential magnetic relay (PMR)

Goodheart-Willcox Publisher

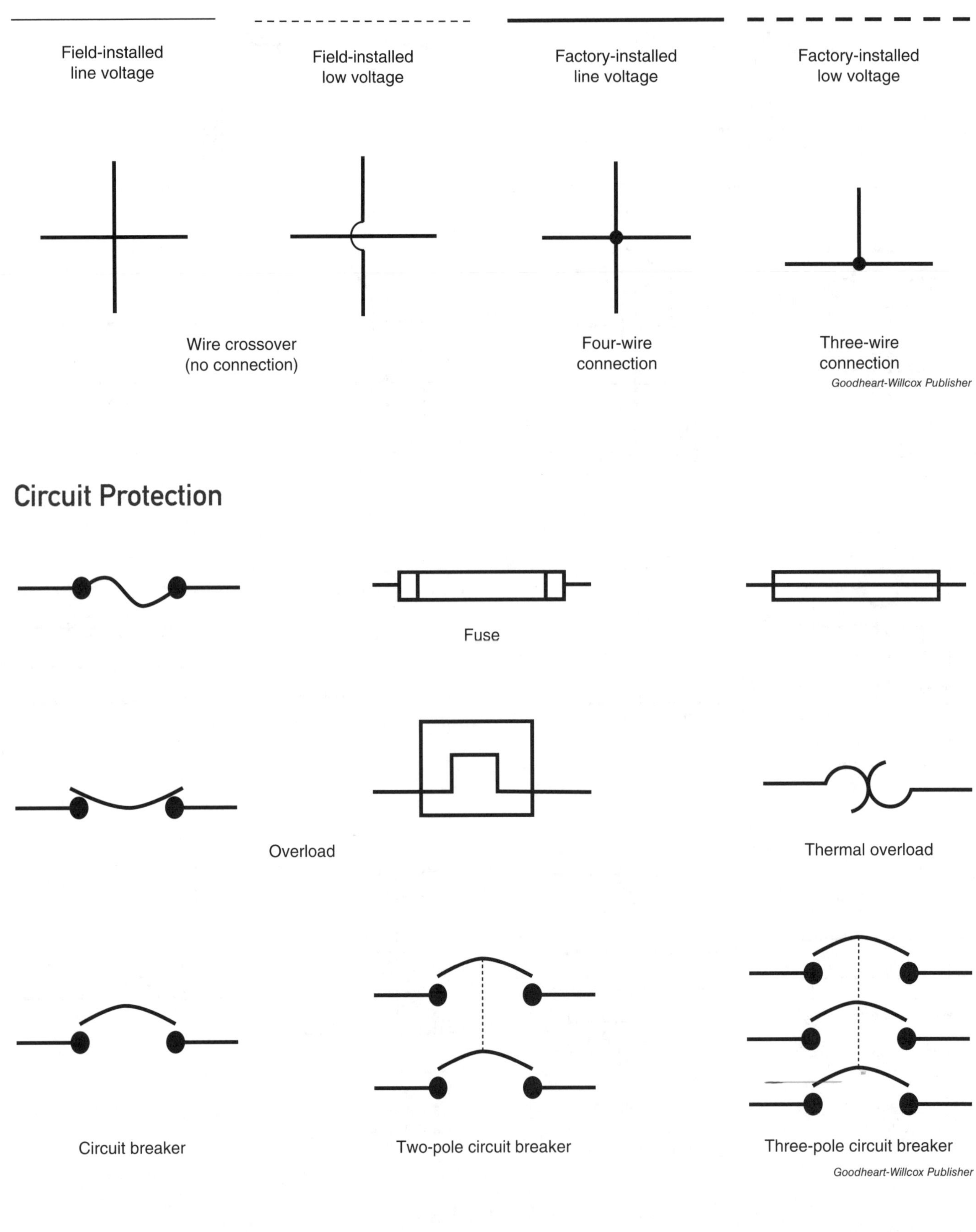

Wiring
Field-installed line voltage
Field-installed low voltage
Factory-installed line voltage
Factory-installed low voltage
Wire crossover (no connection)
Four-wire connection
Three-wire connection
Goodheart-Willcox Publisher
Circuit Protection
Fuse
Overload
Thermal overload
Circuit breaker
Two-pole circuit breaker
Three-pole circuit breaker
Goodheart-Willcox Publisher

Power Supplies and Grounds

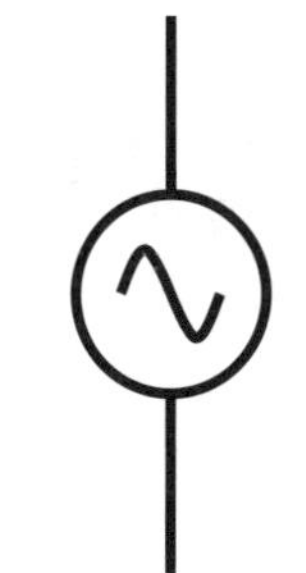

AC power supply

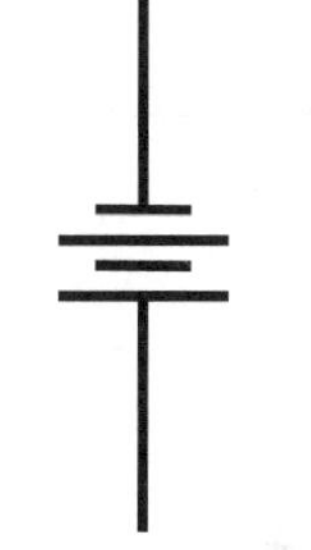

DC power supply

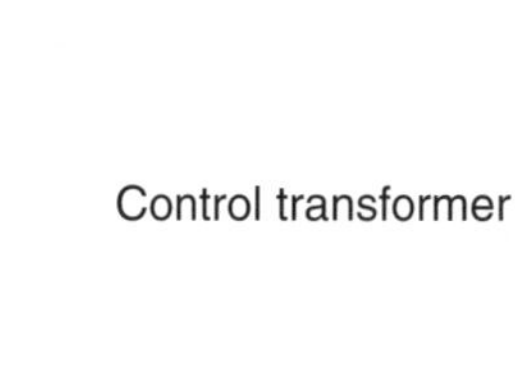

Control transformer

Chassis ground

Earth ground

Goodheart-Willcox Publisher

Coils

Relay/holding coil

Solenoid coil

Goodheart-Willcox Publisher

Resistors and Heaters

Resistor

HTR

Heater

CH

Crankcase heater

Defrost heater

Goodheart-Willcox Publisher

Electronics

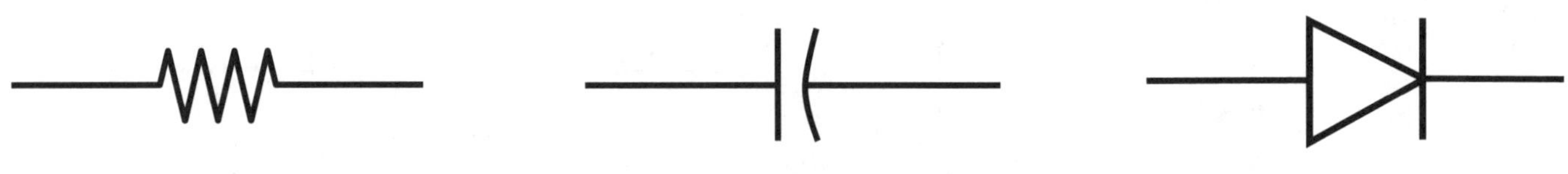

Resistor

Capacitor

Diode

Adjustable resistor

Variable resistor

Thermistor

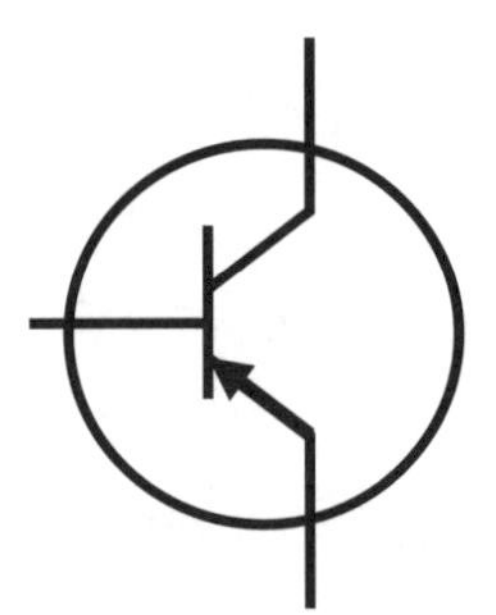

PNP transistor

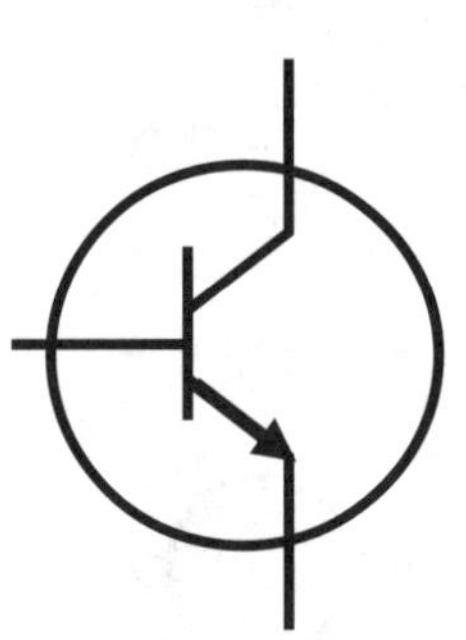

NPN transistor

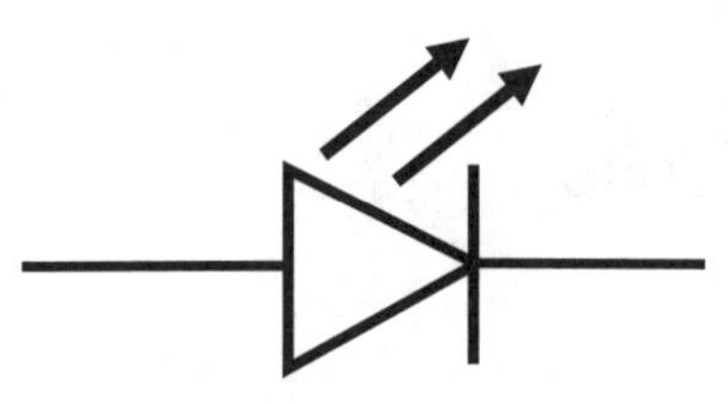

Light emitting diode (LED)

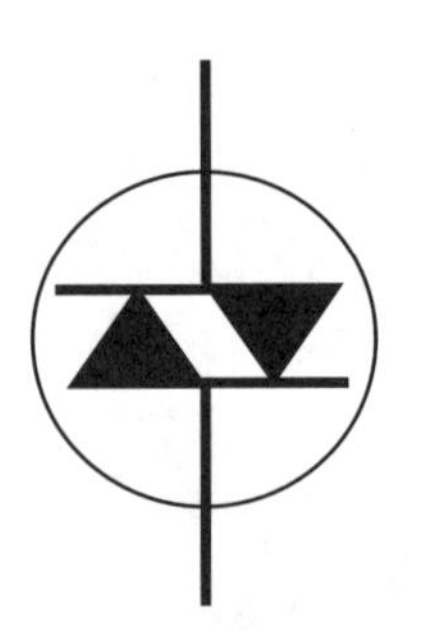

Bilateral switch

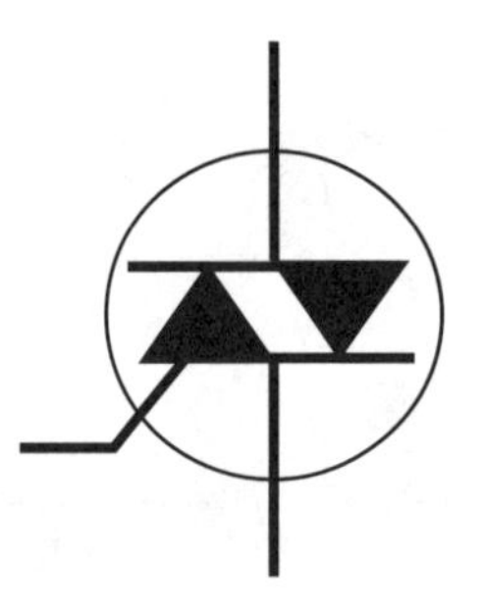

Triac

Inductor (coil)

Goodheart-Willcox Publisher

Piping Symbols

Valves

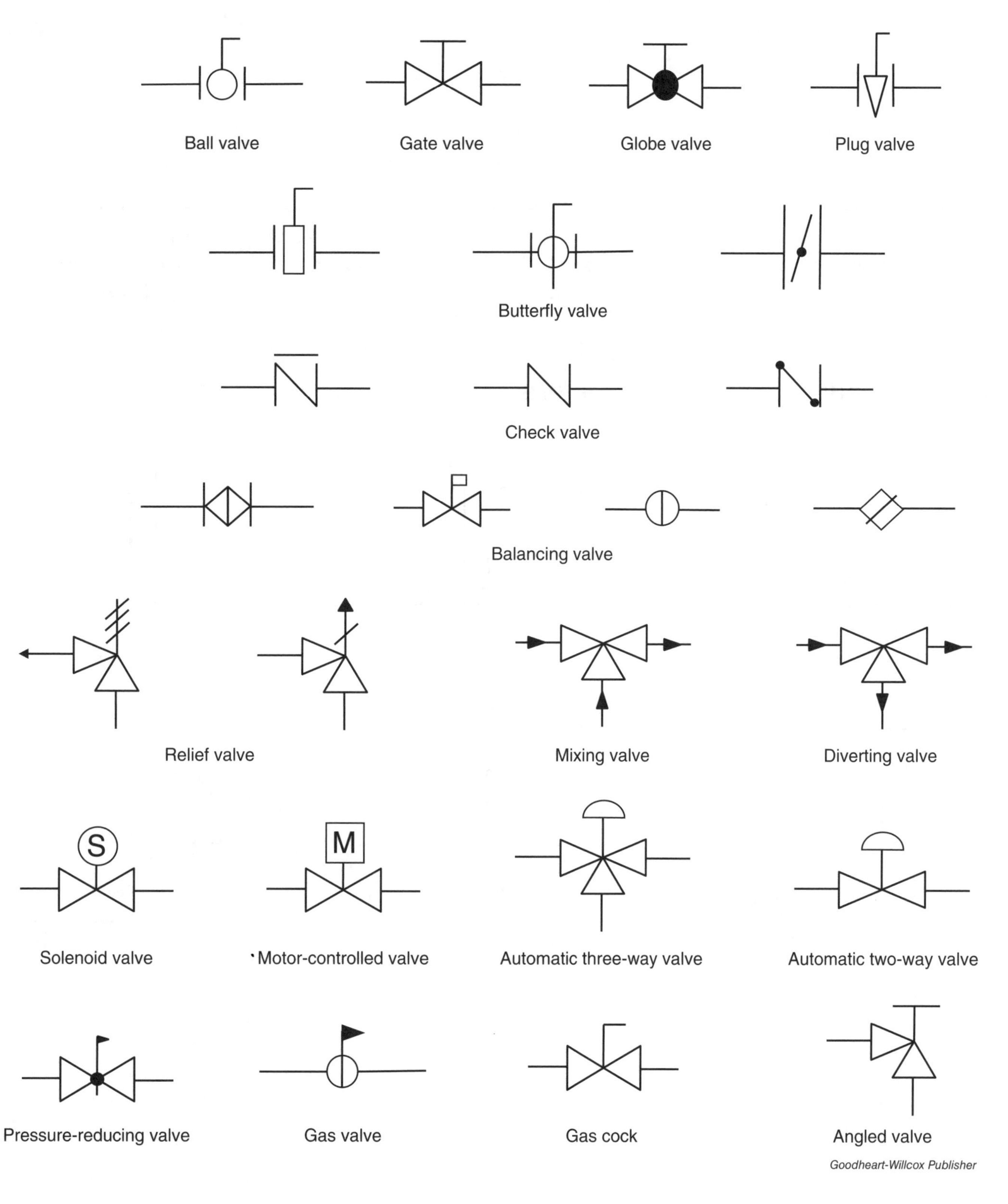

Goodheart-Willcox Publisher

Fittings

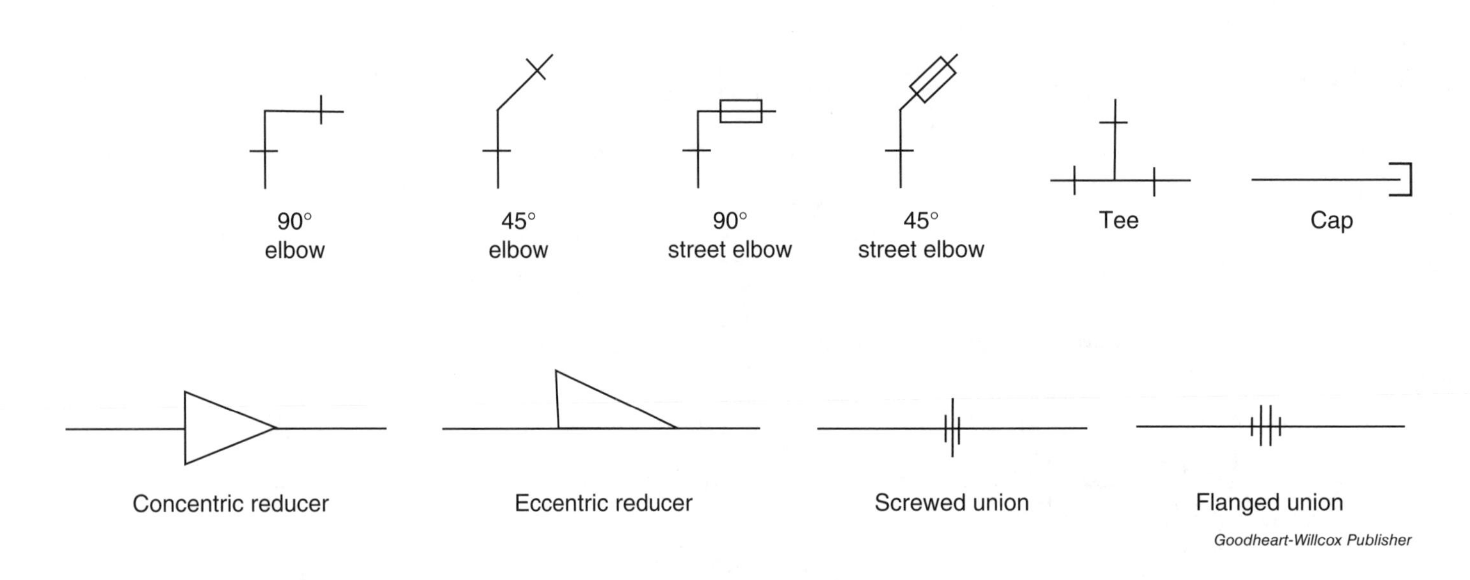

Goodheart-Willcox Publisher

Switches, Gauges, and Thermometers

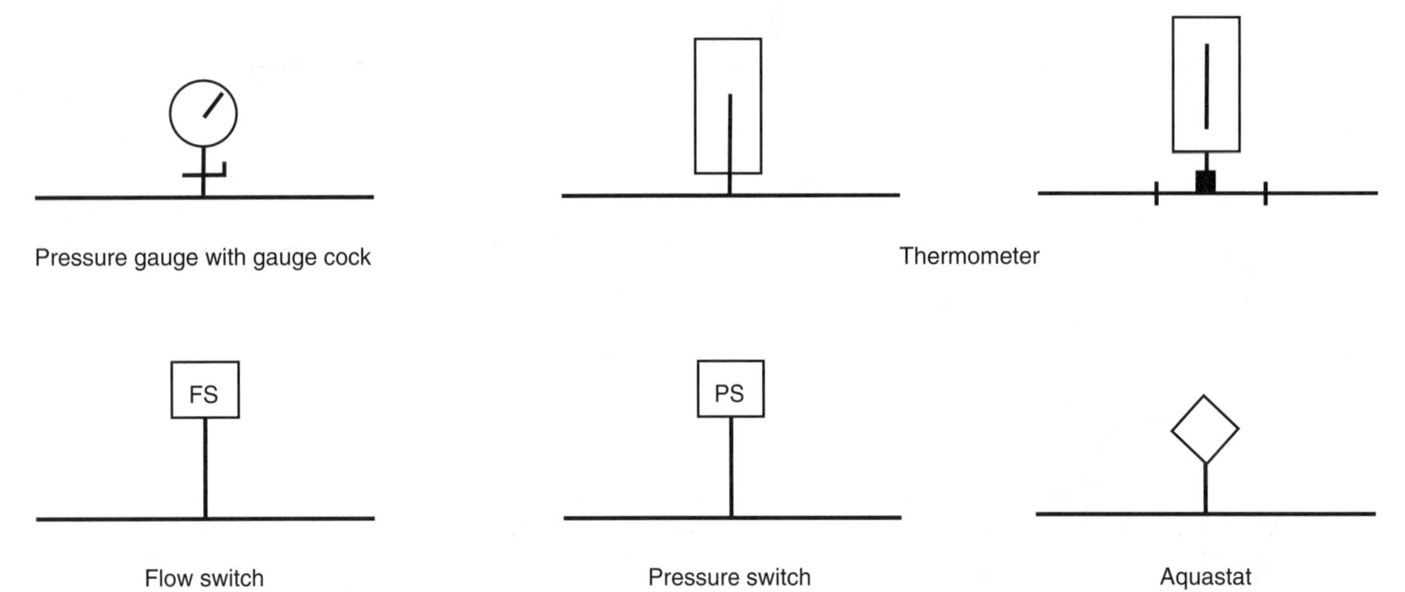

Goodheart-Willcox Publisher

Drains, Strainers, and Cleanouts

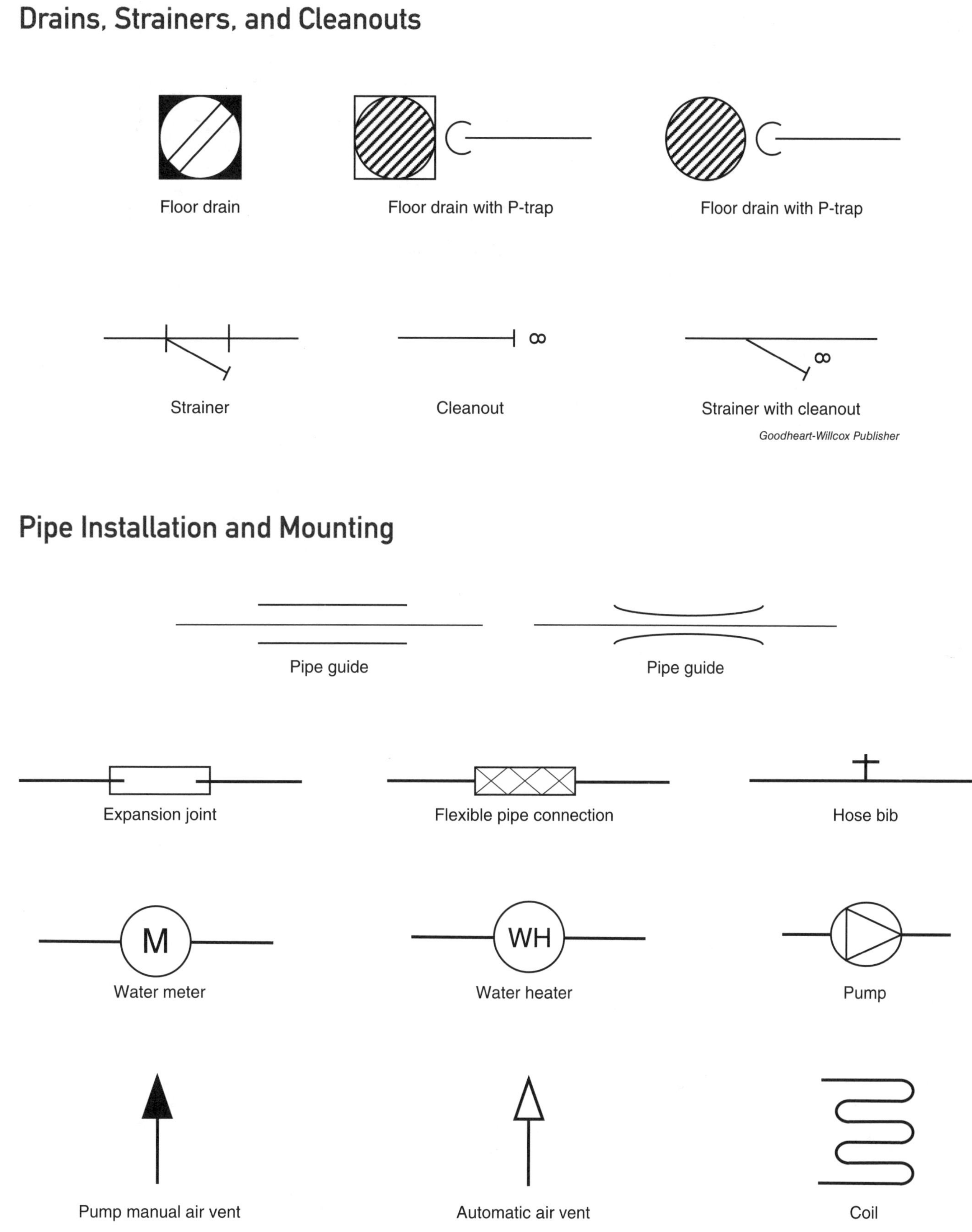

Directional Symbols

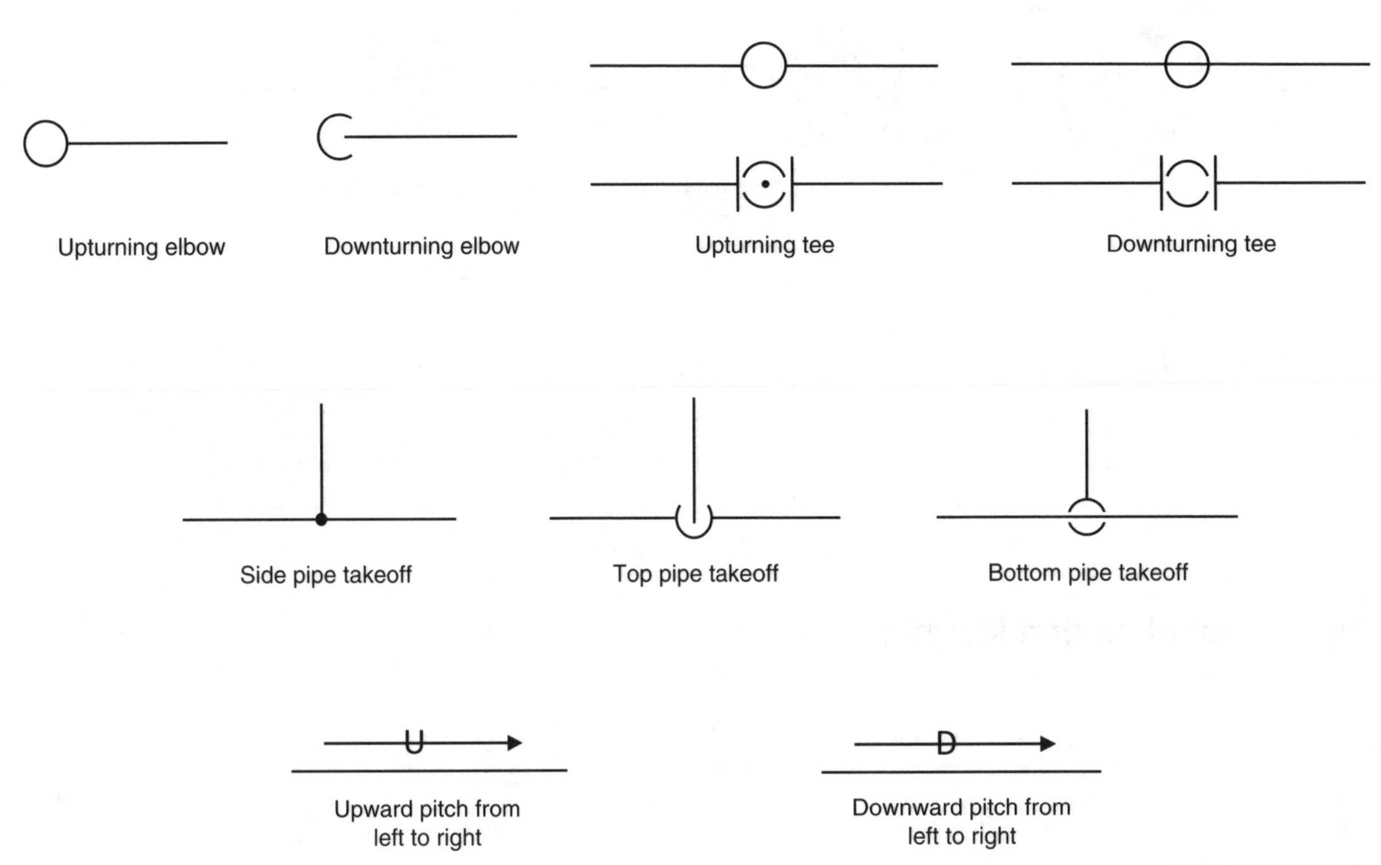

Goodheart-Willcox Publisher

Duct Symbols

Duct Reading Symbols

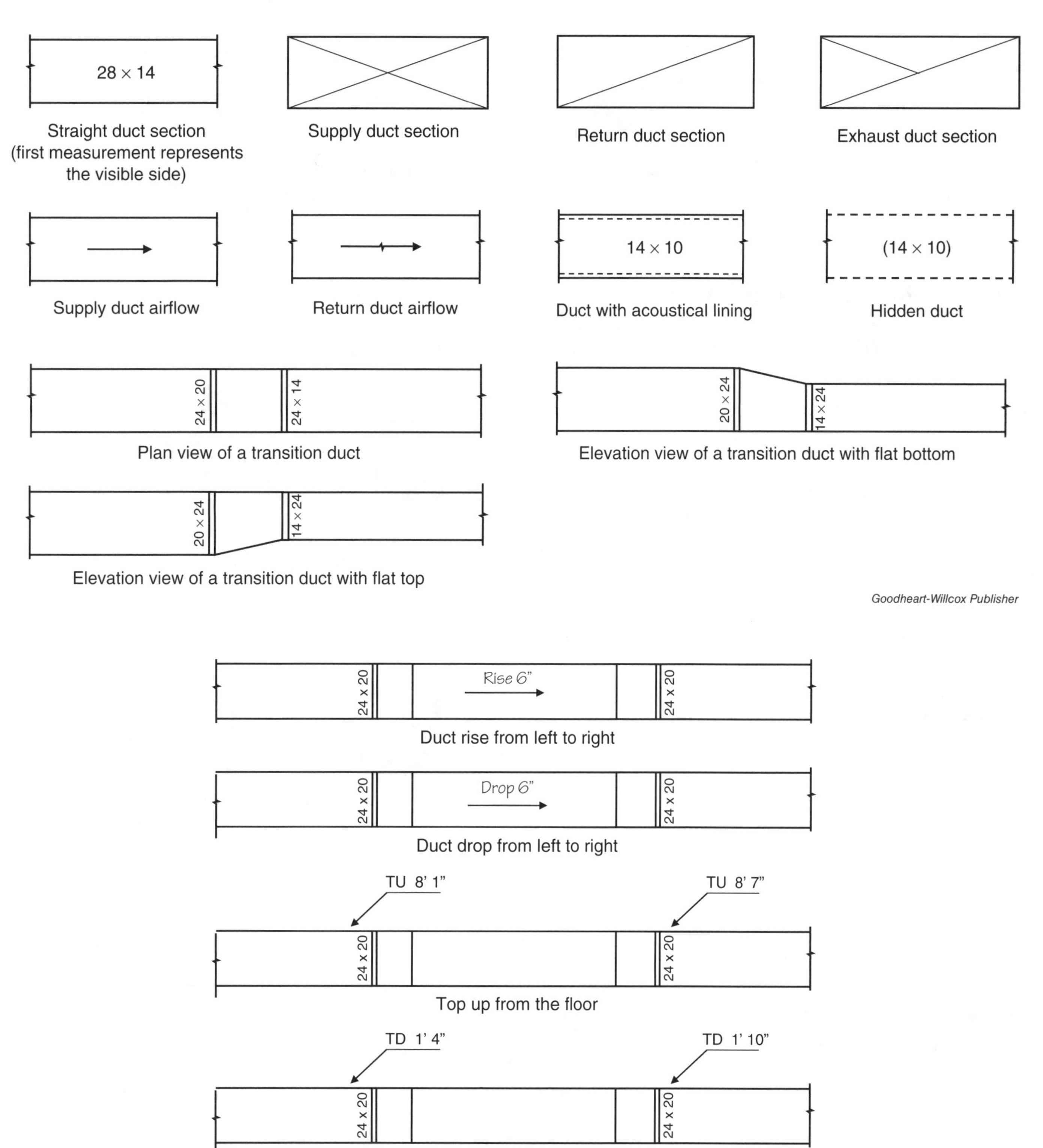

Goodheart-Willcox Publisher

Goodheart-Willcox Publisher

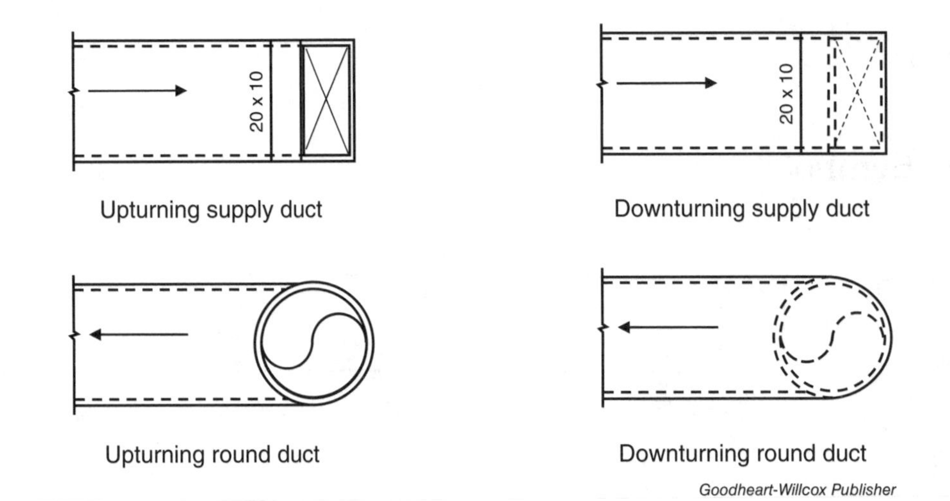

Upturning supply duct

Downturning supply duct

Upturning round duct

Downturning round duct

Goodheart-Willcox Publisher

Supply Registers and Louvers

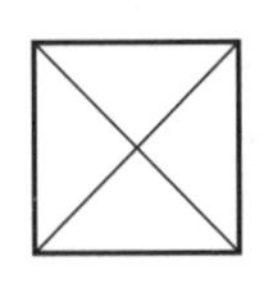
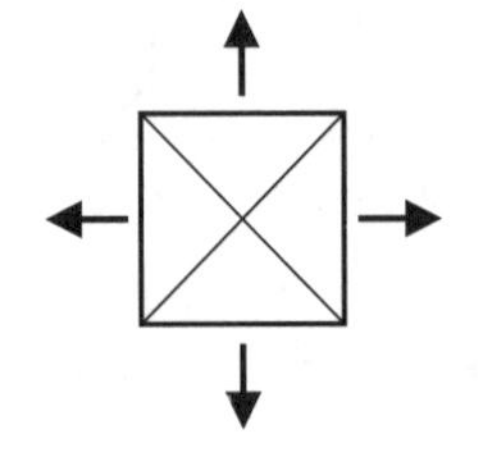

Four-way register

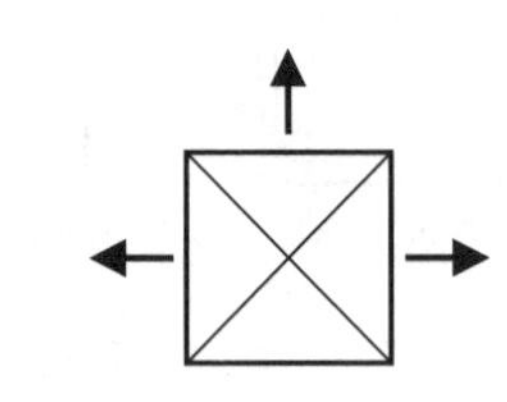

Three-way register

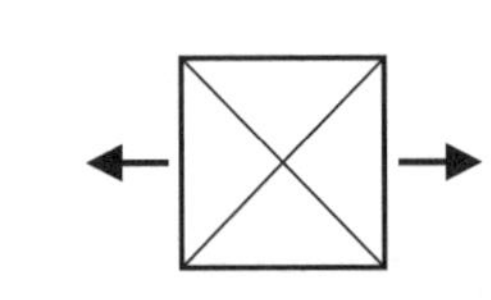

Two-way opposing direction register

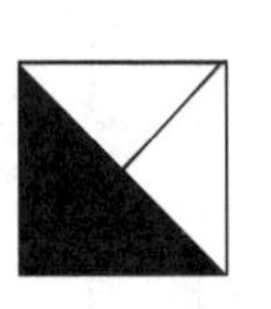
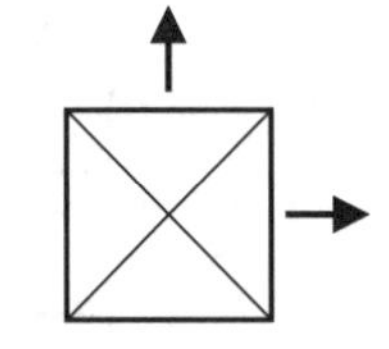

Two-way corner register

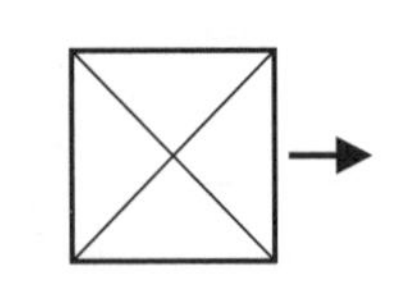

One-way wall register

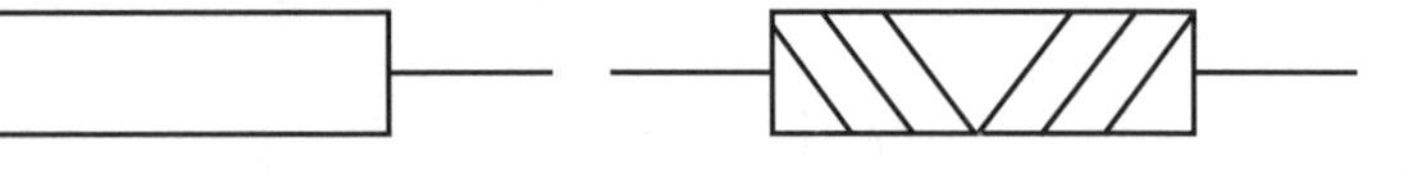

Wall supply louver

Wall supply register

Goodheart-Willcox Publisher

Return Air Symbols

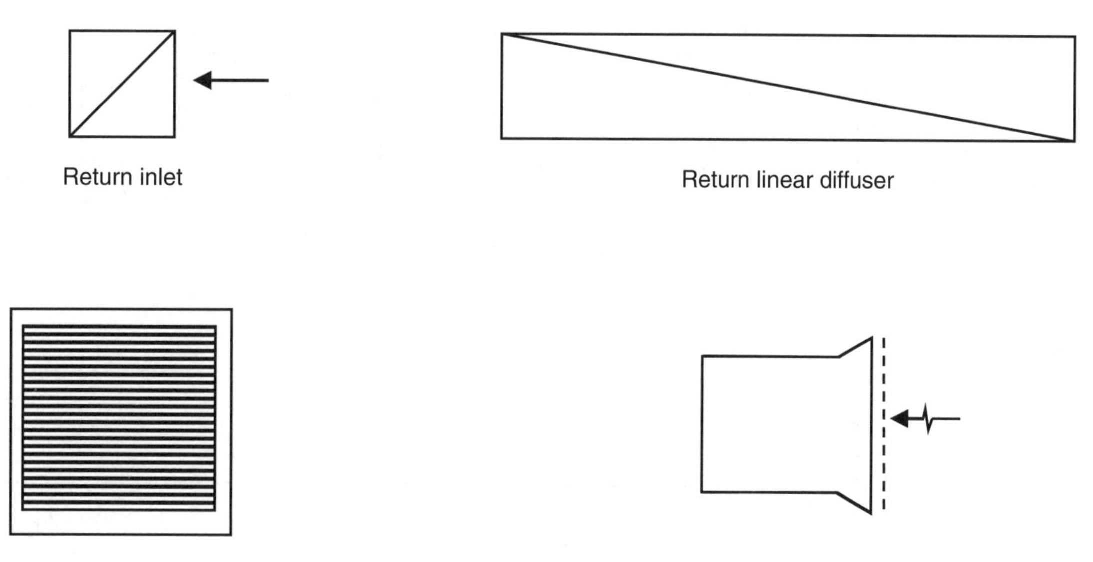

Goodheart-Willcox Publisher

Wall and Door Grilles and Louvers

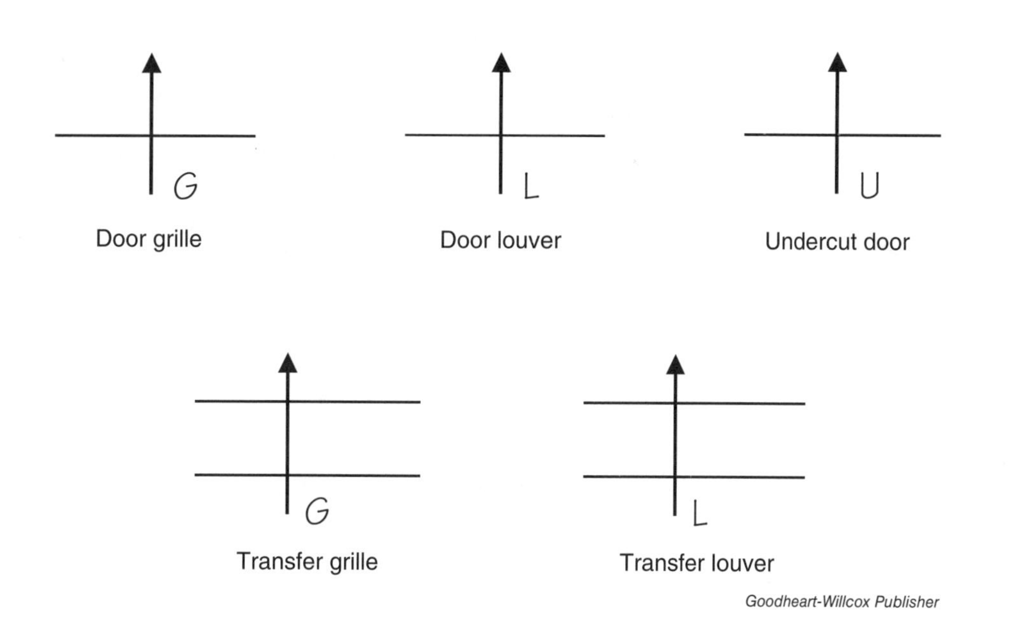

Goodheart-Willcox Publisher

Dampers

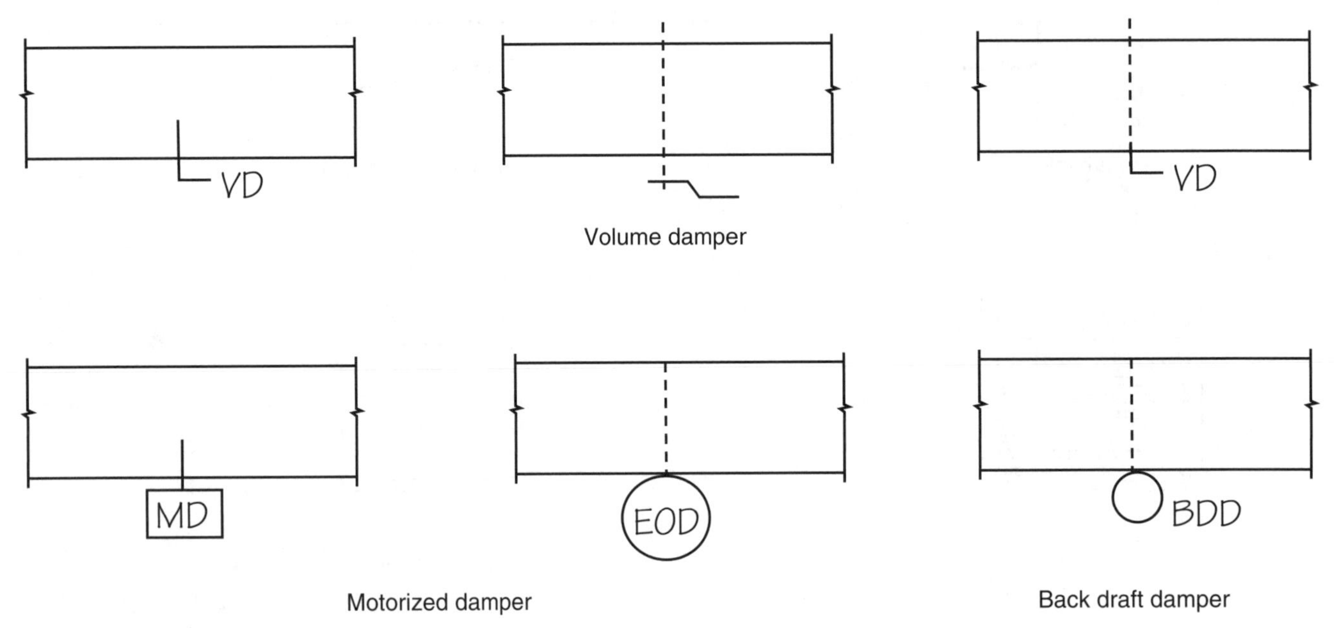

Volume damper

Motorized damper

Back draft damper

Goodheart-Willcox Publisher

Fire and Smoke Dampers

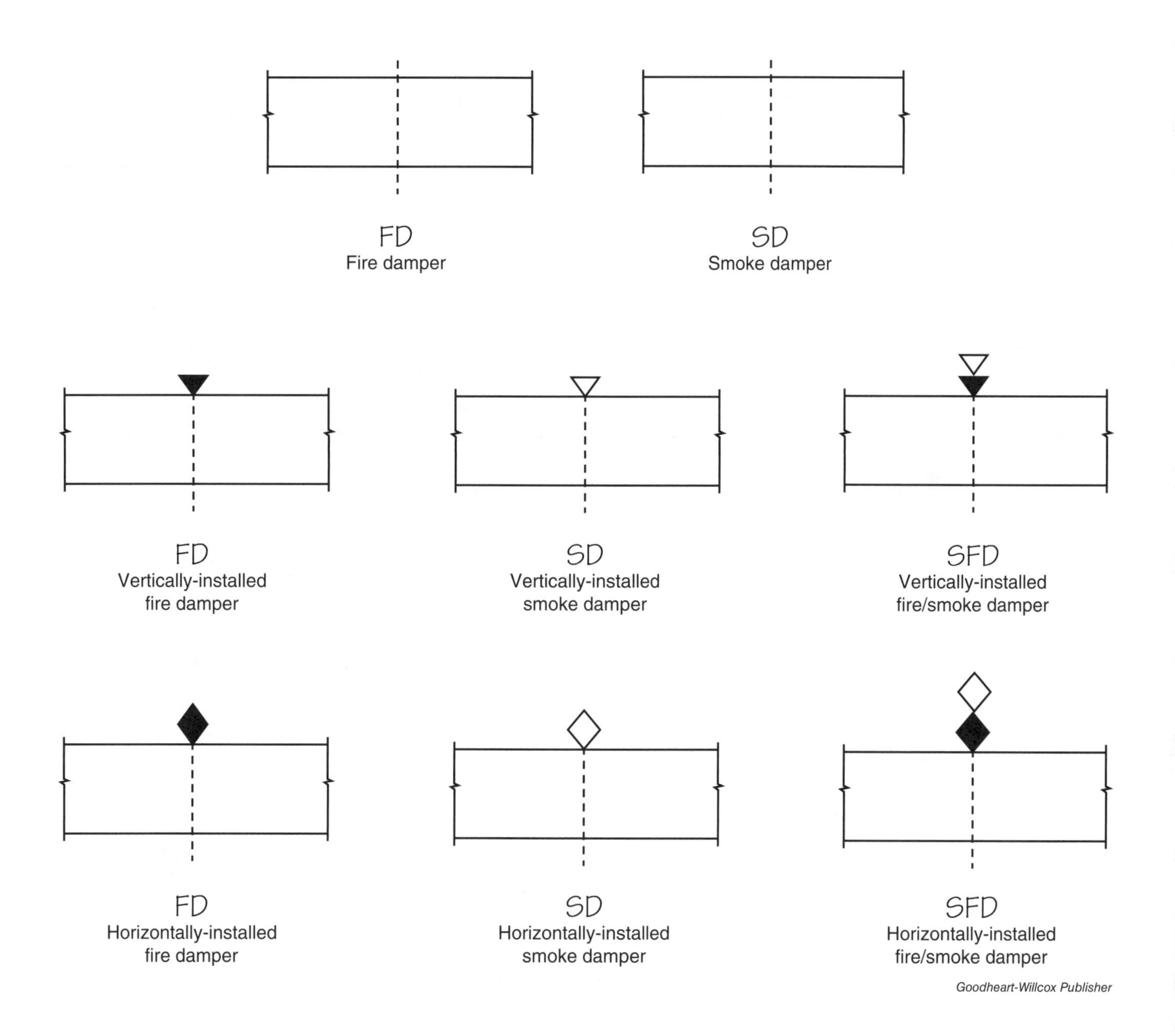

Goodheart-Willcox Publisher

Duct Accessories and Components

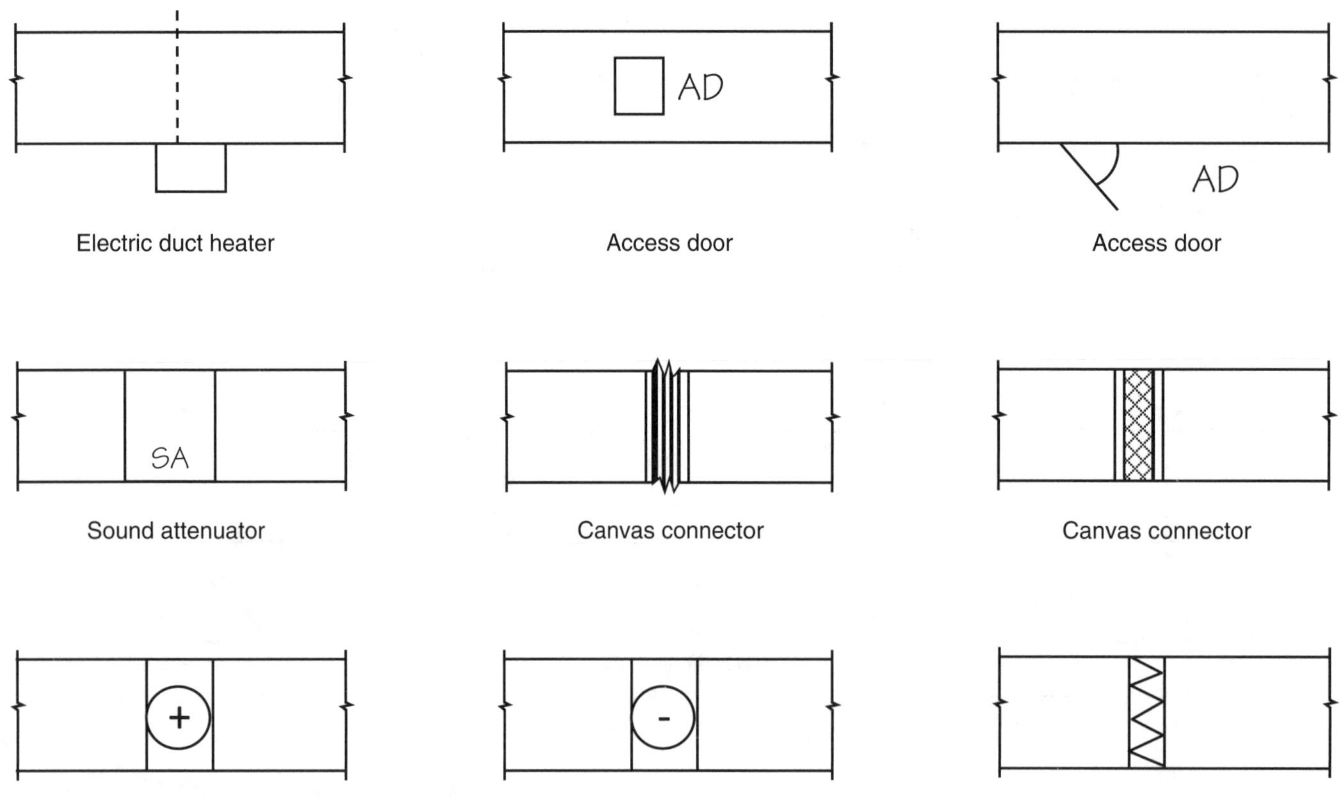

Glossary

A

addend. The numbers being added together to obtain the sum. (2)

air handler. The indoor portion of a split air-conditioning system that houses the evaporator coil, blower, and metering device. (8)

airflow. The amount of air per unit of time that flows through a component. Airflow is measured in cubic feet per minute (cfm). (2)

aquastat. A component that opens or closes one or more sets of electrical contacts when the sensed water temperature reaches a predetermined level. (7)

architect's scale. A tool used to measure distances on a print using a specific scale. This scale is most often used in the HVACR field. (4)

area. The amount of space within a boundary. Area is measured in square units. (3)

arithmetic mean. The average of a particular set of numbers. (3)

auxiliary drain pan. A pan placed under an air handler to eliminate water damage due to an overflow. (8)

B

balancing valve. A type of flow-control valve that sets a desired flow through a particular branch in a multi-circuit piping arrangement. (7)

bar slip. A modified strip that prevents a slip and duct from bending and flexing during blower startup and shutdown. (8)

bid. A detailed document that is prepared based on a specific project plan to be completed exactly as presented. A bid is commonly submitted on governmental or larger commercial projects. (12)

bid overlapping. A problem when multiple trades bid on their portion of a construction project. (12)

borrowing. The process of subtracting 1 from a place value and adding 10 to the next lower place value. (2)

brake horsepower (bhp). The amount of energy used to achieve a desired output. (11)

branch runout. A type of duct run that carries air from a takeoff fitting to a boot. (5)

building setback. Used to identify the required distance between a property line and any structures on that property, or the distance between two property structures. (8)

C

C-clamp. Used for securing a section of threaded rod to a structural steel beam. (8)

centi. A prefix that indicates one thousandth. (3)

check valve. A type of directional valve that allows free movement of fluid in only one direction. (7)

circumference. The distance around a circle. (3)

clevis-type hanger. A three-piece, three-hanging device used to support piping material. (8)

concentric reducer. A reducer that has inline inlet and outlet ports. (7)

condensate. Moisture that is removed from an air-conditioning system. (8)

Note: The number in parentheses following each definition indicates the chapter in which the term can be found.

condensate drain line. A drain line that carries condensation away from an air-conditioning system. (8)

condensate pump. Used to pump condensate from a location that does not include a floor drain. (8)

condensate trap. Installed in condensate lines to help prevent air from flowing up through the drain line. (8)

condensing unit. A component of an air-conditioning system that includes a condenser and compressor. (8)

construction number. An industry-accepted reference to reference a specific fitting. A construction number is a number and letter combination. (5)

contact. The portion of a switch that closes to complete an electrical circuit. (9)

converging transition. A transition through which air flows from the larger size of the piece to the smaller size. (5)

cross-breaking. A process of creating small bends between the diagonal corners on each side of a duct section. (8)

cube. A number tripled by itself. (2)

D

damper. A door that opens and closes to control the airflow through a particular duct system. (5)

defrost mode. A form of supplementary heat provided to increase the temperature of circulated air. (10)

denominator. The bottom number of a fraction. (2)

detail. A exploded view of a particular building component that requires more uncommon assembly. (1)

difference. The final answer derived from the subtraction process. (2)

diffuser. An air side component that introduces air to the space in a manner that facilitates effective air mixing and blending while reducing air stratification. (5)

diverging transition. A transition through which air flows from the smaller size of the piece to the larger size. (5)

dividend. The number being divided into during the division process. (2)

divisor. The number that divides the dividend during the division process. (2)

drive. A strip of metal formed in a C shape that is used to join sections of a duct system. (8)

duct. A round or rectangle pipe used to carry conditioned air. (5)

duct chart. A chart that provides information used to make accurate estimates for duct section fabrication and installation. (6)

E

easy elbow. An elbow that has a width, W, dimension that is smaller than the height, H, dimension. (6)

eccentric reducer. A reducer that has offset inlet and outlet ports. (7)

elbow. A duct fitting connected between two sections of ductwork to change the direction of airflow. (5)

electrical plan. A plan that shows electrical components of a print, including wiring, fixtures, lighting plans, and ceiling plans. (1)

electrical wiring diagram. A representation of an electrical circuit. This diagram shows how electrical loads, switches, and other devices are connected in an appliance or system. (9)

elevation view. The side view of a structure, which shows walls, windows, doors, and other vertical construction elements. (1)

engineer's scale. A tool used to measure distances on a print using a specific scale, normally in whole numbers. (4)

equipment specification. Includes information on a piece of equipment, including the product description, manufacturer, and model. (1)

estimate. A rough approximation of the amount a contractor is willing to accept to perform a given project. (12)

estimator. An individual responsible for weighing many factors to produce a quoted price for a project. (12)

exponent. A superscripted number that indicates the power to which a number is raised. (2)

F

factor. The numbers that are multiplied together to find the product. (2)

fitting. A section of ductwork used to change the direction or size of a duct run. (5)

flanged union. A two-piece connector that is bolted together rather than threaded together. (7)

floor plan. A scaled drawing that shows the view from above of a structure. (1)

flow switch. Detects fluid flow in a pipe. (7)

foundation plan. A plan view drawing of a structure's foundation and foundational elements, such as partition walls, doors, and plumbing fixtures. (1)

framing plan. A plan that shows the details of structure and support elements. Framing plans are used by carpenters during the construction process. (1)

G

gage. A measurement of thickness as it relates to sheet metal. (6)

grille. A framed set of adjustable slates that cover an opening in a duct, doorway, or other construction element. (5)

grouping. The process of combining sets of numbers that add up to easily manipulated values. (2)

H

hard elbow. An elbow that has a width, W, dimension that is larger than the height, H, dimension. (6)

heat pump. A component in a refrigeration system that adds or removes heat from a conditioned space by using mechanical energy. (10)

horsepower (hp). A unit of measure used to describe the size of a motor. (11)

hot water heating system. A heating system that provides heat to an occupied space through the use of boilers and a piping arrangement. (10)

I

improper fraction. A fraction that has a numerator larger than the denominator. (2)

inverse. The interchange of the numerator and denominator of a fraction. (2)

K

kilo. A prefix that indicates one thousand. (3)

L

ladder diagram. A representation of the function or operation of a circuit. (9)

legend. The portion of a print that indicates what each symbol in a drawing represents. (1)

limit switch. A normally closed switch that opens its contacts if the temperature in an appliance gets too high. (10)

lockout relay. A device that prevents a system's compressor from operating if the head pressure reaches an unsafe level. (10)

louver. A framed set of fixed slates that cover an opening in a duct, doorway, or other construction element. A louver can direct air to a particular path. (5)

lowest common denominator. The lowest number that one or more denominators can be changed into in order to make the denominators the same. (2)

lowest common multiple. The lowest number that can be divided by another number without a remainder. (2)

M

material list. A list of items or materials required to complete each project area. (1)

mechanical plan. A project drawing that provides information about mechanical equipment and HVACR systems. (1)

metric system. A standard system that uses the foot as a standard for measuring length. Also known as the *system international (SI) system*. (3)

milli. A prefix that indicates one hundredth. (3)

mixed number. A combination of a whole number and a fraction. (2)

mixing valve. A three-port valve that blends hot and cold water. This valve is usually electronically or electromechanically controlled. (7)

modulate. A control term meaning to increase or decrease the intensity of operations. Examples include variable refrigerant flow or some thermostats controlling multistage heating. (1)

N

normally closed (NC). A set of contacts that are closed in their de-energized state. (9)

normally open (NO). A set of contacts that are open in their de-energized state. (9)

numeracy. The ability to work effectively with numbers. (2)

numerator. The top number of a fraction. (2)

O

offset. A duct fitting used when otherwise straight ductwork needs to bend around an object or obstruction. (5)

overhead. The non-labor expenses paid by a company to operate the business. (12)

overrun. The actual cost to complete a project is more than 10% of the project's original estimated cost. (12)

P

percentage. A portion of a whole that is compared to a base value of 100. (2)

perimeter. The distance around the outside of a space or object. (3)

pi (π). A relationship between the circumference and diameter of a circle that does not change. Pi is closely estimated as 3.14159. (3)

pictorial wiring diagram. A simplified version of a circuit that shows the location of electrical components and electrical connections. (9)

pipe guide. A component that allows a pipe to move in one direction. A pipe guide facilitates thermal expansion or contraction of a pipe. (7)

pitch. The downward slope in the direction of water flow. (8)

plan. A type of drawing that shows features of a structure from above. (1)

plan view. An aerial view of a structure, which shows floors, room layouts, wall locations, and other horizontal construction elements. (1)

plenum. A chamber that facilitates treated air from an air-conditioning unit to a primary supply trunk or primary return trunk. (5)

plumbing plan. A plan that illustrates plumbing components of a print, including hot and cold water systems, sewage disposal, and fixture locations. (1)

pole. The number of poles indicates how many power lines can be open or closed at the same time. (9)

pressure-reducing valve. A type of pressure control valve that reduces the valve's inlet pressure to the desired pressure at the outlet. (7)

pressure switch. Detects when the sensed pressure reaches a predetermined level. (7)

primary return trunk. A duct system component that carries all the treated air to the air handler. (5)

primary supply trunk. A duct system component that carries all treated air by the air handler and distributes it to the network of supply ducts. (5)

print. A graphical representation of an architect's or engineer's design that is used to convey the ideas to a skilled worker performing the construction work. (1)

product. The result of numbers, or factors, multiplied together. (2)

profit. A monetary gain that is received when a service is provided at a price point that is higher than the cost of producing the service. (12)

project cost. The sum of all the costs associated with executing a particular job or project. (12)

proposal. A document that presents a contractor's vision for a project that is intended to sell a client on their plan. (12)

Q

quote. A formal offer to perform a specific task or job at a stated price within a specified time frame. (12)

quotient. The final answer derived from the division process. (2)

R

radius. The measurement of distance from the center of a circle to the outer edge. (3)

ratio. The proportional relationship between two quantities. (2)

reducer. A one-piece, two-port pipe fitting used to change the size of the pipe. (7)

reducing. Lowering the numerator and denominator of a fraction to the lowest terms while maintaining the same ratio between the fractions. (2)

refrigerant trap. Responsible for helping return oil to the compressor of a system. (8)

register. A combination of an adjustable grille and damper. A register has the ability to control the direction of airflow and the volume of air that passes through it. (5)

relay. An electrical switch that is controlled by an external signal. (9)

remainder. The quantity that is left over after dividing two numbers. (2)

return air duct. A duct that carries air from a conditioned space to an air handler. (5)

reversing valve. A valve used on a heat pump to change the operation from heating to cooling. (10)

rollout switch. A normally closed switch that opens its contacts if a gas flame rolls out of an appliance. (10)

root. A number that, when multiplied by itself a given number of times, yields a product equal to the original number. (2)

rounding. The process of altering a number to a higher or lower value for more convenient but less exact calculations. Also called *rounding off*. (2)

S

safety float switch. A normally closed device that shuts down a system if the level of water in the auxiliary drain pan reaches a predetermined level. (8)

scale. A proportion to a size in which a print is drawn. (4)

schedule. Lists required project components and provides information such as the item, size, and number required. (1, 11)

schematic diagram. A detailed illustration of the wiring in a piece of equipment or device. The diagram shows the interconnections of wiring and operation of the circuit. (9)

screwed union. A three-piece union where the male and female ends of the union are secured, either by threaded or soldered connections, to the pipe being joined. (7)

secondary supply trunk. A duct system component that carries a portion of the treated air to a portion of the system's supply ducts. (5)

section. A cross-section drawing that shows a structure as if it were cut through and provides a view of components not seen from the outside. (1)

sequence of operations. A step-by-step procedure that describes how the various components of an HVACR system operate in a particular order. (9)

sheet metal. A class of materials often used to fabricate duct systems. Common sheet metals include hot-rolled mild steel, galvanized steel, stainless steel, and aluminum. (6)

site plan. A drawing that shows the structure, the surrounding area or plot, and various details. Also called *plot plan*. (1)

slip. A strip of metal formed in an S-shape that is used to join sections of a duct system. (8)

solenoid. An electromagnetic device that converts current to mechanical movement by producing a magnetic field. (9)

sound trap. A component that reduces noise transmissions through ducts. (11)

split air-conditioning system. An air-conditioning system that divides its system into two separate locations: a condensing unit outdoors and an air handler indoors. (8)

square. A number multiplied by itself. (2)

square elbow. An elbow that has equal height and width measurements. (6)

static pressure. The resistance to airflow caused by air moving through a duct or other channel. (11)

structural plan. A drawing or set of drawings that provide information on how a building is constructed.

sum. The final answer derived from the addition process. Also called the *total*. (2)

supply air duct. A duct that returns treated air back to the conditioned space. (5)

switch. A device used to open or close a connection in an electrical circuit. (9)

T

tag. A symbol used to indicate a specific location on a print where a device or component is to be located. (11)

takeoff fitting. A fitting that connects a branch runout to the main supply duct. (5)

tape measure. A measuring tool with a retractable ruler that can be set at a specific length to transfer measurements from one place to another. (4)

tee fitting. A fitting used to direct ductwork in different directions at once. (5)

thermostat. The operational and control interface between an air-conditioning system and the system operator. (9)

throw. The number of throws indicates how many possible current paths a switch can control. (9)

time-delay relay. A relay that operates with a delay period between the time the relay's coil is energized and the time the relay's contacts change position. (10)

total equivalent length (TEL). The length of each fitting that is added to the length of the duct run. (6)

total labor cost. The total cost of all aspects of the project that require manpower. (12)

transfer grille. A grille assembly that allows free airflow between two adjacent spaces, helping to equalize space pressure. (5)

transition. A duct fitting used to connect duct sections of different sizes. (5)

turning vane. Helps reduce air turbulence within a duct. (5)

U

underrun. The actual cost to complete a project is 90% of the project's original estimated cost. (12)

union. Connects sections of piping material that may need to be disconnected at a later point. (7)

US Customary system. A standard system that uses measurement units, such as the inch, the gallon, and the pound. Also called the *inch-pound (IP) system*. (3)

V

valve. A system component that is used to control air or fluid through a pipe or duct. (7)

variable frequency drive (VFD). A control device that varies the frequency of power supplied to a motor to modulate the motor's speed. (11)

variance. A document prepared by a manufacturer that states an installation is acceptable and does not affect a system's operation, effectiveness, or efficiency. (8)

vent switch. A switch that prevents the gas valve from being energized unless appliance venting is established. (10)

vicinity map. A visual guide that shows the location of the project in respect to the area in which it is located. (1)

volume. The measurement of space that occupies a three-dimensional object. (3)

W

weighted average. An in-depth average calculation where each value is multiplied by a factor that is proportional. (3)

Index

Note: Page numbers followed by *f* indicate figures.

B

C

D

E

K

L

M

Q

R

S

T

U

V

W

X

Y